ENCYCLOPEDIA OF BIOPROCESS TECHNOLOGY

ENCYCLOPEDIA OF BIOPROCESS TECHNOLOGY

Vol. 4

INDUSTRIAL BIOPROCESS TECHNOLOGY

By

Dr. Arvind N. Shukla

School of Studies of Zoology & Biotechnology

Vikram University

Ujjain

(India)

DISCOVERY PUBLISHING HOUSE PVT. LTD.

NEW DELHI-110 002

Published by:
Tilak Wasan
DISCOVERY PUBLISHING HOUSE PVT. LTD.
4383/4A, Ansari Road, Darya Ganj
New Delhi-110 002 (India)
Phone: +91-11-23279245, 43596064-65
Fax : +91-11-23253475
E-mail: parul.wasan@gmail.com
discoverypublishinghouse@gmail.com
web : www.discoverypublishinggroup.com

***First Edition:* 2012**

ISBN: 978-93-5056-026-6 (Set)
978-93-5056-027-3 (Vol. 1)
978-93-5056-028-0 (Vol. 2)
978-93-5056-029-7 (Vol. 3)
978-93-5056-030-3 (Vol. 4)
978-93-5056-031-0 (Vol. 5)

Encyclopedia of Bioprocess Technology

Printed at:
Shree Balaji Art Press
Delhi

Preface

The present title **"Encyclopedia of Bioprocess Technology"** has been written for undergraduate, post-graduate students and those engaged in pharmaceutical, environmental biotechnology, industrial and clinical research. Actually the explosion of new technologies with their vast potential has brought with it the need for emerging scientists to equip themselves and their laboratories with a whole array of new expertise. With the high discriminating power of the DNA systems has come high potential in evidentiary terms, high profile status for many investigations, and not least, a high degree of professional scruting of evidence produced by such technology.

This book is addressed to all who are curious and who want to understand fundamental biological processes that ensure our survival. The person who might be interested is probably someone just starting research who wants to learn about aspects of regulation, outside his chosen field. It could become a supplementary text for graduate students and post-graduate in biochemistry and molecular biology.

To make the work more comprehensive and informative, the author has consulted many authoritative books, research journals, abstracts, monographs etc., so there can be no claim to originality except in the manner of treatment.

The author express his thanks to his friends and colleagues whose continue inspirations have initiated him to bring out this book.

The author expresses her gratitude to Mr. Wasan and staff of M/s Discovery Publishing House Pvt. Ltd. for their whole hearted co-operation in the publication of this book.

The author will remain sincerely responsible for any shortcomings of the book and be grateful to the readers for their suggestions and constructive criticism for the continuous betterment of the book. He takes this opportunity to appeal to the readers to send their suggestions straightaway to this Publisher.

Arvind N. Shukla

CONTENTS

CELL MEASUREMENTS

A technique for measuring physical or chemical characteristics of single cells (or other biological particles) is called *cytometry*. Such measurements are usually performed with the aid of a microscope, but there are severe limitations both in the number of cells that can be investigated microscopically and in the number of features that can be observed simultaneously. In flow cytometry (which avoids these limitations), the cells pass the measuring volume in a fluid stream, preferably in a single file, and in a short time. The first flow cytometers, developed in the 1940s, were based on the Coulter principle, that is, the detection of changes in the electrical conductivity or impedance of a small saline-filled orifice as cells pass through. The history involved leads most people to think of flow cytometers as instruments that measure optical signals from cells. These optical signals result from light being scattered by the cells (differentially, according to cell type and status) and the fluorescent light emitted by different cell compartments when excited by light of a certain spectral range.

The cellular parameters detected can be characterized as intrinsic or extrinsic, depending on whether their measurement requires use of reagents or probes. A cytometer consists of five main components:

1. *Illumination optics*. Usually consisting of an arc lamp (high-pressure mercury) or a laser (argon ion) combined with crossed cylindrical lenses to focus the light.
2. *Flow cell*. A device for focusing the cells in a linear file by hydrodynamics. The core stream with the cells travels in a laminar flow to the illumination point with the aid of a sheath stream of clean water. The main types are stream-in-air systems and variations thereof passing through a (flat-sided) quartz cuvette.
3. *Collection optics*. Usually consisting of a combination of optical filters for spectral separation, generally set perpendicular to the illuminating beam and in four spectral ranges (green, 510–540 nm; yellow, 560–580 nm; orange, 605–635 nm; and red, 650 nm and above).

4. *Detector electronics.* These consist of photodiodes or photomultiplier tubes (PMTs). The particles of interest in biotechnology, which are frequently very small, (e.g., bacteria), emit only a small number of photons, typically about 1,000 in passage through the detection chamber. Therefore, an appropriate combination of high-power illumination optics and sensitive detectors is necessary for the application of flow cytometry in biotechnology.
5. *Data analysis.* The first flow cytometers displayed and stored the data in multichannel pulse height analyzers. Now the data are stored as standard list mode files in PCs, ready for statistical analysis and sophisticated visualization.

The result of a flow cytometric investigation is a frequency distribution, that is, a histogram. In the channels of this histogram, cells are counted according to the light intensity detected, which corresponds to the relative size or quantity of a specific cellular compartment or component, such as the DNA content. Measuring rates up to 10,000 cells s^{-1} are possible, although about 100 cells s^{-1} is more usual. Flow cytometry is not recommended for measuring the absolute content or concentration of cell compartments. However, it is the best method for classifying cell types with different physiological properties and for selecting and quantifying subpopulations in a cell ensemble. If subpopulations of (vital) cells are to be chemically analyzed or cultivated further, the information provided by flow cytometric measurement can be used for selecting appropriate subpopulations by an electrical or mechanical sorting process. Depending on the type of flow cell and the type of illumination involved, three basic constructions are used:

1. Stream-in-air
2. Contained flow chamber
3. Dark-field illumination

The devices range from large constructions for research purposes to small and inexpensive laptop machines on the way to an on-line approach. The development of new flow cytometric methods is often accomplished using special modular constructions ("cyto mutt" comp.). A further development is the slit-scanning procedure, which allows the shape of the signal generated by the cell passing through the illumination beam to be used for gathering additional information. In this way, morphological features and the distribution of cellular compartments can be recorded. Special demands have led to appropriate modifications. The distribution of aquatic microorganisms can be monitored in the size range of 0.1 to 2000 μm by such a device, for instance. Flow cytometers are best used to answer questions concerning the dynamics of subpopulations in a cell population, for example, cells in a mammalian tissue that have a higher DNA content than that of G1-phase cells. Thus, in medicine, flow cytometry was first applied in oncology, where it was quickly (and frequently) used in cancer diagnosis.

These results generated a solid basis for mathematical modeling of cell-cycle regulation in eukaryotic organisms. Using immunofluorescence and multiparameter analysis, flow cytometry became a successful, and standard, method in laboratory hematology and clinical immunology in the 1980s. Several commercial devices are now available, constructed specifically for medical purposes. An excellent and comprehensive documentation of the method of flow cytometry is given by Shapiro. The use of flow cytometry in biotechnology, and especially in bacteriology was for a long time not as successful as it was in medicine. After early, inspiring investigations by Bailey et al., Slater et al., and Hutter and Eipel further development seems to have been delayed. There are probably three major reasons for this:

1. The high cost of the method
2. The small size of the cells investigated (an *Escherichia coli* cell has only 10^{-3} to 10^{-4} times the size or DNA content, and consequently the potential fluorescence energy gain, of a mammalian cell)
3. The belief of most biotechnologists that analysis of physiological state distributions do not have an important role to play in understanding the dynamics of bioprocesses or for running them safely and economically

The first two handicaps have now been overcome because powerful and inexpensive microcomputers have considerably reduced the cost ofthe technique. Flow cytometry has also benefited from new methods of data analysis and mining, such as neural network analysis, which allow a comprehensive evaluation ofthe rich and complex information gained by flow cytometry. Furthermore, new staining methods combined with improved illumination, detection, and amplification have made flow cytometric analysis highly reliable, even for bacteria and microplankton as small as 1 μm in diameter. Nevertheless, the key to more widespread use of flow cytometry in biotechnology is more general acknowledgment of the importance of a segregated model concept. Toward this end, Fredrickson et al. proved that the statistics, the distribution of cell populations, and the dynamics of bioprocesses are inseparably connected. Moreover, Munch et al. demonstrated, with a sophisticated approach to flow cytometry, that knowledge concerning the distribution of states in the cell cycle is essential for deeper understanding of the growth dynamics of yeast and the metabolism of a biomass.

PROKARYOTIC CELLS

Bacteria are the most common example of prokaryotic cells. They exist as dynamic and diverse populations in nature. Even the cellular heterogeneity within a pure bacterial culture is far greater than previously assumed. To understand bacterial ecology in nature and in bioprocesses, new techniques are required for answering questions that could not be solved by traditional methods alone. Now more than ever, flow cytometry has been recognized as a valuable asset for microbiologists and is perceived to be an important and indispensable tool for investigations on the microenvironmental level.

Bacterial flow cytometry enables visualization of cell states and allows the analyst to follow growth, death, replication, cell division, metabolism, and surface phenomena, greatly enhancing the ability to understand and control cell physiology. A great assortment of fluorescent stains is commercially available; their numbers have rapidly increased recently. It is not possible to give an overview of their application in this article, but Lloyd provides a fine introduction to this topic. Generally, however, it should be realized that these techniques need to be handled more carefully for use with bacteria than with eukaryotes to avoid unclear results from artifacts arising either from sampling or from analytical errors.

Multiplication

All survival strategies should be programmed, above all, to safeguard the genome for the future. Results obtained by flow cytometry allow conclusions to be drawn and tested about how the physiological state of cells is connected with survival strategies under changing environmental conditions. Consequently, the most obvious targets for bacterial staining are the nucleic acids. Steen and Boye and Skarstad et al. presented a model for calculating the relative rates of DNA synthesis per cell from studies based on *E. coli* strains. By means of this model and flow cytometric evaluation of the DNA distribution, the generation time (τ) and the initiation and termination of the replication and post-replication periods (*C* and *D*, respectively) can be estimated. Features of replication (e.g., the number of initiation origins) seem to be dependent both on the genetically determined disposition ofthe bacterial strain analyzed and the growth conditions leading to the initial cell mass. In unperturbed bacterial cultures with high growth rates, several replication forks may be initiated in each cell, if the growth conditions are optimal.

Then, after treatment with antibiotics, *E. coli* cells will contain two, four, or eight fully replicated chromosomes. In contrast, bacteria growing slowly, at doubling rates of less than 1 per hour, exhibit an increase in the pre-replication period as well as in the *C* and *D* periods. This effect can be amplified by further decreasing the growth rate using chemostat cultures under limited growth conditions. Besides the considerations just mentioned, it must be stressed that during limited, and even unrestricted, growth, the behavior of individual cells differs. As pointed out by Akerlund et al., each cell goes through different discontinuous processes in its life (e.g., the cell cycle). They are therefore different both in size and metabolic activity. This is generally true, except in steady-state conditions, to some extent, where only the distribution of the cells and the cell size are time invariant). Considering these results, it is clear that key events in cell physiology, including initiation of replication and cell division, are tightly bound to the microenvironmental conditions in the surroundings of the bacterial cell.

Obtaining information about the replication behavior of biotechnologically useful bacterial strains is, therefore, a crucial pre-requisite for controlling industrial bioprocesses. This knowledge could also be used for investigations into the population dynamics of marine bacteria. Because of fluctuating climatic influences, these organisms must constantly adapt to changing conditions and move between states of growth and starvation. In such cases, major alterations in the level of other cellular components, such as lipids and proteins, may be observed in addition to changes in the nucleic acid contents.

Physiological State

Progress in bioprocess engineering depends ultimately on the level of understanding and control of the physiological state of the bacterial population being exploited. Process efficiency is strongly sensitive to changes in the cell state. Characterization of these states via fluorescence monitoring, evaluation ofthe data obtained, and subsequent regulation ofthe process regime by controlling the surrounding natural or artificial conditions is one ofthe most ambitious tasks in the near future for biotechnology. Flow cytometry appears ideally suited for determination of such parameters as cell concentration and size as well as for visualizing intracellular performance in industrial processes promoted by bacteria. Using this technique, the state of the cellular system can be precisely monitored, allowing optimization of process efficiency in production or degradation of the target substances. Although bioprocess control on the cellular level is not being widely practiced as yet, some processes are now being adjusted by flow cytometry.

One main limitation seems to be the high content of particulate constituents in most industrial media, because discriminating between microorganisms and noise signals from media is not a trivial undertaking. Microorganisms respond to an unbalanced supply of nutrients or deviations from optimum physical factors by increasing the synthesis of intermediates. This phenomenon is called overflow metabolism. The metabolites overproduced can be excreted or accumulated intracellularly. Flow cytometric methods have been established for detecting intracellular accumulation of polyhydroxybutyrate (PHB), a typical overflow metabolite in many bacterial strains. PHB serves as an energy and carbon reserve and has received attention as a thermoplastic and biologically degradable polymer. Furthermore, researchers are searching for novel bacterial strains that may accumulate novel types of polyhydroxyacid (PHA) possessing better physical and mechanical qualities. Some bacterial yield up to 90% of their dry weight in this material. The inclusion bodies so formed alter both the size of cells and their light- scattering behavior.

The effect of different cultivation conditions on PHB formation has been investigated in *Ralstonia eutropha* and recombinant *E. coli* transformed with PHA-synthesis genes. In these cases, cell-sorting technologies can be used to isolate highly efficient strains at a high speed. Flow cytometry can also be used to quantify heterogeneity of PHB production and accumulation, as found in *Methylobacterium rhodesianum* cultures. These investigations are based on the idea that growth and product formation (overflow metabolism) are coupled to specific cell states, which are an expression of maturation of individuals and are connected to stages in the proliferation cycle. It has been found that under growth-limiting conditions, the cells first go though the DNA replication program, thereby safeguarding the genetic information by doubling the chromosome content. Doing this, the organisms maintain the chance of restarting multiplication as a forward strategy of survival if better conditions arise. Cells only lay down PHB as an energy reserve in this kind of situation. Introduction of foreign genes into bacterial cells enables synthesis of desirable products such as proteins, lipids, and a wide variety of other biologically active compounds. Normally, the expression of these genes alters the normal pattern of interaction and synthetic activities in the host cell.

It may be observed that highly active cloned gene expression or extremely large plasmid content leads to a markedly slower growth rate in recombinant cells. Generally, however, some of the recombinant cells revert during prolonged cultivation, either (in the case of plasmids), through defective partitioning during cell division, or through changes in plasmid structure. By this process, a subpopulation of nonrecombinant cells develops that has a higher specific growth rate. In chemostat experiments, the recombinant cells are usually replaced by this phenomenon, and the plasmid-free population comes to dominate the dynamics of the reactor. It is generally believed that the production of heterologous products is metabolically harmful to the cell. Occasionally, it has been proposed that the decrease in the growth rate seen is more likely to be a result of deleterious and injurious interaction between the expressed product and some component of the cell. Flow cytometry can be used to detect segregational instability of a plasmid-containing expression system as well as to estimate the physiological state of the cells involved. There have only been a few biotechnological studies investigating the stability of recombinant cells on near-industrial scales. By way of illustration, however, the method has been used to examine recombinant *E. coli* cells forming protein inclusion bodies via wide-angle light-scattering measurements. Fluorescence-activated cell sorting has also been found to be capable of isolating gramicidin S hyperproducing mutant cells of *Bacillus brevis* from wild-type cells.

Furthermore, flow cytometry can be used to study the progress of gene expression in a wide range of bacterial cell systems. The expression of foreign genes in bacterial cells or the stability of transformed plasmids can generally be determined by the use of bioreporter genes. These systems allow flow cytometric monitoring of the gene expression even when the gene product is difficult to assay. The promoter- less reporter gene is placed under the expression control of the promoter of the foreign gene. A typical example is use of the *lacZ* gene from *E. coli,* one of the most widely used bioreporter systems. The *lacZ* gene encodes the enzyme β-galactosidase, which cleaves a fluorogenic product from specially designed substrates. The emission ofthe resulting fluorescence is proportional to the amount of the enzyme in the cell, corresponding to the expression of the target gene. Consequently, the β-galactosidase-positive cells exhibit a level of fluorescence that is measurable for each cell by flow cytometry.

However, a primary problem in single-cell β-galactosidase assays is leakage of the intracellular fluorescent marker, although a number of methods have been developed to prevent this movement. Other bioreporter systems encode detectable markers such as green fluorescent proteins or surface antigens for monitoring microbial performance. There have also been attempts to use polymerase chain reaction (PCR) in DNA amplification methods for sensitive detection of specific innate or foreign DNA or RNA sequences inside intact cells. Some of the population seem to remain unbroken after such treatment, allowing subsequent flow cytometric assessment. For example, PCR can be used to detect plasmid-encoded gene sequences or poorly expressed mRNA in individual bacterial cells.

Bacterial Physiological Processes

In contrast to biotechnologically processes, bacterial cells in nature live neither in pure cultures nor in surroundings of constant temperature, pH, or nutrient availability. Many of the (microscopically) visible bacterial cells are inactive or dying. Every change in the habitat is potentially a source of stress, which results in a change in growth rates. The sum ofthe cellular responses to stress is a factor associated with the survival strategy. Responses to stress take place at two distinct cellular levels: first, activation of existing enzymes and transport processes and, second, gene expression. The biochemical bacterial stress response can be precisely determined using flow cytometry, by analyzing changes in the type of duplication, cellular pH, membrane potential, and the amount and kind of storage products.

The physiological links between the environment and the genetic systems with respect to survival of stress situations are often mediated by expression of proteins involved in a range of tolerance mechanisms. It is commonly accepted that specific genetic programs exist for prolonging the survival of non- growing bacteria exposed to starvation or stress. Bacterial programmed cell death seems to be mostly related to population development, because the lysed cells are usually essential as sources of nutrient compounds, allowing completion of the developmental cycle of the remaining, living population. To date, knowledge ofthe connections between bacterial processes and bacterial community structure in nature is inadequate and restricted by procedural flaws. Most commonly used methods for the detection and enumeration of bacteria have serious limitations because of their incapacity to allow certain bacteria to grow.

The reason may be that the bacteria concerned need symbiotic partners, anaerobic microenvironments, other uncommon conditions, or unknown growth factors. However, for a variety of reasons, quantifying the bacteria present in the investigated system, independently of growth, is a fundamental requirement for almost all bacteriological studies. Living cells must be distinguished from dormant cells (induced by starvation) as well as from viable (but not culturable) and truly dead bacteria. To characterize the degree of viability of bacteria, flow cytometric detection of basic cell functions such as reproductive activity, metabolic activity, and membrane integrity is an invaluable tool. Reliable flow cytometric methods for viability assessment that circumvent the need for culture were largely developed by Porter et al. Investigations associated with the state of dormancy in bacteria, in which the cells survived for extended periods of time without growth or multiplication, have been performed by Kell et al.

Cationic lipophilic viability dyes should be handled with care in such studies, because they are very often extruded in an energy-dependent manner. There are several indications that a transport system is involved. Excellent correlations can regularly be observed between viability of the cell and the ability to translocate the dye to the cell's surroundings. A wide variety of bacteria are able to metabolize xenobiotica as sources of carbon and energy for growth and to decontaminate polluted ecosystems in this way. However, the effects these toxic substances have on the artificially introduced bacterial species should be considered in such biotechnologically forced processes. Above certain, critical concentrations the chemicals become toxic, the bacteria are poisoned, the bioremediation process slows down and finally stops. It is well known that most of such substances are membrane active and burden the cell energetically. Consequently, cellular changes in the energetic state and the form ofthe membrane potential must be analyzed with the solute transport, ATP synthesis, and pH homeostasis.

To control and optimize such processes, information is necessary about death characteristics, which depend on the physiological state of the cells. Flow cytometric assessment of the membrane potential allows a rapid and differentiated analysis of the ecotoxic potential of pollutants toward bacterial cells and detection of the rapidly changing physiological behavior of the cell in growth phase-dependent variations of the membrane potential. Very little information exists regarding the members of bacterial communities with respect to morphology and taxonomy that are responsible for the decontamination process. There is also little information about differences in the physiological state of specific strains under certain conditions. For this reason, methods must be developed that relate specific cellular parameters to taxonomic classes. The first such investigations of this type were done by Herrmann et al.

In medical science, situations also arise in which living and dead cells must be distinguished. In this way, flow cytometry can be used in the assessment of biocidal drugs, offering rapid assays for prokaryote drug susceptibility. The aim is to use the technique in the clinic for rapid susceptibility testing as an aid for choice of therapy. Common methods for investigating resistance to antibiotics and disinfectants and their effects on bacterial strains are based on estimating the minimum growth inhibitory concentration (MIC) of the compound or the decrease in colony-forming units after exposure. These conventional techniques need at least 24 to 48 h to get information about the action of antibiotics.

Flow cytometry is much faster and, furthermore, gives considerable advantages in drug research for investigating the mode of action of novel antibiotics. The susceptibility of bacterial cells to antimicrobial agents may be so profoundly influenced by age, growth, and metabolism to be heterogeneous within a population. The permeability of the bacterial membrane changes during the life cycle, for instance. Recently, it has become apparent that the permeation of lipophilic substances through the envelope of Gram-negative bacteria has great potential in the development of new chemotherapeutic agents. Generally, using flow cytometry, key drug-induced changes in light-scattering behavior, uptake of vitality stains, DNA ploidy, and protein patterns can be estimated.

Population Complexity

Bacterial ecology requires the application of new techniques to help address problems that cannot be solved by traditional methods alone. Efforts to demonstrate the value of flow cytometry in identifying different species within complex populations have relied on immunofluorescence-based methods such as the application of antibodies, lectins, or rRNA-targeted oligonucleotide probes. The methods chosen must be applicable to as wide a range as possible of different organisms. These techniques have also found uses in areas such as quality control of water and foodstuffs as well as in soil and water ecology. Today, fluorescently labeled oligonucleotide probes are most commonly used for monitoring specific strains. The probes are complementary to group-specific regions of the highly conserved multicopy 16S or 23S ribosomal RNA molecules and are bound using in situ hybridization. In this manner, the bacterial strain of interest can be detected without separating it from the surrounding microflora and contaminating debris.

Furthermore, comparative sequencing and the design of nucleic acid hybridization probes have served to establish an alterated phylogenetic framework. Differentiation between artificial cultures is possible if the members have contrasting proportions or absolute amounts of guanine–cytosine or adenine–thymine. Estimations of guanine–cytosine content vary from nearly 20% up to 80%. Using dyes with different binding preferences to DNA and double fluorescence excitation, composition analysis seems to be workable. Another way of differentiating mixed populations is the genetic insertion of a fluorescent system to split populations with special physiological capacities.

Nutrition

The use of fluorescent labels is accepted as a successful and sensitive approach for detecting bacteria in food. Lactic acid bacteria are fermentative Gram-positive organisms that are widely used as starter cultures for manufacturing a great assortment of foods and drinks, such as cheese, yogurt, and sausages. The quality of a starter culture is highly important. It must be taken into account that industrial fermentations may involve stressful conditions that can affect bacterial gene expression, arrest cell multiplication, interfere with metabolic activity and survival potential, and lead, possibly, to cell death. Studying responses of lactic bacteria to stresses such as freezing and drying, with the aim of monitoring, controlling, and promoting appropriate responses in starter cultures is a typical requirement of flow cytometric investigations in food industries. The method enables determination of cell counts, cell viability, and the effects of stress on the bacterial physiological state.

The method can suffer from interference by particulate materials, but multiparameter counting based both on light scattering and fluorescence measurements generally enables separation of bacteria from background material. There are many situations where the total bacterial concentration is the only measurement that is important. Bacteria are a principal cause of food spoilage. Contamination by harmful organisms can cause severe human food-borne diseases, either caused by the bacteria themselves or the toxins released by the bacteria. The time required for conventional tests (plate-count techniques) can lead to substantial delays (of 24 to 72 h), which is a serious disadvantage.

Direct epifluorescent filter enumeration is another commonly used technique that allows microscopic counting of bacteria retained on a filter. Over the past 10 to 15 years, this method has been used extensively for studies on the survival of bacteria in food and for estimating biomass in drinks. However, results suggest that flow cytometry also has significant potential for the detection of pathogenic microorganisms in the food industry. Using this method, pathogenic bacterial cells can be detected by applying fluorescently labeled monoclonal antibodies, for rapidly detecting *Salmonella* or *Listeria*, for example. Accurate detection has been demonstrated down to levels of below 10^4 cells/ mL or even 1 cell/mL after preenrichment.

Cellular Characteristics

The function of bacterial strains in certain human or animal diseases is increasingly being assessed using flow cytometry. For instance, specific cellular characteristics of bacteria involved in skin diseases are being studied by these techniques. Furthermore, flow cytometry offers a rapid method for characterizing anaerobic bacteria present in human fecal suspensions using specific physical and biochemical features. Attempts have been made to detect bacterial toxins via fluorescently labeled antibodies bound to polystyrene microbeads in stool specimens. Moreover, flow cytometric investigations have been used to identify antisera produced against bacteria capable of specifically binding to surfaces ofthe target bacteria. The resurgence of tuberculosis has caused considerable effort to be focused on developing rapid methods for inhibiting growth of the causal organism by a variety of agents. Also, in animal disease investigations, there is growing interest in the capacity of fish-pathogenic bacteria to survive long-term starvation (in a nonculturable mode) in seawater. Knowledge of how long pathogenic bacteria preserve their infective capacity after being released to the environment is clearly of great practical importance.

In conjunction with other biochemical and genetical methods, flow cytometry has been used as a rapid method for testing the effects of shear stress, extreme temperatures, action of chemical compounds, sonication, and electroporation on bacterial vitality. Bioluminescent *E. coli* harboring lux genes have been used for detecting environmental pollutants, for example. Furthermore, fluorescence-activated cell sorting is often the best available method for obtaining purified samples from natural environments in order to apply molecular techniques or for extracting or enriching samples for detecting low numbers of specific bacteria, such as pathogenic or genetically engineered microorganisms.

The function of bacterial cell surfaces has been investigated in various ways, such as through the interaction of bacterial cells with electrically charged surfaces, macrophages, or other cells through electrostatic interaction. The investigated phenomena may be mediated by bacterial polysaccharides, which are habitually associated with the outer surface ofthe bacterium. These polysaccharides form a unique class of polyelectrolytes, possessing antigenic properties, for which a lot of fluorescent monoclonal or polyclonal antibodies are available.

SACCHAROMYCES CEREVISIAE

Of all the biotechnologically important microorganisms, the yeast *Saccharomyces cerevisiae* has been the most extensively investigated using flow cytometry to date. As baker's yeast and producer (along with certain related species) of beer and wine, this yeast is of great economic significance. Nevertheless, the chief reasons for the exceptional interest for flow cytometry in this yeast is that it is a widely accepted model organism for eukaryotic cells, because of its ease of cultivation and its well-developed genetics. Investigations of other yeast genera have been very rare and have not been done systematically.

Cyclic Behaviour

Basic research into yeast growth and cell cycling by flow cytometry began in the late 1970s with one-dimensional analysis of cell protein content (by FITC-staining) and cell DNA content (using mithramycin staining) by Gilbert et al., Slater, and Hutter. Both common batch-growth phenomena and the typical diauxic growth pattern of *S. cerevisiae* were investigated. These authors elucidated basic rules concerning the timing of DNA synthesis in cells growing at different rates. The first experimental evidence supporting the hypothesis that the eukaryotic G0/G1 phase is the most variable phase of the cell cycle, and that the duration ofthe S phase is nearly constant and is not greatly influenced by the cell's environmental conditions, was generated by these flow cytometrically monitored experiments. In the 1980s, Alberghina et al. and others introduced two-dimensional approaches, involving a light-scattering signal and a fluorescence signal.

In particular, methods for the determination of fluorescently stained single-cell protein were added by Scheper et al. Furthermore, sharper detection of the DNA content and, therefore, of the cell cycle state distribution was also made possible by the introduction of more specific staining methods such as propidium iodide (excitation at 488 nm) and DAPI (excitation in the near UV at 351 nm). The theory that DNA replication of *S. cerevisiae* is initiated when a cell has reached a certain critical size was proved experimentally by these investigations. Asymmetric cell division, subcellular characteristics, and numerous features of the intracellular distribution of cellular components were also shown to be detectable by the application of slit-scanning data acquisition. A double flow cytometric tag allows the progress of the cell cycle to be tracked in the different cohorts of mother and newborn daughter cells during balanced exponential growth.

Physiological State

Substantial progress also came from bioprocess type investigations. Because *S. cerevisiae* populations tend toward autosynchronization under certain process conditions in a continuously stirred tank reactor (CSTR), the changing distribution of states in the cell cycle can be determined by flow cytometry and correlated with the physiological performance of the population. It was demonstrated in this way that the type and rate of substrate utilization and cell compartment synthesis are dependent on the stage of the cell proliferation cycle. In a key paper from 1992, Munch et al. proved on the basis of a comprehensive use of flow cytometric methods that understanding cell cycle behavior is essential for the characterization of growth dynamics in bioprocesses. Using Calcofluor White M2R, the chromosomal and the mitochondrial DNA content in the yeast cells could be separately monitored, and the aging process in the mother cells of the budding yeast could be recorded. Changing respiration coefficients, combined with changes in substrate fluxes and cell processes, such as the storage and mobilization of carbohydrates and the excretion of ethanol, can now be correlated with different stages in the cell cycle. A further important cell component, with cell-cycle sparking and bulk membrane functions (3-β-hydroxysterol), became recordable by staining with FITC covalently coupled to the amino group containing polyenemacrolide antibiotic, Nystatin A1.

Role in Fermentation

Demands of the brewing industry for ever-faster production, without prejudicing the quality of the product, have made flow cytometry more and more significant in this classical branch of biotechnology. The use of cylindrical cone-shaped brewing reactors with volumes of at least 160 m^2 requires increasingly abundant information concerning yeast physiology. Close monitoring of changes in the distribution of DNA, neutral lipid, and 3-β-hydroxysterol contents in *Saccharomyces* cells during propagation, fermentation, and storage enables time-saving process control. DNA distribution monitoring has shown that only cells with a single chromosome content produce ethanol and CO_2 efficiently. On the basis of flow cytometrical investigations, the optimum point for controlling the wort and oxygen supply can be accurately determined. Double-staining techniques prove that cells with a high content of neutral lipids and 3-β-hydroxysterols have a high survival capacity. Only cells with such features are able to generate vigorous subsequent fermentation. However, additional measurements for determining the proportions of living cells or detecting bacterial infections are only sporadically performed in the brewing industry by flow cytometry.

Cellular Characteristics

Wittrup and Bailey use flow cytometry to monitor the β-galactosidase activity in *S. cerevisiae* by measuring the fluorescence decay of the fluorogenic substrate resorufin-β-D-galactopyranoside. With this method, the expression of foreign protein, coupled to the β-galactosidase gene, could be measured, and a segregated model mapping β-galactosidase activity as a function of plasmid copy number was presented. Another method for detecting this protein was a quantitative immunofluorescence procedure developed by Eitzman et al. These authors demonstrated that the activity of the enzyme is correlated with the stage in the cell cycle that individual cells have reached.

Peroxisomes are inducible organelles that may occupy large proportions ofthe cell volume when yeasts are growing on methanol-containing media. *Hansenula polymorpha,* a methylotrophic yeast, was investigated via side-scattered light (which was found to depend on cell volume, morphology, and structure) and FITC retention (which was largely dependent on the vacuole.) A wild strain and a strain partially repressed by glucose were compared. Furthermore, it has been shown that the viability of the pathogenic *Candida* yeasts can be determined by measuring the membrane potential and that of viable blastospores using tetrabromofluorescein.

CELLS OF OTHER ORGANISMS

Cells of Aquatic Organisms

In aquatic science, flow cytometry can make an important contribution to ecological and physiological studies of natural populations of microbial plankton. It has been increasingly used to analyze natural communities of aquatic microorganisms because of its sensitivity and quantification capacity. To understand the processes that influence the behavior of any planktonic species, information is required on both the variability between individuals and groups and the major environmental factors. Flow cytometry was introduced in freshwater and marine sciences at the beginning of the 1980s. However, natural samples of aquatic particles are generally very heterogeneous; there are often immense differences in concentrations and cell sizes (from 0.2 to more than 20,000 μm in diameter) as well as in cell types. Additionally, they may be contaminated by inorganic particles, and artifacts may be generated that can make the data difficult to interpret. However, the application of flow cytometry to planktology is no longer hampered by shortcomings of commercial instruments.

High-resolution and high-speed sorters are available for rapid and precise estimation of cell numbers, differentiation of cell types, and other pertinent problems. Neural network analysis provides an innovative approach for examining taxonomic properties and the state of the marine food web, including its involvement in the surrounding environment. The prokaryotic fraction of natural marine communities is composed of both heterotrophic and autotrophic organisms. In spite of their very small sizes (0.2 to 2 μm), even the oceanic picoplanktonic cells are now easy to identify. Major groups of photosynthetic prokaryotes can be discriminated by flow cytometry because of their different and, compared to other algae, unusual pigment compositions and their forward light-scatter signal. The heterotrophic organisms are mostly bacteria. Their lack of pigments makes them less easy than autotrophs to analyze by fluorescence methods. Nevertheless, flow cytometry has been used to evaluate kinetic constants of nutrient uptake in relation to cell quantity and, additionally, to characterize individual parameters of the marine bacteria.

Frequently, total counts of these bacteria include a large fraction of non-nucleoid-containing bacteria (ghosts), which may be cell residues of virus-lysed bacteria or remains of protozoan grazing. Nevertheless, simultaneously measured DNA content allows further discrimination between autotrophic and heterotrophic components of the picoplankton. In some cases, it has been demonstrated that the cell cycle of autotrophic planktonic prokaryotes progresses in phase with the daily light cycle, in contrast to cyanobacterial populations, which apparently do not adjust their circadian timekeeping. Eukaryotic algae exhibit a strong natural autofluorescence emanating from chlorophyll or other pigments such as phycobiliproteins. The simultaneous measurement of DNA and chlorophyll fluorescence, together with cell size, is necessary to distinguish subpopulations of phytoplankton.

Flow cytometry allows the quantification of changes in cellular pigmentation (e.g., chlorophyll and phycoerythrin concentrations) of divergent phytoplankton subpopulations. Such alterations arise in response to nonsteady light or nutrient regimes, such as may be encountered along a depth profile. Information on the physiological status or response induced by different, artificial treatments can be easily assessed by flow cytometry. Biomass production in phytoplankton and its dependence on environmental conditions can be detected via measurement of the activities of various enzymes such as esterases (FDA) or nitrate reductase (by immunolabeling). The recent development of taxon-specific probes should increase the applicability of flow cytometry for rapid identification of cultured phytopico- and nanoplanktonic strains, especially those that lack taxonomically useful morphological features. Moreover, rRNA probes may provide new information on species that are not amenable to culture in the laboratory.

Furthermore, contemporary research has shown that protozoa are important in the dynamics of the aquatic system. Grazing by heterotrophic protozoa has now been accepted as the most important factor responsible for limiting the numbers of bacterial cells in aquatic ecosystems and maintaining nutrient availability. There is evidence suggesting that bioremediation processes involving bacteria may be strongly inhibited in the presence of phagotrophic protozoa, for instance. Sometimes, microbial population growth rates are difficult to ascertain because ubiquitious grazers remove cells as quickly as they are produced. These taxonomically heterogeneous organisms (including ciliates, amoebae, and a variety of flagellates) can graze voraciously on phytoplanktonic and bacterial cells that form the foundation of the marine food web.

Discrimination has been achieved between prey cells and a dinoflagellate through DNA staining and isolation of the eukaryotic nuclei. Furthermore, differentiation of phagocytosic activity within populations of protozoa is measurable using fluorescently labeled microbeads. Some investigations have been performed on viruses, regarded as ubiquitious, and biologically active members of marine and freshwater microbial communities, in which the viruses maybe responsible for the destruction of a considerable proportion of the bacterial and cyanobacterial populations. In some cases, virus and host support each other in a symbiotic manner, in contrast to the observed annihilation of alternative host populations within the microbial community.

Plant Cells

Application of flow cytometry and cell sorting procedures to study higher plant cells required methods for the production of single-cell suspensions. The need to probe the performance of protoplasts was the main motivation for developing methods for analyzing plant cells. This also required a series of modifications to the flow cytometric instrumentation, because plant cells were sometimes found to be much larger than those of other living organisms. Flow cytometry has been accepted as a rapid and trustworthy technique for genome analysis; for obtaining information about the size, ploidy, and aneuploidy changes during plant evolution; and for the differentiation and state of replication of plant cells. It provides an accurate method for determining the proportions of cells in the G1, S, and G2/M stages of the cell cycle and for calculating cell cycle times. Consequently, cell-cycle-dependent events such as the expression of distinct proteins can be observed in synchronized cell cultures. For example, the expression of a *Nicotiana tabacum* cyclin, a homologue to the *cdc2* gene, disappears in the G1 and S phases of the cell cycle, but is present in the G2/M phase.

Flow karyotyping from mitotic cells and chromosome sorting provide opportunities for gene mapping and the construction of chromosome-enriched DNA libraries. Thus, the analysis and sorting of plant chromosomes is of considerable economic interest. Also, the occurrence and extent of polyploidization, which often accompanies differentiation of plant cells, is important for understanding the regulation of gene expression in differentiated tissues. Furthermore, in situ hybridization methods allow accurate quantification of copy numbers of repetitive DNA sequences on different chromosomes within a species. Examination of intra- and interspecific variation of DNA content can be important in plant hybridization and genetic manipulation programs. Therefore, the method is widely used in molecular cytogenetic research. For example, it can be used to search for foreign chromosomes added to a species, as in the introduction of resistance genes into wheat from the germplasm of rye. Moreover, one of the major problems of using wheat-rye addition lines is their genetic instability.

The absence of the easily lost rye chromosomes must be detected, but counting the chromosomes by methods other than flow cytometry is tiresome and time consuming. For some properties it is necessary to obtain data concerning distinct compartments within a plant cell, and a fundamental goal is appraisal of the physiological behavior of plant cell protoplasts. For this, there are a number of flow cytometric methods for assessing different intracellular components, such as reserve metabolites as well as enzyme activities. Changes in the cytosolic pH using common pH-specific probes have also been shown using these methods. Indeed, it has been proved that analysis of the quantity and quality of cellular fluorescence can yield important information concerning tissue status and derivation.

Furthermore, plant flow cytometry can be a powerful tool in the examination of symbiotic systems. For instance, *Azolla filiculoides* harbors phototrophic, aerobically grown cyanobacteria in vegetative cells and nitrogen-fixing anaerobic cyanobacteria in heterocysts. Some of these differentiated tissues have been investigated regarding the internal distribution of the blue-green algae and the capacities of the various tissues for nitrogen fixation using DNA estimation. Recently, the green fluorescent protein (GFP) from the jellyfish *Aequora victoria* has been used as a powerful new vital reporter in diverse plant cells.

In higher plant protoplast studies, it can be used to monitor gene expression, signal transduction, co-transfection, transformation, protein trafficking and localization, protein–protein interactions, and cell separation and purification. In plant cell systems, the bright green fluorescence of GFP is visible even in the presence of the red chlorophyll autofluorescence. GFP is very stable in plant cells and shows little photobleaching. Viable cells, marked with GFP, can be recovered after fluorescence-activated cell sorting.

Animal Cells

Flow cytometry is widely exploited in the study of spermatogenesis and routinely applied to estimate viability. It is also used to determine the chromosome content and sex of the offspring. It is a valuable and adaptable tool for research into the genesis of germ cells in both normal and perturbed situations, and it can be used in the control of reproduction, domestic animal breeding, and wildlife conservation. Furthermore, cytogenetic investigation of eukaryotic organisms is important for understanding evolutionary processes and essential for testing evolutionary models. Improved knowledge of the morphology of germ cells, and the motility and fertility of spermatozoa enhance the efficiency of pre-insemination technology. Also, functional heterogeneity can be easily detected, as in the assessment of mitochondrial lesions, which are connected to subfertility.

The abundance of live cells, altered sperm morphology, and chromatin structure and the resistance of membrane-bound enzymes and membrane permeability to environmental stress factors can also be calculated from flow cytometric data. In contrast to other methods, the productivity of the flow cytometric separation of the X or Y chromosome-containing germ cells (using highly efficient staining and sorting methods) yields highly purified samples that can be used for sex control of domestic animals. More accurate still is slit-scan technology, which allows the whole X or Y germ cell to be spatially visualized along the direction of flow. This is important, because the intensity of fluorescence detected is affected by the flat ovoid head shape and the compactness of the chromatin. Thus, the orientation of the germ cell during passage through the laser beam, detectable by slit-scan flow cytometry, is critical for a high efficiency sorting procedure. Another range of applications in flow cytometry lies in improving the control of antibody production. Generally, hybridoma cell lines serve as sufficient carriers.

Optimizing the productivity of hybridoma cell lines depends partly on developing suitable methods for screening and selecting highly productive cultures and on understanding the regulation of antibody production. Surface-antibody immuno-fluorescence can be used to select high yielding cells by sorting and to gauge the efficacy of the whole population by cytometric cell analysis. Furthermore, information can be obtained about the stability of expression, the yield, and the integrity of the antibodies obtained. Antigen production can be improved using cytostatically growth-arrested cells, because they do not need to consume cellular resources for biomass production. Besides the hybridoma lines, other cell lines (often from insects) have been systematically developed for somatic cell genetics, expression of transfected genes, and synthesis of hormone- inducible proteins. Toward this end, flow cytometric-based studies on animal cell systems can be used for developing convenient and widely applicable procedures for gene cloning.

There are, for instance, protocols for DNA transfection and subsequent selection of rare transfectants by sorting. For such investigations, the green fluorescent protein (GFP) may be inserted. Several new GFP variants have been generated that are brighter or have altered excitation spectra, facilitating the monitoring of the marked cell lines. Viruses have a particularly important biotechnological role in the expression of both foreign prokaryotic and eukaryotic genes in host cells. However, only a few virus-cell systems have been analyzed by flow cytometry until now, though the method is competent for broad applications in detecting and quantifying virus-infected cells. Viruses can be assayed in animal cell cultures by in situ hybridization using fluorescently marked oligonucleotides and quantitatively estimated by flow cytometry. Characterization of the accessible virus host systems offers information about their capability for providing adequate quantities of virus outgrowth. The process of productive and nonproductive virus infection in animal cell lines can also be investigated, for example, by in vivo analysis of promoter activity regarding transcription of the virus gene. The staging of the virus can be estimated using a virus reporter gene, whose product is, for instance, immunofluorescently stainable. Another possibility is to insert an insect luciferase gene, encoding a protein with characteristic light-emission properties that are measurable by flow cytometry.

High sensitivity and rapidity of the reaction (in combination with the luminescence, which does not naturally occur in most ofthe cellular systems used) makes the method valuable for flow cytometric estimation of virus production. It is also possible to get information about the reproductive capabilities of the virus-infected and uninfected host systems. Differences can be analyzed between highly susceptible and less susceptible cells. Furthermore, virus-induced DNA synthesis may be monitored by assisted determination of BrdUrd incorporation, using monoclonal antibodies. Another biological field where the prospects offered by flow cytometry seem extremely encouraging, besides clinical diagnostics, is pharmacology. Before the application of a drug is allowed for human use, the mode of action at cellular and subcellular levels must be clear.

The experimental conditions must be as close as possible to physiological drug–cell interactions within living target cells. Model systems have been developed to study the operation of curative drug systems. On-line flow cytometry is the most sensitive and fastest method available, and it causes the least possible harm to living cells. Additionally, the method is appropriate for investigating the mode of action of other biologically active substances. In phytopathological sciences, for instance, insect cell lines are very suitable target systems for studying the cytotoxic potential of microbe-produced cytostatic agents.

2

BIOANALYTICAL TOOLS

In recent years, various bioanalytical systems for improved bioprocess monitoring have been developed and investigated. The application of these systems in pilot facilities and production plants can provide the information needed to better understand all types of bioprocesses, including cultivations, biotransformations, and downstream operations. In addition, better documentation of the processes, especially important for the production of pharmaceuticals, can be obtained. However, most of these sensors are invasive, and thus there is the chance of impacting the cultivation process (e.g., contamination) and causing complications in the processes. For this reason, more and more research groups are involved in the development and application of noninvasive bioanalytical tools, which can be interfaced to the bioprocess without causing problems such as contamination or changes in cell metabolism.

Optical sensors offer tremendous advantages for bioprocess monitoring. They can be interfaced noninvasively to the bioprocess via glass fiberoptics. The information can be transported via fiberoptics without interferences caused by electromagnetic fields over long distances, and signals can be obtained instantaneously, because spectro-optical systems have extremely short response times. Fluorescence sensors have been investigated intensively during the past 15 years for different applications in biotechnology, especially biomass estimation, reactor characterization, metabolic studies, and general bioprocess monitoring. These so-called fluorosensors were developed mainly for monitoring of the reduced form of nicotine adenine dinucleotide (NADH) and its phosphorylated form (NADPH). This coenzyme is present in all living cells and is involved in different metabolic reactions; its level is extremely sensitive to environmental conditions.

MONITORING NADH AND NADPH

The NAD(P)H pool gives information about the metabolic state of the cells. In 1957, Duysens and Amesz first showed that relative NAD(P)H measurements could be performed in living cells using spectrofluorometric techniques. UV light of 340 to 360 nm was used to excite fluorescence of NAD(P)H in cells in a suspension. The NAD(P)H fluorescence was monitored at 460 nm while metabolic experiments were performed. The first spectrometer device for biotechnological application was designed by Harrison and Chance in 1970. It could be used in glass bioreactors. The excitation light was guided through a quartz window into the bioreactor, and the fluorescence was collected at an angle of 60° through a second quartz window. The penetration depth of the light was only a few centimeters at low cell densities. The sensitivity decreased as the cell density increased. However, it was possible to monitor aerobic and anaerobic transitions in *Klebsiella aerogenes*. Zabriskie and Humphrey used this technique in 1978 for biomass estimation in different cultivation experiments (e.g., *Saccharomyces cerevisiae, Streptomyces* spp., and *Thermoactinomyces* spp.). Based on these experiments, different smaller miniaturized process fluorometers were developed that could be interfaced to a bioreactor via a standard port. This port is used to guide the excitation light into the bioreactor and to collect the backward (180°) fluorescence.

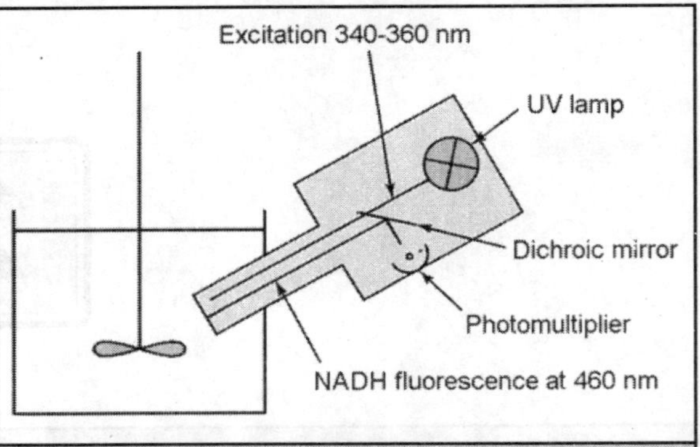

Figure 2.1 Principle setup of miniaturized spectrofluorometer.

In 1981, Beyeler et al. developed a miniaturized fluorosensor in an open-end detection mode. This design has the advantage of eliminating the inner filter effect by monitoring the surface fluorescence at the observation window. All these sensors shared two disadvantages. First, they were limited to detection of NAD(P)H fluorescence, it was only possible to excite fluorophores in the range of 340 to 360 nm and to detect the fluorescence intensities at 450 to 460 nm. And they also lacked the ability to distinguish between different fluorophores that might interfere in this region. In order to obtain more information, a flexible two-channel fluorometer was developed by Scheper and Schugerl. In this device, the light of the UV lamp was focused out the optical filters and lenses onto a quartz fiber bundle. This fiberoptic was used to guide the light into the bioreactor. The backward fluorescence light was collected via the same fiber bundle and split into two bundles, passing interference filters that selected for different wavelength bands. Thus, it was possible to monitor simultaneously different fluorescence wavelengths.

SIGNIFICANCE

Single-wavelength fluorosensor devices have been used to monitor different cultivation processes, including suspended as well as immobilized organisms. Investigations in which the NAD(P)H fluorescence of these cultivations was monitored had one of two goals: on-line determination of biomass concentration or on-line detection of metabolic changes. A significant advantage of using fluorescence for biomass concentration monitoring is that only living cells are detected. To use culture fluorescence as an indicator for biomass concentration, it is necessary that the culture fluorescence per cell is constant during the cultivation process. In this case, a linear correlation between fluorescence and the biomass concentration could be expected. However, the intracellular NAD(P)H pool changes in a cell during the batch experiment. Thus, it is necessary to linearize the corresponding biomass concentration and fluorescence data to obtain a calibration plot. One problem that must be overcome is that all process conditions affecting the culture fluorescence intensity cause problems with the biomass determination.

Table 2.1 Application of fluorescence monitoring for different cultivation processes and organisms

Suspended	
	Saccharomyces cerevisiae
	Penicillium chrysogenum
	Cephalosporium acremonium
	Escherichia coli (incl. recombinant)
	Zymomonas mobilis
	Bacillus licheniformis
	Clostridium acetobutolyticum
	Spodoptera frugiperda
	Baby hamster kidney cells
	Wastewater treatment
Immobilized	
	Cephalosporium acremonium
	Saccharomyces cerevisiae
	Zymomonas mobilis

The two most common interferences are gas bubbles and the presence of compounds with fluorescence spectra that overlap that of NAD(P)H. Interestingly, fouling of observation windows was not observed. Many investigators have reported efficient biomass concentration determination via culture fluorescence monitoring. Examples include cultivations of *Methylomonas mucosa, Zymomonas moblis* in synthetic medium, and *Pseudomonas putida*. In all these applications, suspended cell cultivations were studied, and it was shown that on-line biomass estimation is also possible in technical media under certain process conditions.

A wide range of metabolic changes have also been studied using NAD(P)H culture fluorescence. For example, Ristroph et al. studied the cultivation of *Candida utilis* under different conditions and developed a feeding strategy for fed-batch cultivation of this organism on the basis of NAD(P)H-dependent culture fluorescence. Einsele reported on different applications of culture fluorescence monitoring and determined the mixing-time behavior of different bioreactors and the substrate uptake rate of different microorganisms during cultivation.

Reardon et al. studied immobilized *Clostridium acebutylium* in a fixed-bed reactor. The cells were immobilized in calcium alginate beads, and the cultivation medium was circulated through the reactor. The growth and productivity of the cells could be monitored under different cultivation conditions. Scheper et al. reported on the application of this technique for the monitoring of immobilized yeast cells.

Table 2.2 Different phenomena studied via fluorescence monitoring

Phenomen studied	*Organism*
Aerobic–anaerobic transition	*Klebsiella aerogenes*
	Saccharomyces cerevisiae
	Candida tropicalis
	Escherichia coli
	C. guilliermondii
Aeration rate	*Penicillium chrysogenum*
Addition of carbon source to starved cells	*Saccharomyces cerevisiae*
	Candida tropicalis
	Escherichia coli
Diauxic growth	*Saccharomyces cerevisiae*
Dilution rate changes	*Saccharomyces cerevisiae*
	Escherichia coli
	Pseudomonas putida
Glycolytic oscillation	*Saccharomyces cerevisiae*
Culture synchrony	*Saccharomyces cerevisiae*
Metabolic shifts	*Thermoactinomyces* sp.
	Clostridium acetobutylicum
Addition of metabolic uncouplers	*Saccharomyces cerevisiae*

FURTHER APPROACHES

Several approaches have been taken in the past 8 years to circumvent the restrictions of single-wavelength NAD(P)H-dependent culture fluorescence monitoring. In 1991, Li et al. reported on a special fluorosensor with five different wavelengths for the monitoring of *Saccharomyces cerevisiae* and *Candida utilis* cultivations. They showed that other wavelengths could be used for biomass concentration estimation and for metabolic studies. Horvath et al. used a fluorescence spectrometer coupled to a bioreactor via quartz glass fibers for the determination of *S. cerevisiae* biomass concentrations. Based on his results, the tryptophan fluorescence was better correlated to the cell mass concentration than was the NAD(P)H fluorescence. A more recent development was described by Chenina in 1993 for the cultivation of yeasts (*Cyathus striatus, Eurotium cristatum,* and *Crinipellis stipitaria*). A special fluorescence spectrometer was developed that can also be interfaced to a bioreactor via light guide cables. Monochromators were used to monitor the fluorescence of wide excitation and emission ranges.

FLUORESCENCE SPECTROMETRE

These fluorescence spectrometers can be used to produce a two-dimensional (2-D) fluorescence spectrum, which represents the emission spectra at different excitation wavelengths. Such a spectrum is described by the three parameters of excitation wavelength, emission wavelength, and fluorescence intensity, which can be shown as an excitation-emission-data matrix. In this matrix, each row represents an emission scan and each column an excitation scan. Different compounds can be identified by their position in the matrix. The most illustrative way to show a 2-D fluorescence spectrum is to use an isometric plot, but contour plots are the most useful because no fluorescence peaks are hidden. Dark shadings in the fluorescence plots indicate high fluorescence intensities.

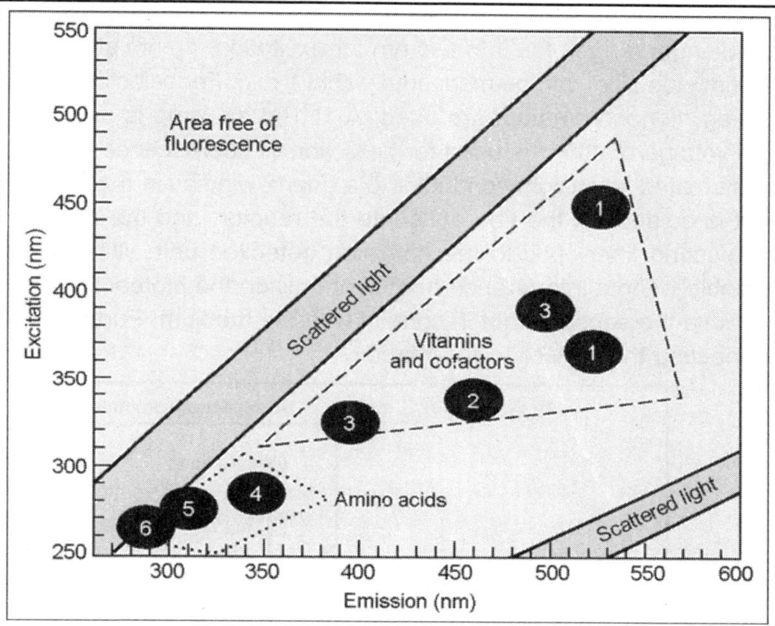

Figure 2.2 Biogenic fluorophores in 2-D fluorescence spectrum. 1=ribovlavin, FAD, FMN; 2=NAD(P)H; 3=pyridoxamine, pyridoxal-5'-phosphate; 4=tryptophan; 5=tyrosine; 6=phenylalanine.

BIOGENIC FLUOROPHORS

The spectral ranges of important biogenic fluorophores, which occur inside the cell and can be detected simultaneously by 2-D fluorescence spectroscopy. Almost all biological processes depend on proteins, which fluoresce because of aromatic amino acids such as tryptophan, tyrosine, and phenylalanine. In another region of the 2-D fluorescence spectrum (λ_{Ex} = 310-480 nm, λ_{Em} = 350-550 nm), vitamins (e.g., pyridoxine [vitamin B-6]), riboflavin [vitamin B-2]), and coenzymes (e.g., NADH, NADPH, FMN, FAD) can be found.

CONTROL OF BIOTECHNOLOGICAL PROCESSES

For on-line monitoring and control of biotechnological processes using 2-D fluorescence, the instrument must be capable of measuring a complete 2-D fluorescence spectrum in a short time (about 1 min), and the time between two measurements should not be longer than 5 to 15 min. This is to ensure that a complete 2-D fluorescence spectrum represents the fluorescence at a certain time and that changes during a cultivation can be monitored in real time.

HITACHI F4500 SPECTROFLUOROMETER

For 2-D fluorescence measurements, fluorescence spectrometers for use in laboratories can be used. One suitable apparatus is the fluorescence spectrophotometer model F4500. This spectrometer allows a very fast scan speed (30,000 nm/min) and an entire 2-D spectrum with a wavelength range of 250 to 550 nm for excitation (10-nm step size) and 260 to 600 nm for emission (20-nm step size) can be measured within 1 min. For selection of excitation and emission wavelength, grating monochromators are used. A 150-W Xe lamp is used for excitation, and a photomultiplier at a voltage of 700 V is used for detection of fluorescence. The spectrometer is connected by a 1-m bifurcated liquid light conductor to a quartz window in a 25-mm port of a bioreactor. The excitation light is guided via the fiberoptic into the reactor, and the backward fluorescent light is guided via the second fiber-optic to the emission detection unit. With this setup, in situ measurements are possible without interference from light outside the bioreactor, and there is no risk of contamination because the sensor is not in contact with the medium. Furthermore, sterilization is possible without connecting the sensor to the reactor.

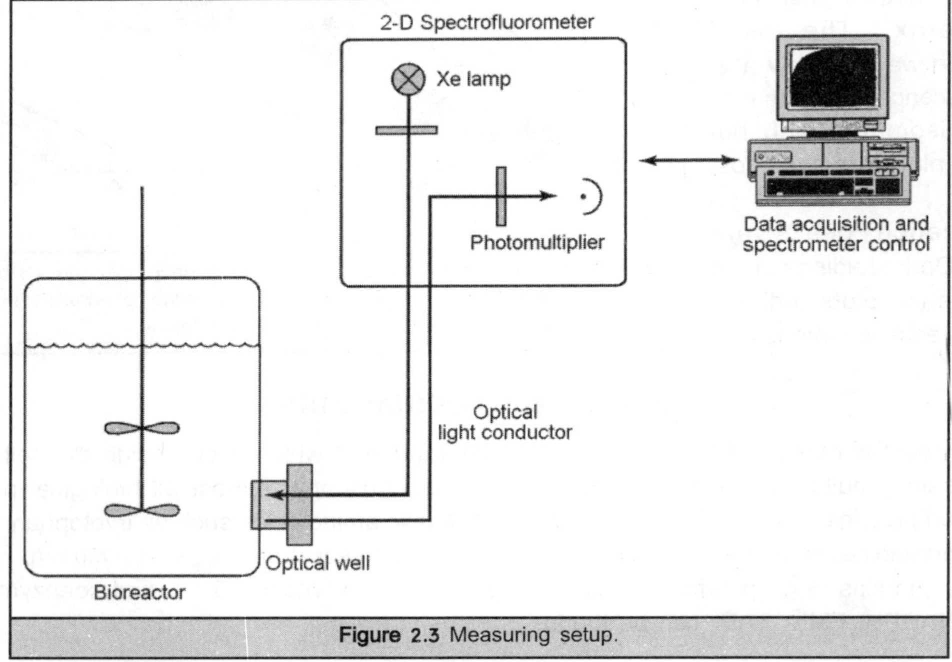

Figure 2.3 Measuring setup.

BIOVIEW

The BioView sensor is a fluorescence sensor optimized for fully automated industrial measurements and applications to monitoring and control of bioprocesses. It is very robust and suitable for harsh environments (e.g., high temperature or moisture) and electromagnetic interference. For data transfer, a single-fiber modem is used, which allows a distance between sensor and computer up to several hundred meters. The optical parts of the BioView are optimized for use in biotechnological processes in that interference filters, rather than spectrometer gratings, are used by the BioView sensor for selection of excitation and emission wavelengths. Filter wheels (one for excitation and another for emission) with 16 filters each are used. These filters can be designed and chosen individually. Usually, measurements in steps of 20 nm are performed in a wavelength range of 270 to 550 nm for excitation and 290 to 590 nm for emission. For the measurement of a complete 2-D fluorescence spectrum, the excitation filter wheel is set to the first filter position. With this excitation filter, the fluorescence spectrum is monitored via the emission filters that switch from filter to filter until a cycle is completed. Afterward, the excitation filter switches to the next position and the next fluorescence spectrum is measured. This procedure is continued until the excitation filter wheel has performed a complete cycle. The complete measurement takes about 1 min, depending on the number of different filter positions and measurements for each filter combination, which can be chosen individually. The connection to the bioreactor is the same as that described for the Hitachi instrument, except that a 2-m long liquid light conductor is used.

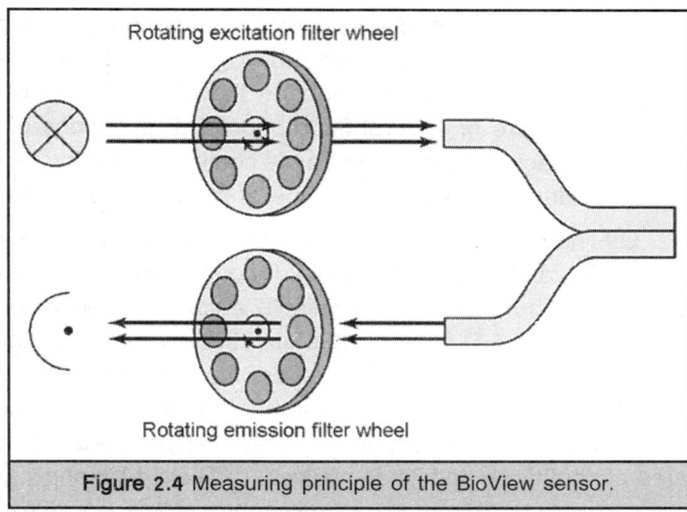

Figure 2.4 Measuring principle of the BioView sensor.

USES

Two-dimensional fluorescence spectroscopy offers the possibility of simultaneous analysis of several fluorophores. The location ofthe peak maximum in a 2-D spectrum characterizes each fluorophore, and the relative fluorescence intensity correlates with its concentration. Despite this conceptual simplicity, the interpretation of 2-D fluorescence spectra is not trivial. Interactions ofthe fluorophores with one another and with other nonfluorescent compounds, bubbles, and other physical influences such as pH, temperature, and viscosity, quenching effect, and changes in morphology of the cells all cause problems in the interpretation of the 2-D spectrum. To make the system user friendly, automated interpretation by artificial intelligence was developed for 2-D fluorescence spectra. The intelligent BioView sensor is able to predict the course of a cultivation if it is taught with a training cultivation under comparable cultivation conditions.

Modern chemometric systems like principle component analysis (PCA) or neural networks are able to consider all influences on the fluorescence signal and lead to a sensor with high capacity, able to measure various cultivation variables. The changes in fluorescence intensities during cultivation processes are shown in difference spectra, which are obtained by subtraction of a 2-D spectrum taken during the cultivation from the one in the beginning. Through comparison to the spectrum of a mixture of pure biogenic fluorophores, such as riboflavin, NADH, pyridoxine, pyridoxamine, pyridoxal-5′-phosphate, and tryptophan, the peaks could be assigned to the expected fluorophores. All changes have a biological origin and offer information about the cell growth and metabolism. Even small changes in the concentration of a single fluorophore can be seen in a complex mixture of several fluorophores. The regions of changes are regarded as the areas of interest. Measurements made nearly continuously allow us to construct the course of various culture variables during a bioprocess. One of the most important is the biomass concentration.

Off-line analysis of counting cells, measuring optical density or weighting dry mass are time consuming and fail if filamenteous fungi are used. On-line tools such as a turbidimeter often cause problems. The search for a reliable automated technique for on-line monitoring of biomass still continues. In a 2-D fluorescence spectrum, various excitation–emission regions can be correlated to biomass concentration. This offers the advantage to change to another region if one cannot be used because of overlapping or other influences. Fluorescence of proteins (because of their high quantum yield, especially in the case of tryptophan), pyridoxine, the area of NAD(P)H, and the area of riboflavine and its derivatives (FAD or FMN) show good correlations with biomass for several organisms (e.g., *Acremonium crysogenum, Claviceps purpurea, Sphingomonas yanoikuyae, Bacillus licheniformis*, different strains of *Escherichia coli, Enterobacter aerogenes, Saccharomyces cerevisiae, Schizosaccharomyces pombe,* mammalian cells or *Tetrahymena thermophila.* For the filamenteous fungi, *Clavicepspurpurea* fluorescence intensity in the region of riboflavin and its derivatives (λ_{Ex} = 450 nm, λ_{Em} = 530 nm) allow on-line monitoring of biomass concentration, which is not possible by optical density or cell count because ofthe mycelial growth. Scattered light is also measured in a 2-D fluorescence spectrum.

These data correlate well with data derived from conventional turbidimetric sensors. Along with the ability to estimate biomass concentration, 2-D fluorescence sensors such as the BioView system offer on-line information about intracellular variables without interference into the process. This is achieved by monitoring changes in the excitation–emission regions associated with the redox-sensitive coenzymes NADH, FAD, and FMN. This approach can be used to study aerobic–anaerobic transitions, the tricarboxylic cycle, effects of uncoupling agents such as 2,4-dinitrophenol (DNP) on oxidative phosphorylation, diauxies, the effects of pulse additions of substrate, and oscillations of a culture. Fluorescent products can be observed by 2-D fluorescence. The productivity of ergot alkaloids could be estimated by 2-D fluorescence spectroscopy. Degradation of fluorescent substrates such as polycyclic aromatic hydrocarbons can be observed. Other extracellular or physical variables such as the microenvironment ofthe cells are captured by 2-D fluorescence. Two-dimensional fluorescence spectroscopy is ideal for a wide variety of organisms and installation uses and can therefore be seen as a multianalyzer that is capable of replacing several other sensors in the future.

3

ANIMAL CELL IN BIOPROCESSING

Animal cell lines are a resource in biotechnological processes. They are used increasingly in molecular biology developments, fermentation technology, the production of diverse health care products, and cell-based screening systems in toxicology and pharmacology. Large-scale technology began with the use of primary monkey kidney cells for the production of poliomyelitis vaccines in the 1950s. After intense debates on safety issues, human diploid cell strains (e.g., WI-38) were accepted for the production of mumps, measles, and rubella vaccines in the 1960s. The next step was taken in 1964 with the commercial production of foot-and-mouth disease virus (FMDV) from baby hamster kidney (BHK 21) cells for veterinary purposes.

After the discovery of interferons and their clinical importance, the growing demand for these compounds led to the search for alternative, easily accessible sources. Interferon was the first licensed product derived from Namalwa cells, a human lymphoblastoid heteroploid cell line, in the late 1970s. Soon after this, hybridoma and monoclonal antibodies (mAb) took over the fast-developing market with their diverse applications in diagnostic and therapy. Although strict safety standards have to be met, the value of animal cell lines is undeniable, and they are indispensable for the industrial production of eukaryotic proteins.

Since the acceptance of cell lines for the production of vaccines, cytokines, mAb, and therapeutic proteins or even the cells as product themselves, constant progress in a broad range of areas has contributed to the improvement of products. Research fields having major impact are analysis and adaptation of suitable cell lines, high-level expression systems, genetically engineered cell lines and products (e.g., protein modifications), scale-up methodology, bioreactor development, and downstream techniques (e.g., purification, quality control). This chapter presents an overview of cell lines used for various applications and assays and gives an insight into different technologies and products. The purification of therapeutic products from natural sources was and still is restricted because of limited supply, and contamination of known or unknown origin presents a serious danger (e.g., HIV in blood products). In addition, bacterial systems are not capable of the post- translational modifications eukaryotic proteins undergo. Animal cell technology has been able to fill this gap and provide proteins of high quality on an industrial scale.

TYPES OF ANIMAL CELLS

Ovary of *Cricetulus griseus*

The anchorage-dependent cell line from Chinese hamster ovary (CHO) was derived from a biopsy of an ovary of an adult Chinese hamster (*Cricetulus griseus*) in 1957. Various derivatives were developed, including nutritionally deficient mutants and transfected cells. The most popular variants are CHO-K1, a proline-requiring mutant, and CHO/dhFr_, which is deficient in dihydrofolate reductase and therefore useful for the selection of transfected cells. Other variants have been adapted to grow in suspension and in serum-free media. Advantages of CHO are their high capacity of amplification and expression of recombinant genes and proteins, the ability to synthesize proteins with similar glycosylation patterns as observed in native proteins, and the ability to grow in large-scale bioreactors. Therefore, they are very suitable for the expression of heterologous proteins by recombinant DNA technology.

Because CHO cells are the most important substrate for the production of therapeutic proteins, new products are developed and licensed continuously. Product status, either in research or as industrial process, is indicated in the tables under Application. Generally, CHO cells are kept in attached culture on tissue culture material or on microcarriers, but they can readily be adapted to grow as single cells or aggregates in suspension. This allows the usage of highly efficient bioreactors such as fluidized bed reactors. Because aggregates can lead to non-homogenous product formation and affect system monitoring, single-cell suspension cultures are often preferred. To achieve this status, media formulations have been developed maintaining single-cell suspension during the production process. Cell lines that have been adapted to grow in suspension do not maintain all characteristics seen in attached cultivation. In a study by Brand et al., it was shown that specific production rates of cells in attached cultures are not necessarily connected to similar rates after adaptation to suspension. This observation illustrates the difficulties of scale-up procedures.

Serum-containing media impede the purification of pharmaceutical products to a great extent, and therefore efforts are directed to establish cultures in serum-reduced, serum-free, or even protein-free media. To facilitate the scale-up of transfected cells, CHO cells preadapted to serum-free and suspension culture have been developed by Sinacore et al. Other aspects important for efficient production are the study and regulation of inhibitory substances such as ammonia that can exert an influence on the glycosylation pattern. A reproducible product consistency is of course a major aim in all animal cell systems used for the manufacture of proteins to be administered as therapeutic agents. The effect of genetic and environmental factors on purity and posttranslational modifications are discussed in "Post-translational Modifications of Recombinant Proteins in Animal Cell Lines."

Kidney Cells of *Mesocricetus auratus*

The attached BHK 21 cell line was derived from five unsexed 1-day-old Syrian hamster (*Mesocricetus auratus*) kidneys in 1961. BHK 21 clone 13 was obtained after single-cell isolation and is now widely used, very often only addressed as BHK or BHK 21. Several variants have been developed that are being adapted to suspension culture or suitable for genetic engineering techniques. Because of their susceptibility to various viruses, they have been extensively used for the production of vaccines. Since the 1960s, BHK cells have played the major role in the production of FMDV vaccine in industrial-scale fermenters (with a capacity of several thousand liters) for veterinary applications.

The experience and know-how gained with this manufacturing system led to the increased use of animal cells and the improvement of process parameters. Like CHO, BHK cells show spontaneous aggregation that can be advantageous for cell retention and recycling systems. Efforts have also been directed toward high-cell-density culture and serum- or protein-free media formulations. The first product approved by the Food and Drug Administration (FDA) from BHK cells cultured in perfusion reactors was the recombinant human factor VIII for the therapy of coagulation disorders. Since then, many other proteins have been produced on laboratory- and industrial-scales using this versatile cell line, ranging from human erythropoietin and antithrombin III to enzymes and a range of vaccines for veterinary or human health applications.

Antibody Secreting Cells

Since Kohler and Milstein discovered the technique to produce immortalized, antibody-secreting cells—hybridomas—the development in this area has had a major impact on the production and application of the secreted proteins (mAb). Therefore, it is not surprising that this tool occupies an important and still growing place in industrial biotechnology because the range of applications is constantly expanding. The advantages of hybridomas and myelomas (i.e., their natural secretory ability and, as lymphocytes, growth in suspension) renders these cell lines ideal for exploitation in biotechnical processes. To optimize the production of mAb, efforts have been undertaken to develop efficient and economic media, including serum- and protein-free formulations, generating high cell densities by using perfusion reactors instead of batch processes, and selection of suitable clones with higher resistance against shear stress in agitated bioreactors.

In addition to the ready-to-use serum-free media that are already commercially available, cultivation of hybridomas is constantly improved by research activities related to specific needs. Examples are cholesterol-free media or the use of genetically engineered cell lines that are glutamine-independent, which reduces the accumulation of inhibitory substances such as ammonia. The ability to produce recombinant mAb contributed to the development of chimeric mAb containing several therapeutic advantages. It is now possible to construct heterohybridoma cell lines where different species can contribute to the variant (e.g., mouse) and constant (e.g., human) region of the secreted mAb. These antibodies are less immunogenic and more effective in therapeutical applications.

The constant region can also be replaced with enzymes or toxins, which is useful in tumor targeting. Bispecific mAb can react with two antigens (e.g., tumor markers and specific toxic drugs) and thereby bind and destroy specific cells only. Myeloma cell lines such as J558 L, NS0, SP2/0, or P3X63Ag.653 are not only malignant fusion partners used for the generation of hybridomas, but are also suitable for the production of recombinant proteins. Once again, it can be attributed to the tremendous new techniques of genetic engineering that opened the door for myeloma-based expression systems leading to a new branch in animal cell technology. These naturally secretory cells can be grown to high cell densities and are robust and relatively easy to transfect. With the use of recombinant myeloma cell lines, raising of antigen-specific B cells as prerequisite for the production of hybridomas and finally mAb can be avoided, and time- as well as cost-consuming optimization procedures can be limited.

Insect Cells

In addition to mammalian cells, insect cell lines for the production of recombinant proteins have been established in the past decade. The techniques used most frequently can be distinguished by the application of different expression systems: the baculovirus expression vector system (BEVS) and the *Drosophila* metallothionein (Mt) promotor system.

Baculovirus and insect cells

Baculoviruses replicate exclusively in invertebrates and are often highly species-specific. The BEVS most commonly used for the expression of heterologous proteins is the *Autographa californica* (alfalfa looper) nuclear polyhedrosis virus (AcNPV), where the gene of choice is inserted into the baculovirus genome under the control of a strong late viral transcription promoter. The recombinant virus is then used to infect lepidopteran insect cells, mainly from *Spodoptera frugiperda* (fall armyworm). These cells express the recombinant protein during the late stages of the infection process. Several human, animal, and plant proteins have been expressed using this system. The insect cell line used most frequently in combination with BEVS is SF 9; the parental cell line, SF 21, was obtained from pupal ovarian tissue of *S. frugiperda*. SF 9 cells are highly susceptible to infection with various baculoviruses.

A main advantage of the baculovirus/insect cell system is its potential to yield milligram quantities of recombinant product. To obtain mammalian cell lines producing reasonable amounts of heterologous proteins, it is not only necessary to invest valuable time on transfection, selection, and optimization, but the resulting cell line might also show a basic constitutive expression activity and is constantly under selective pressure.

Cell of Drosophila metallothionein

In the *Drosophila* Mt. system, the inducible Mt. promotor is used to control the expression of recombinant proteins. This system offers the benefit of a well-known host and genomic structure with a tightly regulated promotor. In contrast to the viral system BEVS, where the production peak is found when the cells already die from the viral infection, *Drosophila* cells and the Mt. expression system can be used continuously. Despite the different expression systems for insect cell lines, development of efficient fermentation techniques and process optimization are similar to the ones already described for mammalian cell types.

High-cell-density reactors to achieve a maximum yield of desired product have been developed by various groups. Cavegn et al. established a system monitoring the production of insulin-receptor tyrosine kinase domain and the soluble part of endothelial leukocyte adhesion molecule as model proteins and achieved up to 3×10^7 cells/mL in a serum-free perfusion bioreactor. Other developments are immobilized systems, such as cultivation on porous microcarrier, or the use of simulated microgravity to reduce hydrodynamic forces, as described by O'Connor et al. applying the rotating-wall vessel. Investigations at the laboratory level indicate interesting areas still to be exploited for biotechnological purposes. The high level of expression of measles virus protein in SF 9 facilitates research into structure, function, and immunogenicity.

Expression of recombinant bovine β-lactoglobulin for the processing of milk proteins to inhibit allergic reactions is another potential application. Because of the ability to express two or more proteins simultaneously in the baculovirus system, analysis of multi-subunit formation (e.g., for bluetongue virus-like particles) is possible. Research activities are also concentrating on other insect host cells to improve the production process and, as an important requirement for clinically administered compounds, obtain a correctly glycosylated and biological active protein. SF 9 cells display only a very simple glycosylation capacity, whereas *Estigmena acrea* cells are able to produce complex glycoforms very similar to their authentic counterparts. Other insect cell lines investigated for the manufacture of recombinant proteins are Schneider 2 (*Drosophila melanogaster*) and Tr 5 (*Trichoplusiani*). Apart from using insect cell lines to facilitate production, they can also be applied in screening or cytotoxicity tests.

Other Types of Cells

Although the cell lines described in the previous section probably contribute to the majority of industrial processes, various other cell types are used as substrate for the production of biologicals. On the one hand, this might be of historical reason. Certain cell lines are well established; the production of vaccines and systems for manufacture have been optimized already. On the other hand, cell lines synthesizing proteins naturally without transformation procedures have clear advantages considering the production process (where a selective pressure to maintain expression is hence not necessary) and acceptance of the final product. Other cell lines have been shown to possess valuable properties such as glycosylation characteristics. This section describes some of these cell lines and their biotechnological usage; although not great in number, they are still significant in the pharmaceutical industry. VERO cells, derived from the kidney of a normal adult African green monkey, are susceptible to a wide range of viruses, including measles, polioviruses, and rubella. These properties led to the early establishment of vaccine production in VERO cells. This now rather classical cell line is still used as capable substrate for virus production.

VERO cells replaced primary monkey kidney cells for the production of oral polio vaccine and are now cultured on microcarriers in large-scale bioreactors. They were also successfully applied for the study of infection with Aujeszky virus in the VERAX system, a perfused fluidized bed with macroporous collagen carriers as microspheres to evaluate the possibility of vaccine production in a high-cell-density system. Baijot et al. established the manufacture of an Aujeszky vaccine using swine testicular cells (NLST) on microcarriers. Although VERO cells are mostly grown as attached cells, they can be adapted to grow in suspension in serum-free medium. The inducible expression of HIV-1 viruslike particles in transfected VERO and COS 7 cells (a SV 40 transformed monkey kidney cell line) was described by Haynes et al. This important development offers an exciting alternative to the use of live virus vectors for the production and evaluation of AIDS vaccines based on noninfectious particles. The human lymphoblastoid cell line Namalwa has been used as substrate for biologicals since the 1970s. The secretion of INF-α being naturally induced in Namalwa cells by treatment with sodium butyrate and addition of sendai virus is well established for industrial manufacturing.

Process optimization and increase of productivity are still continuing and achieved by various modifications, such as treatment with tetramethyl urea or sequential induction with sendai virus. Namalwa KJM 1 was derived from Namalwa and adapted to grow in serum-free medium. Miyaji et al. describe the efficient expression of recombinant human INF-β in these cells in high-cell-density perfusion culture. The comparison of recombinant pro-urokinase expression in Namalwa KJM 1 and CHO cells revealed a partly cleaved and thereby inactive form secreted in CHO cells due to proteases present in the supernatant. Because the protein secreted by Namalwa KJM 1 showed a single-chain form, this indicates that these cells are more suitable for the production of protease susceptible heterologous proteins.

The carcinoembryonic antigen (CEA) is a high molecular weight glycoprotein identified as tumor-associated marker. For the purpose of early clinical diagnosis, therapy response, and recurrences, it is important to be able to detect CEA in monitored patients. Several cell lines synthesize CEA and have been used in various large-scale production systems. mAb raised against this antigen can improve clinical detection assays and cancer specificity. Other examples of cell lines used for the production of naturally secreted proteins are HUH 6 (fibronectin) and RPMI 8866 (CD 23). A list of all cell lines currently used in biotechnology is clearly beyond the scope of this article. The constant improvements in this field, such as transformation of cell types without losing specific properties or differentiation characteristics (e.g., hepatocytes, neurons, keratinocytes) and optimization of cell lines for fermentation processes, will certainly lead to a variety of options for animal cell biotechnologists.

MICROCARRIER TECHNIQUES

Animal cell lines have to be classified according to their ability to grow in suspension or be anchorage-dependent. For the large-scale production of biologicals, suspension cells are clearly the substrate of choice. With the development of microcarriers initiated by van Wezel 1967, it is now possible to culture attached cells in bioreactors very similar to suspension cells. Growth of animal cells on microcarriers or immobilized systems facilitates production processes and allows high yield of cells, heterologous proteins, or viruses. Microcarrier techniques are especially important for the handling of primary cell types or cell strains because these are, with only a few exceptions, mostly anchorage-dependent. The microcarrier technology offers several advantages: providing a high surface-to-volume ratio; allowing manipulation during culture without interrupting the fermentation process, and facilitating scale-up possibilities.

A further advantage is the stabilization of cells in culture, which applies to both attached and suspension cells, thereby permitting prolonged culture periods. Macroporous microcarriers have proved to be suitable for the protection of cells from shear forces and have been applied successfully for the immobilization of adherent and suspension cells. The ability of some cell lines to form aggregates is another feature used for increased productivity in perfusion bioreactors. However, aggregation has to be evaluated in a case-to-case study because it can result in inhomogeneous expression of proteins and might impair monitoring. Finally, three-dimensional culture conditions can simulate the natural environment of cells, including cell-cell interaction and differentiation processes that are not always possible to achieve in two-dimensional systems. An example is the establishment of a bioreactor that provides high, almost tissue-like cell densities and thereby simulates cell-cell interactions.

RECOMBINANT DNA TECHNOLOGY

Recombinant DNA technology opened the field for the production of an unlimited variety of clinically relevant therapeutics in various host cell lines. The majority of proteins applied undergo more or less extensive posttranslational modifications in their natural counterparts, including mainly glycosylation, but also phosporylation, carboxylation, and signal-peptide processing. These modifications affect the biological activity of the heterologous protein by being involved in solubility, pharmacokinetics, antigeneicity, circulating half-life, secretion, protein folding, oligomer assembly, and susceptibility to proteolytic attack. Choosing a host cell line able to fulfill these criteria is, therefore, important. Regulatory authorities such as the FDA demand a comprehensive analysis of carbohydrate structures and consistency in the production process. Because prokaryotes are not capable of performing posttranslational modifications observed in eukaryotic cells, they cannot be utilized for the production of recombinant proteins, where bioactivity relies heavily on correct glycosylation, for example.

These unique advantages of animal cell technology inspired research directed toward overcoming process or regulatory difficulties. Research has increasingly revealed already substantial differences in the glycosylation of recombinant proteins using several host cell lines or even transgenic animals. Variations can also stem from different environmental factors, such as media or bioreactor configuration, or the internal cell status at the time of production. A few examples illustrate the importance of choosing a host system able to produce heterologous proteins with appropriate characteristics for clinical applications. Although CHO and BHK lack certain functional glycosyl-transferases found in human cells, they still represent the most favorable host cell lines for large-scale manufacture of protein-based pharmaceuticals such as erythropoietin or tissue-type plasminogen activator (t-PA). Studies on product consistency revealed an independence of glycosylation pattern by applying varied process parameters and showed that glycosylation was identical to natural sources.

Constant improvement in the methods available for the analysis of carbohydrate structures allow a detailed pattern identification and have been used to detect species-specific oligosaccharide variations of recombinant human IFN-γ produced in CHO and SF cells and transgenic mice. The glycosylation status of recombinant proteins synthesized in baculovirus-infected insect cells can vary quite considerably, as shown by Ogonah et al. *Estigmena* cells are capable of producing complex oligosaccharides, whereas the still more frequently used SF 9 cell line is restricted to the performance of only simple glycosylation reactions. These findings demonstrate the importance of investigating and identifying an animal cell system best suited for the production of therapeutics.

Even shortcomings, such as the lack of glycosyl-transferases in CHO cells, can be overcome by transfection of appropriate enzymes, indicating the potential for future processes.

BIOPRODUCT TECHNOLOGY

Several examples given in the preceding sections already point toward activities undertaken to optimize animal cell technology and bio-products of interest. Cell lines other than the well-established hamster lines CHO and BHK present valuable alternatives, for example, the efficient production of recombinant proteins. Concomitant with an increasing understanding of media formulations, bioreactor design, and processes, efficiency and quality are still subject to improvement. Tailored cell lines can be selected by appropriate cloning strategies, using either naturally occurring variants or conferring specific characteristics by recombinant DNA methods. The development of vectors with a predictable integration site for foreign DNA into the host genomic structure, including efficient expression systems, can overcome low-productivity problems. An example for enhanced productivity is the oncogene-activated system applied by Ternya et al., where cells are co-transfected with the *ras* oncogene. Recombinant DNA technology gives new directions for the design of therapeutics with improved or even novel characteristics compared to their natural counterparts (e.g., pharmacokinetic properties). Ahern et al. investigated recombinant variants of tPA produced by site- directed mutagenesis and identified proteins with increased fibrinolytic activity in vitro and decreased clearance in vivo. New technologies, such as the encapsulation of cells, genetically engineered or as primary isolates, and transplantation into patients with various disorders or illnesses, could provide a cellular therapy tool for long-term treatment or even cure. Other approaches attempt to simulate the three-dimensional microenvironment for constructing organotypical systems, such as artificial liver support devices or drug metabolism studies. Investigations into the involvement of cytokines, extracellular matrixes, and cell-cell interactions are required to optimize organ models and lead to successful applications.

CONCLUDING REMARKS

Animal cell lines are an established tool in biotechnology and will be used extensively in a variety of emerging areas, such as tissue engineering, organ replacement, or drug discovery. A vast amount of knowledge and experience has accumulated concerning the cell lines described in this article, and research into improved industrial large-scale use is still on going. Safety concerns using continuous cell lines with a tumorigenic potential are addressed frequently. As a result of numerous efforts made in response to recommendations by regulatory bodies and the World Health Organization, cell lines and derived products are now a prosperous factor in biotechnology. Selecting a cell line for the manufacture of pharmaceuticals is dependent on the protein that has to be produced. Posttranslational modifications, secretion, and purification are aspects that have to be considered, but species (e.g., human or rodent origin) and regulatory issues are equally important. The examples described in this article give some indications on cell line suitability, but research as well as industrial production is improving rapidly in this area. New sources (i.e., cell lines and methods) are being established continuously and should therefore be closely studied before choosing an appropriate cell substrate.

BIOPROCESSING OF FOOD

Foods and beverages are processed to increase their safety and shelf life, to favorably modify their sensory and nutritional quality and composition, and to improve ease and reliability of preparation by consumers. Foods must be protected from contamination and stored and transported safely and without adversely affecting quality. Food process engineers employ engineering principles and methods to identify, select, design, develop, and "debug" processes, equipment, systems, and plants used to accomplish these tasks. Food process engineers also supervise the start-up and operation of food processing systems and plants; devise related control, maintenance, and sanitation systems; devise ways to eliminate hazards and minimize waste; and help develop new food products and ways to upgrade existing processes.

In carrying out these tasks, engineers must account for the biological activity and chemical instability of foods and for the ability of microorganisms and pests to invade and grow in or on foods. Many different types of food products are produced. The processes used are extremely diverse and too numerous to cover in detail. Therefore, this article examines only the engineering aspects of a limited number of processes or operations. Many of these are used to preserve foods. Others are used to separate food components, and still others are used to change the shape, size, and texture of foods. Food processing frequently involves heat transfer. This article briefly examines such transfer. Individual operations may produce effects that fall into several processing categories. Thus drying, which separates water from other food ingredients, is used mainly to preserve foods. Drying often changes the texture of foods and involves transfer of heat from hot air to moist solids. It also involves transfer of moisture (mass) within solids and from solids to air and transfers of momentum, which affects airflow.

TEMPERATURE IN FOOD PROCESSING

Food process engineers are frequently asked to model heat transfer, devise ways to control or improve such transfer, and calculate heating and cooling loads and sizes of equipment needed to provide desired amounts of heating or cooling. These tasks are accomplished by means of methods widely employed by other engineers. However, in using these methods, food process engineers must account for the physical, heat transfer, and sensory properties of foods, and for their temperature sensitivity. Food components rapidly decompose in undesirable ways at high temperatures. The highest product temperature reached during food processing is probably 250°C, the end-of-roast temperature for very darkly roasted coffee. Maximum tolerable temperatures for many foods are much lower.

Higher processing media temperatures maybe used (e.g., gas temperatures ranging up to 500°C are used in roasting coffee). For economic reasons, foods are rarely cooled to temperatures lower than –40 °C. Foods usually are either solid or complex composites (e.g., moist multicellular tissue, porous materials, stiff doughs) that act like solids. Heat transfer in such foods is modeled by means of partial differential equations based on Fourier's laws of conduction. Modeling also requires specification of boundary conditions based on surface temperatures, surface heat transfer coefficients, or surface rates of heat transfer.

Solutions to partial differential equations for conductive heat transfer and many boundary conditions of practical interest can be found in the work of Carslaw and Jaeger and other standard texts. Similar solutions for partial differential equations and boundary conditions describing diffusive mass transfer equations are provided by Crank. Many pairs of heat transfer and mass transfer equations and boundary conditions are formally equivalent. Therefore equation solutions for diffusive mass transfer often can be readily converted into solutions that apply for conductive heat transfer and vice versa. Heat transfer calculations for solid foods may be complicated by a need to account for heats generated by ongoing biological processes, surface evaporation, anisotropic conduction, and internal vaporization and condensation. Often boundary conditions are spatially nonuniform, or they may vary with time during food processing.

The thermal conductivities of foods may be difficult to predict. Nevertheless, most problems relating to heat transfer in solid foods can usually be solved fairly easily by means of computer-based numerical methods. Food engineers also deal with heat transfer that is due to natural convection in liquid-filled containers, in pore spaces in stacks of heat-generating, respiring produce, and in the air that surrounds solid foods discharged from heating devices. In addition to widely encountered problems involving forced convection of heat in heat exchangers, evaporators, and condensers, they deal with convective heat exchange between streams of air or water and solid foods lined up one behind another in rows on shelves.

These problems are often solved empirically. Food engineers working on aseptic preservation and ohmic heating processes are now attempting to precisely predict heat transfer to individual solids in groups of solids suspended in heated, flowing fluid. Food engineers also deal with prediction of temperature changes during microwave and dielectric heating. In doing so they must account for attenuation and focusing of energy as electromagnetic radiation passes through foods, refraction and reflection of such radiation at food surfaces, energy distribution patterns in food-containing microwave cavities, and temperature-induced changes in the energy absorption characteristics of foods.

PRESERVATION TECHNOLOGY

Drying, chilling, freezing, heating-induced inactivation of microorganisms and enzymes, application of bactericides, storage in controlled atmospheres, solute-induced water activity or pH reduction, and combinations of some of these processes are used to preserve foods.

Drying

Drying, the evaporative removal of water from moist solids or solidifiable solutions, may well be the earliest method of preserving foods and still serves that purpose today. It is also used to harden foods that have been shaped in a moist state, to reduce food weight, and volume, and to reduce moisture contents to facilitate processing. Drying preserves foods because biological activity (e.g., microbial growth) radically slows or stops and chemical reactants become immobile when little water is present. Fresh fruits and vegetables usually contain 5-10 kg water/kg dry solids; stable, dried vegetables, 0.03-0.1 kg water/kg dry solids; and stable, sugar-rich dried fruits, 0.20-0.39 kg water/kg dry solids. The water content of freshly harvested grains is up to 0.22-0.43 kg water/kg dry solids; to provide safe storage, this must be reduced to 0.11-0.15 kg water/kg dry solids.

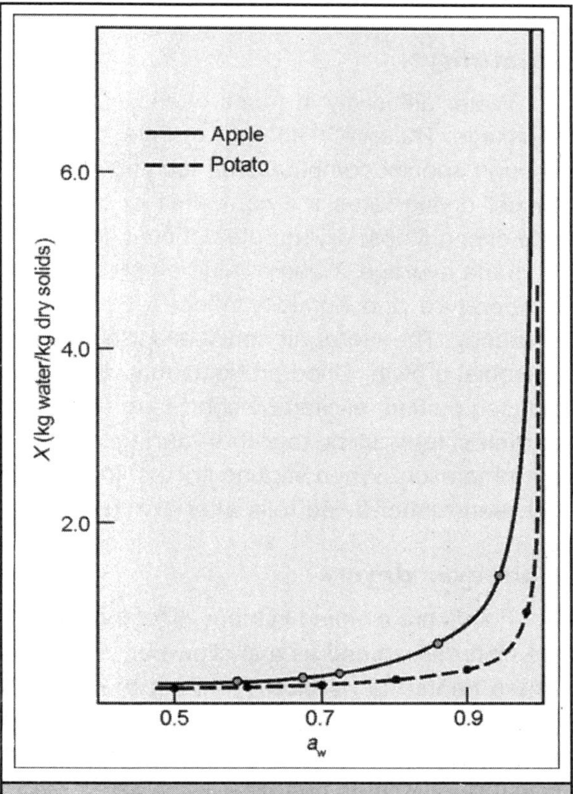

Figure 4.1 Moisture sorption isotherms for apples and potatoes.

Hot Air Drying

Food drying most frequently is carried out by transferring heat to moist food particles from flowing hot air. This causes food moisture to evaporate and move away with the air. Consequently, the air moisture content (humidity) rises, and the air temperature drops proportionately. Relative humidity (RH) is the partial pressure of water in air divided by the vapor pressure of pure water at the same temperature. Water activity (a_w) is a measure of the chemical activity of water. It is also the equilibrium RH for a food at a given moisture content.

Food moisture sorption isotherms, are constant-temperature plots of a_w versus a food's dry-basis moisture content, X. Moisture sorption isotherms shift downward moderately as temperature rises. The minimum X obtainable by drying is X_E, the sorption isotherm X at which a_w at the air temperature equals RH. After start-up transients have decayed, food temperatures during drying depend on a_w at the food's surface. As long as the surface a_w is greater than 0.9, the food temperature will be no more than 1°C higher than the air wet-bulb temperature; if air conditions are almost constant, the drying rate will be almost constant. Constant-rate drying does not last very long for foods. Surface a_w depends on X_S, X at the surface. As drying removes water from the surface, water diffusing to the surface from the food's interior partially replaces it. Because water diffuses slowly in foods, X_S very rapidly drops enough to drive a_w below 0.9. After this, a_w drops sharply as X_S decreases, the food temperature rises and progressively approaches the air temperature, and the drying rate progressively decreases.

Drying rates are strongly influenced by heat transfer and airflow rates as long as the air temperature is significantly higher than the food temperature. As food temperature approaches the air temperature, drying is more and more controlled by water diffusion in the food. Near the end of drying, diffusion completely controls drying rates. The air temperature sets the food temperature. Hence, it affects diffusion and drying rates. Enough airflow must be provided to remove evaporated moisture fast enough to prevent X_E from rising excessively. Other than that, airflow no longer affects drying.

Humidity

Water diffusivity in foods often decreases markedly as drying proceeds. Drying also causes shrinkage. Theoretical analysis of water diffusion in drying foods is difficult because of these changes. Though special computational techniques can be used to deal with these complications, in most cases, drying rates are correlated empirically rather than theoretically predicted. After constant-rate drying stops, drying rates at constant air conditions are often proportional to $(X_A - X_E)^n$, where X_A is the average X and n is an empirical constant, which ranges between 0.7 and 2.4. Drying air temperature and humidity affect food quality in addition to affecting drying rates and thermal efficiency. Therefore, air conditions are adjusted to minimize quality loss and to prevent in-process microbial growth. Dried products may not rehydrate well. This can be partially remedied by vapor-induced puffing, elicited by short-term use of high air temperatures late in the drying process. Wet particles may stick together during drying. This property can be used to provide desired agglomeration. When sticking occurs and is unwelcome, particles are broken apart by passing them between rubber-faced rolls after drying.

Conveyer dryers

Foods are air-dried in many different types of equipment. In conveyor dryers, moist food particles are uniformly spread as a shallow bed on a perforated belt and conveyed through a series of stages where heated air is blown through the bed and belt. Different air temperatures, humidities, and flow rates are often used in the stages. To improve uniformity of drying, up-flow is used in some stages and downflow in others. In conveyor-based drying of vegetables, 60-85% of the food's moisture content is removed in the first 1.5-2 h of drying. An additional 2-8 h usually is needed to complete the process. Air temperatures between 66 and 120°C are used in early stages of drying. Lower air temperatures (32-52°C) are used later, at the end of drying. Quite different conditions are used for some foods. Low temperatures are used during first stages of gelatin strip drying to prevent melting. Quick-cooking rice is dried in less than 15 min using temperatures up to 200°C. Temperature around 175°C may be used to puff diced potatoes at late stages of drying. In tunnel dryers, hot air passes over shallow beds of food particles conveyed through a rectangular tunnel on racked trays mounted on carts. Airflows in the direction of cart movement or counter to that direction. Clearances between the carts and the tunnel are small, to minimize by-passing of air. Periodically, a freshly loaded cart is pushed into the tunnel and a cart containing dried material simultaneously leaves.

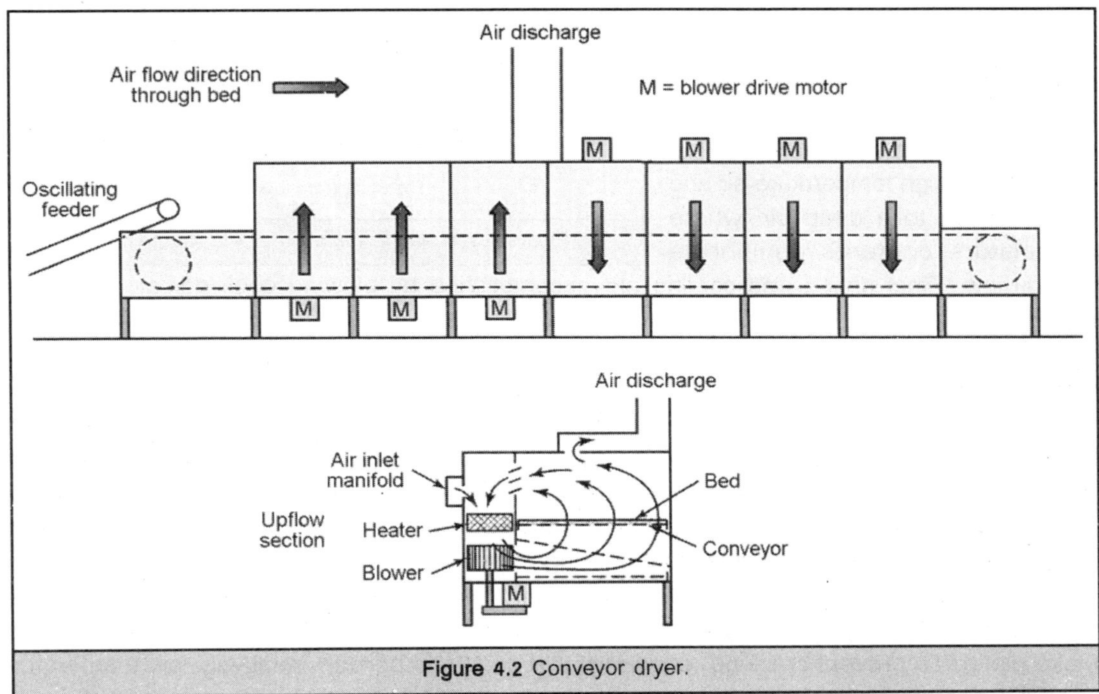

Figure 4.2 Conveyor dryer.

Drying tunnels often are used in series. Cocurrent drying with high inlet-temperature air is followed by countercurrent drying with lower temperature air. Conveyor dryers are used instead of tunnel dryers wherever feasible because tray loading is burdensome and conveyor dryers permit automatic loading and discharging. In addition, drying conditions are more readily controlled in conveyor dryers.

Bin drying

Bin dryers in which air passes through a deep bed of particles may be used to complete the drying of partly dried vegetables discharged from shallow-bed dryers. Grain is often directly dried in deep beds in bins, where air is introduced through a false floor or ducts near the floor. Ambient air may be used if its relative humidity is less than 0.55, or it may be heated to 11°C above ambient. In such dryers, moisture desorption waves and accompanying temperature-change waves move upward through the grain bed. Most of the time, the exit air temperature equals the initial grain temperature and the exit air RH equals the initial grain a_w. Design engineers select airflow rates that provide drying rapid enough to prevent mold growth in moist region: 0.007 m^3 air/(m^3 grain s) may be used for relatively dry feed; 0.08 m^3 air/(m^3 grain s) may be needed for very moist feed. In other grain dryers, air flows across descending beds of grain that are 0.2-0.4 m thick. In these cases, airflow rates of 0.7–2.7 m^3 air/(m^3 grain s) are used. Inlet air temperatures up to 82°C are used for animal feeds. Less than 54°C is used for grain that is to be milled and processed, and less than 45°C is used for seed grain. Grain is cooled with cool, dry air after such drying. Grain may be partly dried with high temperature air and then transferred to a deep bin where local moisture contents equilibrate without airflow. Final drying and cooling are accomplished by blowing cool, dry air through the grain.

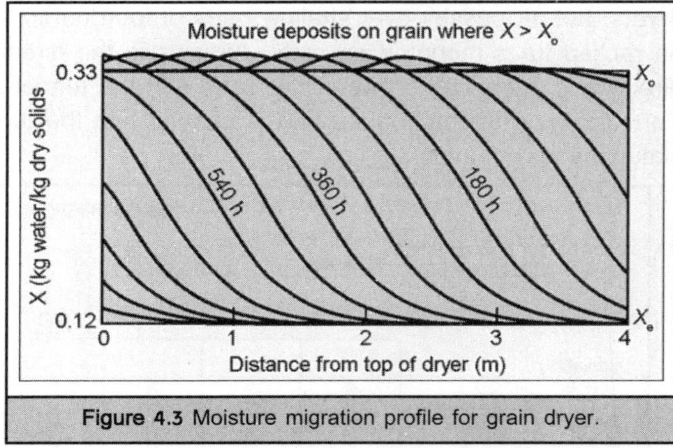

Figure 4.3 Moisture migration profile for grain dryer.

Pasta dryers

Special dryers are used to dry pasta. Freshly extruded pasta contains roughly 0.45 kg water/kg dry solids. That moisture content is reduced roughly 0.13 kg water/kg dry solids to set pasta's shape and preserve it. Long goods (e.g., spaghetti) are draped over moving rods and carried through the dryer system. Short goods (e.g., macaroni) are conveyed through dryers by simpler means. To prevent adjacent pieces of pasta from sticking together, the pieces are pre-dried using 63°C, 65% RH air. They leave the predryer and enter a large, final dryer where very humid air is used. Zones in which active drying occurs are followed by zones in which internal equilibration of moisture can take place. To prevent cracking (checking) due to stress buildup, relatively low temperatures (e.g., 50°C) and very long drying times (up to 18 h) were and often still are used for long goods. Now, based on better modeling of moisture diffusion and stress development, temperatures up to 110°C and drying times of 5 h are frequently used for long goods. Short goods are less susceptible to cracking and can now be dried in 2 h. Use of higher temperatures also reduces microbial growth.

Spray dryers

Spray dryers are used to dry liquid foods and pumpable food slurries. Such foods are converted into drops with mean diameters ranging between 50 and 250 μm by nozzles or rotary atomizers and sprayed into a large chamber. The drops enter at high velocity but soon decelerate. Hot air is blown in and flows in the same direction as the descending drops. Inlet air temperatures as high as 270°C may be used; often they range between 160 and 200°C. Outlet air temperatures usually range between 80 and 110°C. Drop temperatures quickly rise to the wet-bulb temperature of the inlet air and then progressively approach the local air temperature as drying proceeds. Because air cools as it moves through the dryer, cocurrent air-drop flow prevents drops from overheating. Dried particles leave the dryer with temperatures 10-20°C below the exit air temperature. Large particles drop out of the airstream. Smaller particles are collected in cyclones. Drying times range from 5 s for small drops to 50 s for large drops. Spray-drying involves exchanges of heat, momentum, and mass between the air and the drops, nonlinear diffusion of moisture within drops, and changes in drop size and shape. Complex computer programs have been developed to account for these processes. Spray-dried foods are rapidly cooled after drying by sucking cool, dry air into solids discharge lines or by drawing cool, dry air through a screen-supported product bed on a vibrating conveyor.

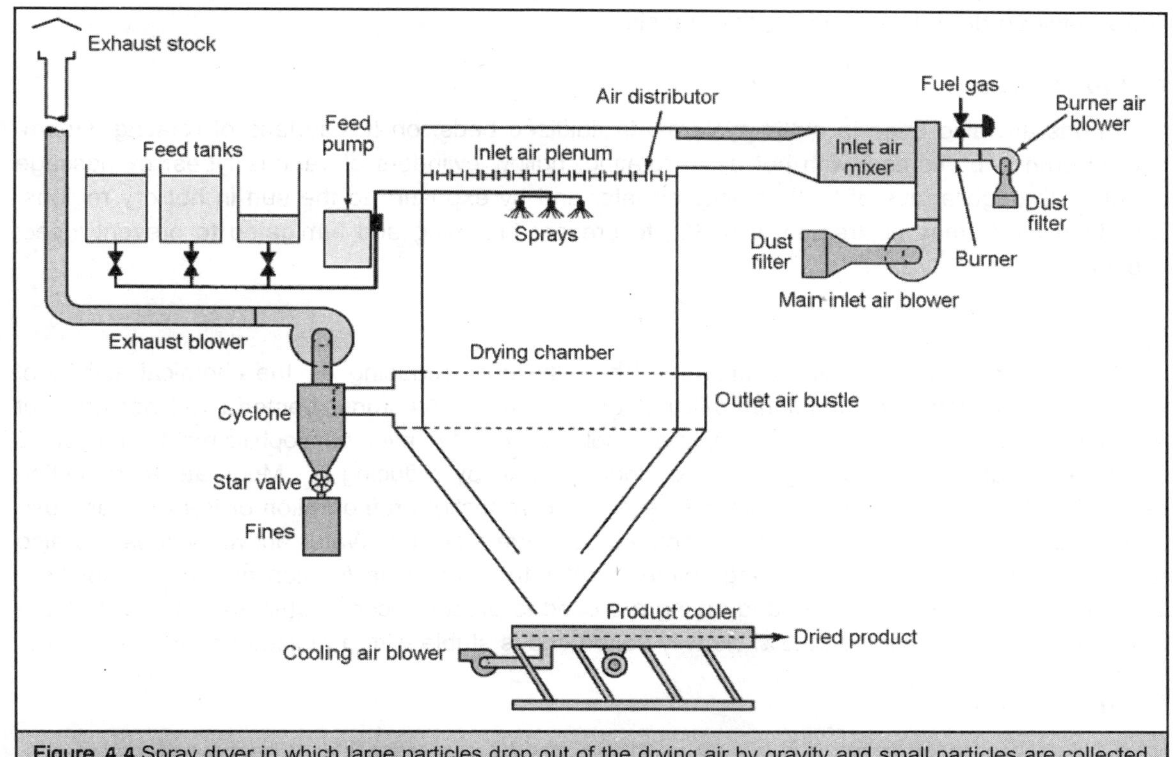

Figure 4.4 Spray dryer in which large particles drop out of the drying air by gravity and small particles are collected in a cyclone.

Freeze-dryers

Freeze-dryers are used to dry beverage extracts, shrimp, soup ingredients, and military and camper rations. In these dryers, moisture sublimes from frozen food in a vacuum chamber and condenses on very cold refrigerated coils. Heat transfers radiantly from heated panels to the surface of the food and then passes by conduction through a porous dry food layer to a sublimation interface. The interface recedes into the food as drying proceeds. Vapor created by sublimation escapes from the food through pores produced by prior sublimation.

Temperatures at the product surface and sublimation interface depend on the radiant panel temperature and on the thermal conductivity, vapor permeability, and thickness of the dry layer. Depending on whether heat conduction or vapor permeation controls the drying rate, radiant panel temperatures are programmed either to prevent product surface overheating or to prevent melting at the sublimation interface. Small, frozen particles are sometimes used to circumvent drying rate retardation caused by low dried-layer permeability. Noncondensibles must be removed before and during freeze-drying. If adequate noncondensibles removal is achieved, chamber pressures in the dryer approach ice's vapor pressure at the condenser temperature. Condenser temperatures of –40°C are frequently used, resulting in absolute pressures around 100 μm Hg. Local chamber pressures in plant-scale freeze-dryers are effected by pressure drops due to high-velocity vapor flows between dryer trays and heating panels.

Other dryers

Foods are also dried in air-lift systems, in fluidized beds, on the surface of rotating, steam-heated drums, by contact with hot air in rotating, hollow cylinders of various types, by passage under or through arrays of high-velocity air jets, and by exposure to the sun in hot dry regions. Sun-dried foods may be treated with SO_2 to prevent browning and fumigated to prevent insect infestation.

a_w Reduction

Microorganism growth can be prevented by sufficiently reducing a_w, the chemical activity of water. Thus, *Clostridium botulinum* will not grow at $a_w < 0.94$, most bacteria will not grow at $a_w < 0.905$, most yeasts at $a_w < 0.88$, and most molds at $a_w < 0.8$; even osmiophilic molds and yeasts will not grow at $a_w < 0.6$. Drying preserves foods in part by reducing a_w. Moderate a_w reduction can be used in combination with other steps (e.g., pH reduction, refrigeration or freezing, and use of antimicrobial agents and fungistats to provide storage stability). Water activity values are also reduced by evaporation or by adding solutes that reduce the mole fraction of water in solution. Different combinations of salting and drying are used to preserve cod. Lightly salted cod is stable when its moisture content is 12.3%; heavily salted cod is stable with a moisture content of 52.4%.

Nonfrozen Storage

Chilled, nonfrozen storage is widely used to extend the storage life of fresh produce, meats, fish, and dairy products. Recommended temperatures and humidities and approximate storage lives for the commercial storage of fruits and vegetables are listed in U.S. government publications and by the American Society of Heating, Refrigeration, and Air-Conditioning Engineers (ASHRAE). These sources also provide heat capacities and heats of respiration for many of the commodities listed. Typical rates of heat influx are also given to permit computation of refrigeration loads for storage facilities. Storage temperatures close to 0°C and humidities ranging between 85 and 95% are frequently recommended; cold-sensitive produce is stored at temperatures between 7 and 13°C. Storage life at these conditions varies widely, ranging from as short as 2 days to as long as 12 months; storage life is less than 1 month for roughly 60% of the foods surveyed. Design and operating considerations for food storage refrigeration systems are treated by ASHRAE. Prechilling before storage is carried out by blowing chilled air on products, by immersing products in chilled water containing chlorine or an approved phenol compound or spraying similar chilled water on them, or by mixing products with ice or slush. Leafy products that conduct heat poorly are often cooled by evaporating a small part of their water content in a vacuum chamber. To minimize weight loss, such products may be prewetted before vacuum cooling.

Storage life

Chilling is combined with use of controlled atmospheres that contain as little as 2-5% O_2 and up to 5% CO_2 to increase the storage life of apples, pears, and cabbage. CO_2-rich controlled atmospheres are also used during long-distance shipment of meat. The rooms or shipping containers used must be sealed and air-tight. Exceeding tolerable CO_2 levels or falling below tolerable O_2 levels damages apples. These levels significantly differ for different varieties of apples. Respiration can be used to reduce O_2 and raise CO_2 levels. Any excess CO_2 produced is absorbed by hydrated lime, ethanolamine (which can be regenerated by heating), water, or less frequently activated charcoal or molecular sieves. To more rapidly adjust O_2 and CO_2 levels, catalytic or noncatalytic combustion of fuel and flushing with nitrogen are used. Multilocale O_2 and CO_2 sensors and loggers have been used to ensure that improper local environments do not develop in controlled-atmosphere storage rooms. The use of selectively permeable wrapping can generate modified atmospheres in packages and improve shelf life during food distribution. Vacuum packing is also used.

Preservation of Moist Food

Freezing, the cooling-induced conversion of most of a food's water content into ice, is used to preserve virtually all types of moist food. It is also used as a step in freeze-concentration and freeze-drying and to produce ice cream and sherbet. Freezing preserves foods by reducing rate constants for chemical reactions, radically reducing reactant mobility, and reducing a_w. The water activity of a frozen food depends solely on its temperature and is independent of the nature of the food involved, as long as it is suitably moist; a_w = 0.908 at –10°C, 0.824 at –20°C, and 0.748 at –30°C. Solid foods are frequently frozen by exposing them to suitably cold streams of refrigerated air. This may be done by passing cold air through a shallow, fluidized bed of small particles or by blowing cold air over larger pieces of food carried on perforated conveyors or mounted on trays in racks. Meat carcasses are frozen bypassing cold air around them while they hang by the hind legs on hooks.

Foods have also been frozen by direct contacting with brine–ice mixtures or by contact with an inert, evaporating refrigerant or cryogenic gas. Packaged foods are often frozen by compressing them between refrigerated plates. Ice cream and sherbets are partially frozen in scraped- surface freezers. Finish freezing (hardening) is carried out in plate freezers. Ice cream pops are frozen in molds conveyed through baths of refrigerated brine. Descriptions of most types of commercial freezing equipment, their specifications, and operating conditions are provided by Postolski and Gruda. Most water in moist foods will freeze, but part (0.17– 0.31 kg water/kg solids, depending on the particular food) is bound to food solutes or solids in nonsolvent form and will not freeze, even at very low temperatures.

Food solutes depress food freezing points. Equilibrium initial freezing points for moist foods range between –0.4 and –3.2°C, and usually lie between –1 and –2°C. Foods frequently supercool 5 or 6 centrigrade degrees below their equilibrium initial freezing point before ice nucleates and freezing starts. Then the temperature rapidly rises to the initial freezing point. As freezing proceeds, solutes concentrate in residual unfrozen solution, further depressing the freezing point. Thus, food freezing occurs over a range of temperatures rather than at a single temperature. Fractional conversions of freezable water to ice, fractions of water that remain unfrozen, and thermal properties of partly frozen foods can be predicted with the aid of the freezing point depression equation. These predicted values can be used in calculating freezing times for foods and temperature profiles for foods undergoing slow to moderately rapid freezing (i.e., freezing in which local phase equilibrium can be assumed without serious error).

Freezing involves incorporation of water in ice crystals and diffusion of water to active growth surfaces on ice crystals. These processes occur at rates slow enough to cause significant local deviations from phase equilibrium during very fast freezing. They also largely determine ice crystal size and shape (but not the total amount of ice formed) during slower freezing. Ice usually grows as dendrites (treelike crystals) when foods freeze. Spacings between ice dendrites tend to be inversely proportional to the square root of the initial freezing rate. Specific surface areas and surface free energies of dendrites are much larger than those of rounded crystals. Therefore ice dendrites tend to convert into rounded ice. Conversion occurs rapidly in agitated slushes but is quite slow in unagitated frozen food, particularly food that is kept well below its initial freezing point.

Specific surface energies are larger for small crystals than for large crystals. Therefore small ice crystals tend to sacrificially melt and cause growth of larger ice crystals. If storage temperatures are not low enough or are allowed to cycle excessively, ice crystal size will increase appreciably after long storage, affecting texture adversely. When unblanched fruits and vegetables are frozen, ice initially nucleates outside cells. If freezing is slow, much water transfers outward from cell interiors and forms extracellular ice before nucleation occurs inside cells. This affects texture adversely. Good practice calls for rapid freezing. Then, intracellular nucleation occurs quickly, most water freezes inside cells, and adverse textural effects are minimized. Preferential growth of ice in extracellular space usually does not occur during slow freezing of blanched vegetables. Fast freezing of meat and fish helps minimize losses of water-holding power and decreases drip following thawing.

Sterilization

Sterilization, pasteurization, and blanching are processes in which foods are heated and held at suitable temperatures long enough to kill the microorganisms and inactivate the enzymes they contain. Commercial sterilization is designed to kill substantially all microorganisms that can grow or spores that can germinate at storage conditions. Because sterilized foods are packed in sealed containers that contain virtually no air, anaerobic organisms are of greatest concern. Destruction of *C. botulinum* spores is often used to evaluate adequacy of sterilization for low-acid foods (i.e., foods with pH > 4.5). Fractional survival levels must be less than 1×10^{-12} for such spores. Adequacy of sterilization for low-acid foods may also be based on destruction of a hardier putrefactive anaerobe called (PA) 3679.

Adequacy of sterilization for foods with pH values less than 4.5 is based on destruction of one or another of the facultative anaerobes, *Bacillus coagulans, B. mascerans,* or *B. polymaxa*. Fractional survival levels for these organisms should be less than 1×10^{-5} or 1×10^{-6}. Sterilization of low-acid foods involves maximum temperatures in the 116–149°C range. Maximum temperatures up to 110°C are used for more acidic foods. Sterilization of canned products is carried out batchwise by steam heating in vertical or horizontal retorts. The retorts must be vented for at least 5 min with steam flow to remove air before steam pressurization and starting a heating cycle. Foods in jars are often sterilized in air-pressurized hot water in retorts. Sterilization based on steam heating is also carried out continuously in retorts with sections separated by pressure locks, in rotary sterilizers (where a rotor causes cans to move through the unit along spiral guide rails), and in hydrostatic sterilizers (where entering and leaving cans are conveyed through a long, steam-heated zone confined between vertical water legs that balance the steam pressure). Headspace is left when food containers are filled. The container walls or tops have elements that flex. These elements and headspace prevent thermal expansion from damaging containers during sterilization. Headspace air and dissolved gases are partly removed from cans, jars, and their contents before sterilization by steam flushing or vacuum evacuation before capping. Thermal exhaustion (i.e., preheating material to expel dissolved gases) is sometimes used instead of or in conjunction with these processes. Thermally processed products are cooled under air pressure with chlorinated cold water immediately after heating. Air pressure is used to prevent bulging of containers or unseating of caps caused by excess residual pressures inside containers. Cooling is continued until products are just warm enough to fully evaporate water clinging to the container surfaces.

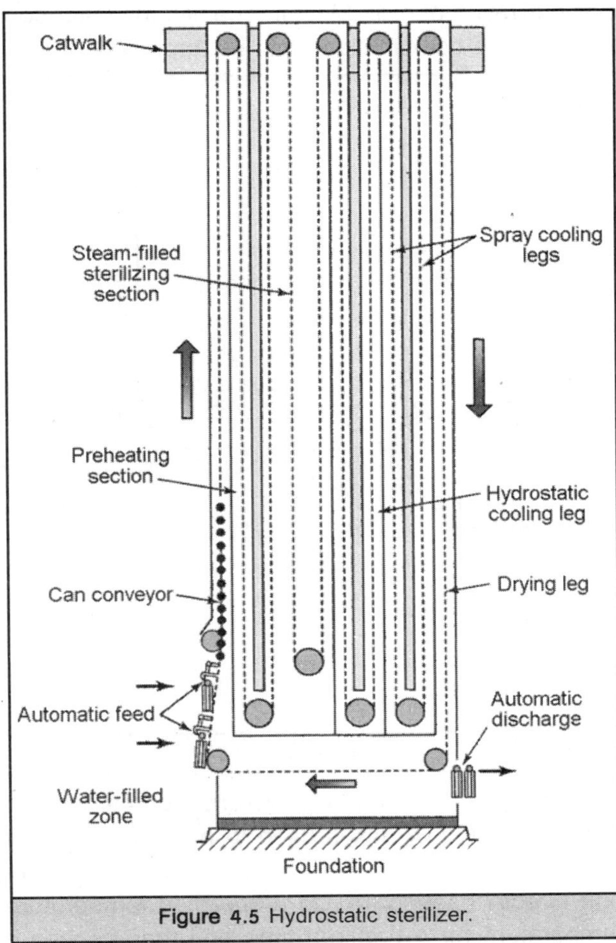

Figure 4.5 Hydrostatic sterilizer.

Pasteurization

Pasteurization, the thermal treatment at temperatures less than 100°C, is used to inactivate enzymes, virtually all pathogenic microbes, and many food spoilage organisms. It is used in combination with protective packaging and other treatments (e.g., chilling, pH, and a_w reduction) that greatly retard subsequent microbial growth. The target organism or enzyme on which adequate inactivation is based depends on the product involved. For milk it is *Coxiella burnetti*, which causes Q fever. For wine or beer it is wild yeasts. For high-acid fruits it is molds and yeasts. For citrus juices it is pectinesterase. Inactivation of pectinesterase also provides more than adequate microbial inactivation. Acceptable holding time-temperature combinations may be specified by government regulations. In the United States, at least 30 min at 62.8°C or at least 15 s at 71.7°C is required for pasteurization of milk. Most frequently, the latter conditions are used. Other, equally effective time–temperature combinations can be approved.

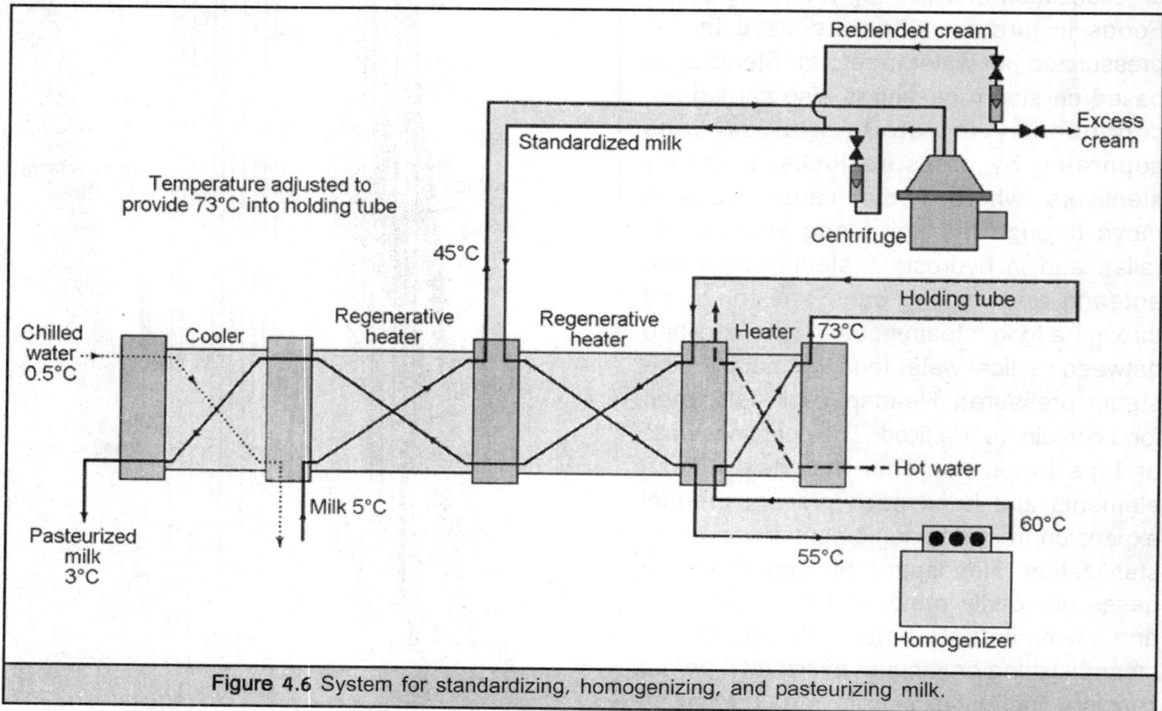

Figure 4.6 System for standardizing, homogenizing, and pasteurizing milk.

Ultrahigh temperature pasteurization carried out at 88.3°C for 1 s or 100°C for 0.01 s is used sometimes. Liquid mixes used to produce ice cream and frozen desserts are pasteurized for at least 24 s at 79.4°C. Milk is pasteurized before it is put in containers. This is done by regeneratively and directly heating milk to a specified temperature in a plate heat exchanger, holding it at that temperature for a required time and then cooling it in the same exchanger. Integrated systems that also provide cream removal by centrifugation and homogenization are often used. Other arrangements for pasteurizing milk are described by Kessler, who also provides correlations for overall heat transfer coefficients versus fluid velocities or Reynolds numbers for different plate–heat exchanger flow arrangements. Simpler plate–heat exchanger setups are used to pasteurize fruit juices. Other foods and beverages are pasteurized in containers. Products in glass jars are conveyed on metal-link belts, heated gradually, held at temperature, and cooled gradually in stages to prevent thermal shock to the glass. Long hot water baths or sprays followed by similar cold water baths or sprays are used for jarred pickles, canned fruit, and bottled beer. Heating by passage through steam at atmospheric pressure is sometimes used in pasteurizing products packed in glass.

Blanching

Blanching, heating in water somewhat below it boiling point or in steam at atmospheric pressure, is used to inactivate enzymes that adversely affect food storability and quality. Sterilization and pasteurization processes usually inactivate enzymes as well as microorganisms; blanching is used primarily before drying and freezing for vegetables and some fruits. Blanching also causes expulsion of gases from foods, softens foods, facilitates cutting and peeling, sets some colors, and provides some microbial inactivation. Unfortunately it also leaches nutrients from foods, which in turn increases the biological oxygen demand (BOD) of effluent blanch water. Less leaching occurs when steam blanching is used and still less when very humid hot air is used.

Aseptic processing

Aseptic processing involves heating liquids, semiliquid products, or liquids containing small particles to sterilization or pasteurization temperatures in heat exchangers or scraped-surface heaters or by injection or infusion of culinary steam and then passing the products through a holding tube. They are then cooled in heat exchangers or, for products heated by steam addition, by vacuum evaporation, and put in sterilized containers. The containers may be cans, jars, or drums that have been heated with superheated culinary steam. They are capped with steam-sterilized covers in chambers filled with super-heated culinary steam. Continuously formed laminated containers or plastic cups are also used. Before they are filled, they are bathed in H_2O_2 and dried. These containers are heat-sealed by fusing together narrow bands of a meltable plastic. Heated product temperatures range from slightly less than 100°C for acidic products to 149°C for low-acid products.

Inactivation kinetics

Thermally based microbial destruction and enzyme inactivation are treated as first-order reactions; that is, destruction or inactivation rates are proportional to the residual concentrations of living microbes or active enzymes involved. Inactivation rates are strongly affected by pH and food composition. If $k(T_p)$ is the destruction rate constant for a particular microbe or enzyme in a food expressed as a function of product temperature T_p and T_p is a function of time *t*, we can write

$$\ln\left(\frac{N}{N_0}\right) = -\int_{t=0}^{t} k(T_p)dt \qquad \ldots(1)$$

where *N* is the residual number of microbes per unit volume or enzyme concentration at *t* and the spot where T_p and $k(T_p)$ are evaluated and N_o is the initial value of *N*. To achieve acceptable destruction or inactivation, the highest N/N_o at any locale in a product for the most critical microbe or enzyme in question must be equal to or less than a required value. For low-acid foods, N/N_o must be less than 10^{-12} for *C. botulinum* spores. For nonmobile canned foods, the greatest N/N_o is assumed to occur at the center of a can's contents. Convection occurs in foods containing free liquid, and the greatest N/N_o is assumed to occur at the slowest heating point. This usually lies along a can's axis and near its bottom, but its locus should be experimentally determined. In the food industry, adequacy of thermal processing is often evaluated by determining whether inactivation is equal to or greater than that provided by treatment for a specified time at standard temperature, 121.1°C (250°F).

Temperature-induced changes in thermal inactivation rates are accounted for in terms of Z, the temperature reduction that will cause $k(T_p)$ to decrease by a factor of 10. A standard *Z* of 10°C (18°F) is often used for *C. botulinum*. Depending on the food and corresponding *Z* involved, acceptable treatment of low-acid foods should provide *C. botulinum* destruction equal to that provided by 1.2-2.4 min isothermal exposure at 121.1°C. In evaluating adequacy of thermal processing, one must account for changes in product temperature that occur during both heating and cooling. Product temperatures predicted from heat transfer analysis are often used in trial-and-error fashion to determine suitable heating and cooling times. These calculations are useful for estimating required processing time, but determination of adequacy of processing is usually based on measured product temperature–time data. This is particularly true for new products, where adequacy of sterilization must be further verified by microbial inactivation tests in inoculated packs.

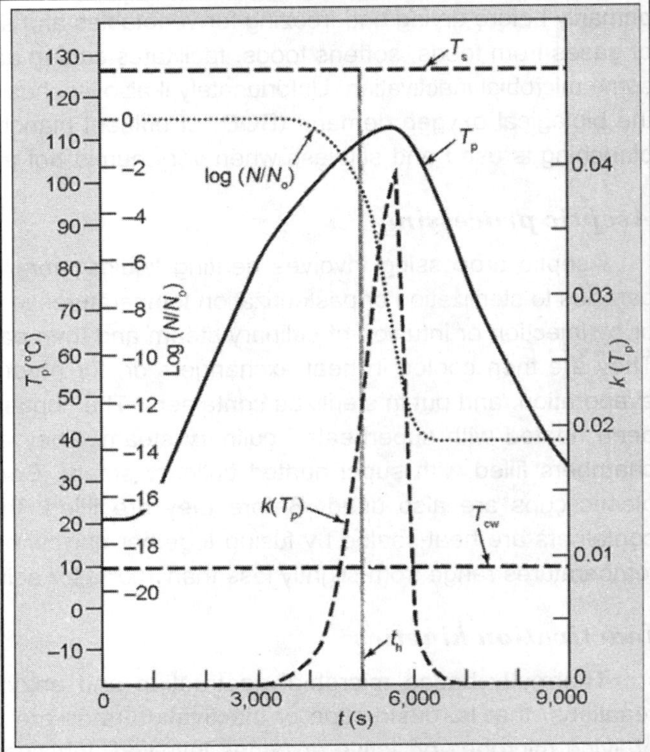

Figure 4.7 T_p, $k(T_p)$, and N/N_o versus t during sterilization of a canned solid food by thermal processing: T_s and T_{cw} are respectively, the steam temperature and the cooling water temperature in the retort, T_p is the temperature at the center of the food, and t_h is the time at which heating ends and cooling begins.

Formula methods and nomographs based on characteristics of semilog plots of unaccomplished temperature change versus time were once widely used to determine adequacy of sterilization and needed thermal processing time. Now, simple computer programs are used for the same purpose. Sterilization causes partial destruction of some nutrients and often adversely affects food flavor and color. Destructions of nutrients and some colors and flavors are first-order or pseudo–first-order processes for which kinetic parameters are sometimes available. New flavors, colors, and textures that affect product quality also develop by processes with more complex kinetics. Effects of product temperature on nutrient, flavor, and color destruction often are framed in terms of activation energies and the Arrhenius equation instead of Z. In evaluating nutrient destruction and quality reduction, fractional retention of the nutrient or attribute being evaluated is averaged over the whole volume of the container.

Sterilization methods and temperature–time conditions that provide adequate pathogen or spoilage agent inactivation but minimize adverse effects should be used. Rates for reactions that destroy nutrients and adversely affect sensory properties increase less sharply than microbial destruction rates as temperature increases. Therefore high-temperature, short-time (HTST) processing often improves nutrient and flavor retention. Heat-resistant enzymes may not be adequately inactivated when microbial destruction is adequate if very high temperatures are used for very short times. Outer regions of containers are often overprocessed and suffer excessive quality degradation when the points that heat most slowly are just adequately processed. Overprocessing can be reduced by providing more uniform heating (e.g., by using aseptic processing and microwave and ohmic heating). Other preservation processes that avoid or greatly limit heating (e.g., radiation preservation, micro- filtration, and high-pressure treatment) have been extensively investigated or, in some cases, are already in use.

Removal of Water

Evaporators

Water and solvents can be removed from liquid foods by means of evaporators. In multiple-effect evaporators, the most commonly used type, vapor generated by steam or vapor heating in an effect provides heating that causes evaporation in a subsequent effect operating at a lower pressure. In *N*-effect evaporators, roughly *N* kg of vapor is generated for each kilogram of steam used. In some cases, a steam ejector is used to compress part of the vapor generated in a second or third effect, and the compressed vapor and motive steam are used to heat the first effect, providing more than *N* kg of vapor per kilogram of steam used. Vapor recompression may also be carried out by large centrifugal compressors. Electrical energy is used to drive the compressor; virtually no steam is used directly. Evaporation, particularly multiple-effect and vapor recompression evaporation, is more thermally efficient than drying. Therefore evaporators are often used to concentrate liquid foods before drying.

Evaporation is also used to concentrate juices, syrups, sauces, fermentation residues, stillage bottoms, and absorption refrigeration solutions. It is also used to induce crystallization and produce glassy supersaturated sugar solutions used to make hard candies. Vacuum evaporation, often combined with single-pass operation in each effect to provide low product holdup, is used to minimize thermal damage to foods. In single-pass evaporators, steam condenses on the outer walls of vertical heat transfer tubes 6–12 m long mounted in a steam chest. Water evaporates from liquid flowing in film form down the inner walls of the tubes. Concentrated liquid and vapor leave through the bottom of the tubes. Entrained liquid is separated from the vapor in a cyclone. High-velocity recirculation in liquid-filled tubes is used to improve heat transfer in evaporators used to concentrate viscous materials such as tomato paste and gelatin and pectin solutions. Liquid superheats as it passes through tubes in such evaporators and flashes as it leaves the tubes.

Cyclones are used to separate entrained liquid from the vapor produced by flashing. Scraped-surface, thin-film evaporators are used for very viscous products (e.g., hard candy melts) and for heat-sensitive products. Natural convection and boiling cause circulation of liquid in evaporators containing short, steam-heated tubes mounted in a wide-diameter assembly (calandria). Vapor separates from the foaming, boiling liquid in a large, vapor-filled chamber above the calandria. Heat transfer coefficients in these evaporators peak at a certain liquid level in the calandria. Multieffect systems containing such evaporators are frequently used to concentrate sugar solutions in sugar refineries. Part of the vapor produced may be withdrawn from effects and used for process heating elsewhere in the refinery. Most food aromas have high relative volatility with respect to water and almost completely vaporize during initial stages of evaporation. Such aromas can be recovered in concentrated form from first-stage vapor streams and used for add-back purposes.

Membrane processes

Pressure-driven flow through permselective membranes is used to gently remove water or aqueous solutions of low molecular weight solutes from liquid food products. Reverse osmosis (i.e., selective removal of water) is used to raise the solids content of maple sap from roughly 2% to 10-12%, to recover water and solutes from waste processing streams, and to concentrate dilute aqueous caffeine extracts. Ultrafiltration (selective removal of water and low molecular weight solutes) is used to concentrate proteins in milk, milk whey, and soy whey and to recover protein from potato fruit water. Protein-rich concentrates obtained from heated milk by ultrafiltration are used to make cheese with reduced loss of milk solutes and reduced use of rennet and starter cultures. Diafiltration (ultrafiltration accompanied by water addition) is used to de-ash gelatin. Pervaporation has been used experimentally to produce aroma-rich concentrates by selectively evaporating water through permselective membranes. Mass balances for retained species and equations governing permeation rates are used in designing membrane-based processes. Design calculations must not neglect the effects of concentration polarization on permeation rates and selectivity.

Freeze concentration

Freezing is sometimes used to gently and selectively remove water from fruit juices, beverage extracts, beer, and wine. Scraped-surface coolers partially freeze the solution being processed, producing small ice crystals. Ostwald ripening (i.e., growth of large crystals induced by sacrificial melting of smaller crystals) is then used to produce large ice crystals, which are separated from residual solution in wash columns or centrifuges. Thermal efficiency and upper concentration limits have been increased, and costs reduced, by means of countercurrent freeze concentration, but the use of freeze concentration remains limited because of its high cost.

Extraction of Liquids from Solids

Solid-liquid extraction is used to recover sugar from sugar beets and sugarcane, oil from oilseeds, juice solutes from fruits, beverage extracts from coffee and tea, protein-rich materials from soybeans, fermentable solutes from grains, and hydrocolloids, gums, and natural food colors from various sources. It is also used to remove unwanted constituents (e.g., caffeine from coffee, bitter compounds from olives, cyanogenic compounds from cassava, sarcoplasmic protein, and undesired flavors from minced fish, and salt from pickles). Extraction also occurs when flavoring agents leach out of charred barrels during the aging of wines and distilled spirits, when solutes leach out of vegetables and fruits during blanching and fluming, and when plasticizers leach out of plastic containers into liquid foods and beverages. Vegetable or animal matter from which solutes are to be extracted is cut or otherwise divided into small particles. Oilseed grits are flaked or extruded as steam-puffed cylindrical particles. This coalesces isolated oil bodies and creates fissures or pores that allow solvent or extract to reach the oil. During extraction, solutes diffuse out of the solid through internal liquid-filled paths and pass into solvent or extract that contracts the solid. Dry materials (e.g., coffee and tea) must imbibe solvent or extract before diffusion can proceed. In rare cases (e.g., extraction of vanilla from vanilla beans with alcohol), imbibition is very slow and is a rate-controlling step. Plasma membranes are denatured to facilitate water-based extraction of solutes from cellular material. Then, pores in cell walls become the main diffusion bottleneck. In-particle diffusivities for water-soluble extractibles are 0.1–0.5 times as large as in water alone. Required extraction times are inversely proportional to the in-solid solute diffusivity and proportional to the particle size squared. The extract carries away the solute, following a contacting path that promotes high solute recovery.

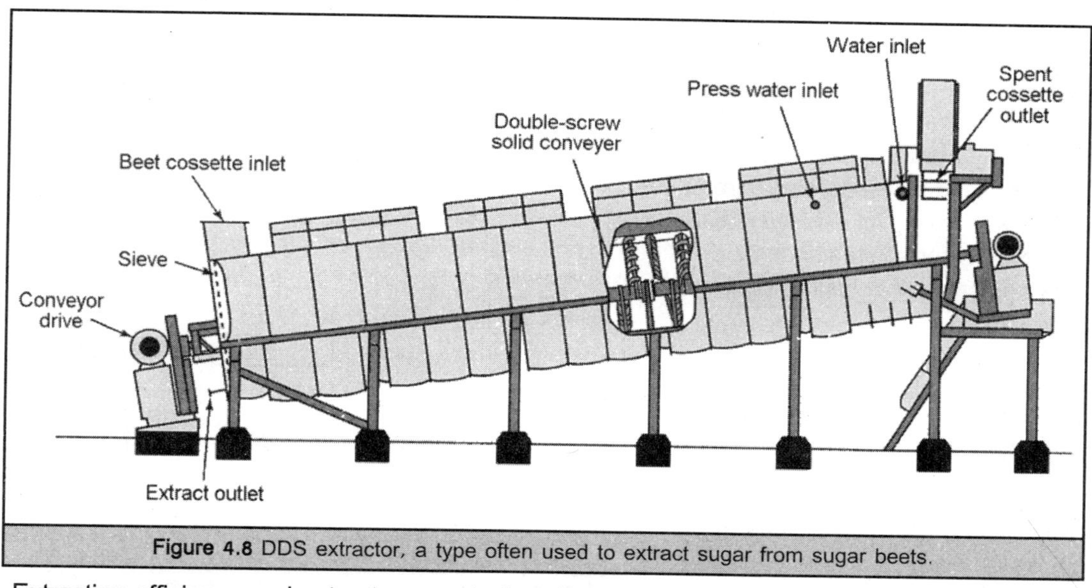

Figure 4.8 DDS extractor, a type often used to extract sugar from sugar beets.

Extraction efficiency and extract concentration depend, in part, on the stripping factor, the solute distribution coefficient multiplied by the ratio of extract to moist solids. High concentration extracts can be obtained by using a low stripping factor (i.e., low solvent flow rates or small amounts of solvent), but efficient extraction can be obtained only if the stripping factor is greater than 1.0. Stripping factors of 1.2-1.4 are often used in water-based extractions. Higher stripping factors (e.g., 2.4) are used for hexane-based extraction of oil from oilseeds. Small-sized particles favor fast extraction but cause excessive flow pressure drop and displacement instability. Large particles extract slowly. Particle thicknesses or diameters in the range of 3-5 mm are a good compromise. Where possible, particle shapes that prevent blinding of surfaces by interparticle contact are used.

Commercial extraction is often carried out in very large, continuous or semicontinuous countercurrent extractors where extract percolates downward through a rising bed of solids or in units where a bed of solids moves horizontally while extract repeatedly percolates through it in a flow arrangement that effectively provides countercurrent contact. In other cases, a cyclically loaded and discharged set of interconnected columns is used to provide nearly countercurrent contacting. Most of these processes can be analyzed and designed by means of solutions of partial differential equations describing diffusive mass transfer in solids or mass transfer analogs of partial differential equations describing conductive heat transfer in solids. Solutes are sometimes produced from insoluble precursors by reactions whose kinetics must be accounted for in designing the extraction process. Extraction processes involving foods often are subject to imperfect contacting (e.g., channeling, unsteady displacement, and axial dispersion). These imperfections must be accounted for in design calculations. Supercritical carbon dioxide is now used as a solvent in decaffeinating coffee and tea and in recovering solutes from hops. Supercritical carbon dioxide is a particularly attractive solvent because it is nontoxic and can be readily removed from foods. Further, its dissolving capacity and selectivity for solutes such as caffeine can be adjusted by adjusting temperature and pressure.

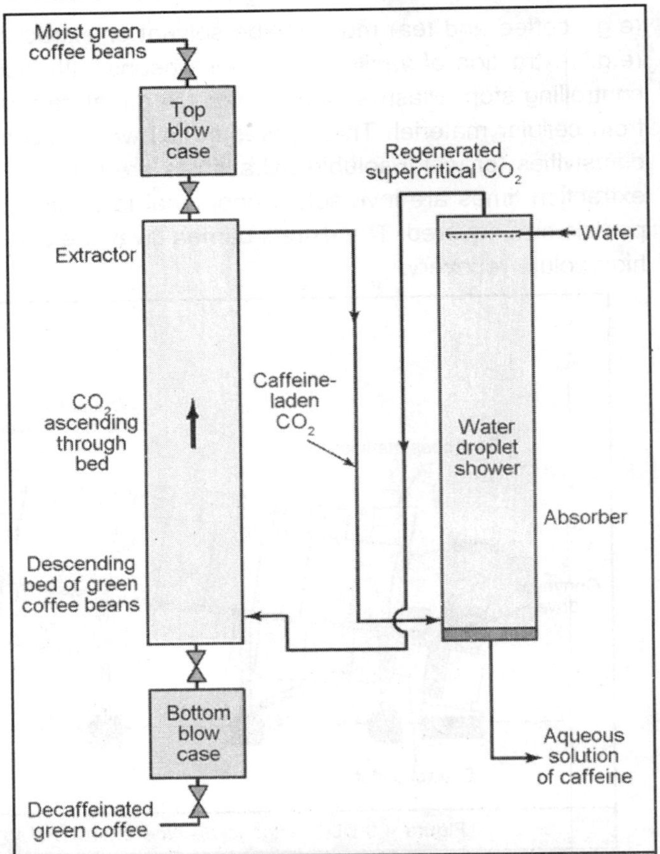

Figure 4.9 System for extraction of caffeine from green coffee by supercritical carbon dioxide.

Separation Techniques

Filtration

Filtration is used to remove solid particles from or to clarify juices, extracts, vegetable and fish oils, fermented beverages, recirculated, cooking oil, flume water, milk, and soy milk. It is also used to separate potato starch from potato fruit water, high-melting fats from vegetable oils in fractionation processes, crystals from mother liquors, and chemically precipitated impurities from sugar juice. The engineering principles that govern filtrations involving foods are very similar to those that apply to the filtration of chemicals, but superior sanitation must be maintained.

A fluid that contains particles is passed through a cloth or other porous medium that retains particles. The particles deposit on the medium, forming a porous cake that thickens as deposition continues. As long as total pressure drop remains constant, total flow resistance increases linearly as the cake thickens. Depending on the feed pumping system used, flow resistance increases cause either decreases in flow rates and solids deposition rates, increased fluid pressure drop, or a combined increase in pressure drop and decrease in flow rate. Filtration is stopped when these changes become excessive, and the filter cake is manually or mechanically removed. Extents of these changes and optimal scheduling of cake removal can be predicted by means of calculations that relate cake depth and flow pressure drop to the initial solids content of the fluid, the amount of filtrate discharged, and the experimentally determined relationships between applied pressure, cake density, specific filtration resistance, and applied pressure. Some food-based filter cakes are highly compressible, and their specific filtration resistance increases markedly as pressure increases. Filter aids (inert, highly porous solids) are sometimes added to filter feeds to reduce cake filtration resistance and compressibility and to improve solids retention.

Cake filtration resistance tends to be inversely proportional to the particle size squared. Therefore chemical agents that cause particles to flocculate (stick together in clumps) are often added to filter feeds. Deep porous media that capture and retain particles within their pores are used for some feeds. Media flow resistance increases as filtration proceeds. Consequently flow pressure drops increase and/or flow rates decrease. The medium is replaced or cleaned by backwashing when these changes become excessive. Unlike cake filtration, where total filtration resistances at constant pressure increase linearly as solids removal progresses, filtration resistance for in-media solids capture is more than linearly proportional to the amount of solid removed. Therefore in- media capture is used only for fluids with light solids loads.

Sterile microfiltration based on use of membrane filters with 0.2-μm pores is used to remove bacteria from beers, thereby eliminating or reducing needs for pasteurization. Ceramic microfilters that can be sterilized and cleaned by back-flushing are used to clarify fruit juices. They are very expensive but will last for 10 years or more if care is taken to avoid heat shock. Beds of particles used in solid–liquid extraction and adsorption and ion-exchange processes act like deep filter cakes. Pressure drops during flows through such beds often can be predicted with the aid of the Ergun equations or Kozeny-Carman equation. In other cases, flow pressure drops are calculated by means of empirically determined permeabilities or specific filtration resistances.

Centrifugation

Particles forced to rotate about an axis at high angular velocity in a centrifuge experience a radially acting force proportional to the radial distance from the axis, the angular velocity squared, and the density difference between the particle and surrounding fluid. At rotational speeds used in commercial centrifuges, particles denser than the fluid move outward and particles lighter than the fluid move inward thousands to tens of thousands times faster than they would sink and rise, respectively, under the influence of gravity. Therefore, centrifuges are used to separate and collect dispersed particles and droplets too small to be cleanly and quickly separated by other means. Centrifuges are used, for example, to separate dispersed oil, microorganisms, and fine biological particles and precipitated curds from aqueous solutions and to remove yeasts from fermented beverages and waste treatment sludges from dischargeable water.

Density differences driving centrifugal settling are usually much smaller for food processing than for chemical processing; the particles involved are more compressible, and, again higher levels of sanitation are required. The particles or droplets involved move outward or inward until they strike a collection surface or material previously deposited on that surface. Deposited material then slides along the collection surface or is mechanically conveyed along it to a conduit or port that permits it to leave the centrifuge. Settled solids may leave through peripherally located nozzles as a dense slurry or leave in even denser form through intermittently opened peripherally located slits. Settled solids may also be knifed or augered out of centrifuges. Clarified liquid and separated oil or oil-rich creams leave through ports located closer to the rotational axis of the centrifuge.

Centrifuges are sized to provide holdup time sufficient to permit complete settling or a desired level of settling. Large numbers of thin, closely spaced, concentric conical disks are mounted in centrifuges to reduce settling distances and times for extremely fine materials (e.g., butterfat droplets, and microorganisms). This greatly increases usable throughput rates for feeds containing such fine material. Filtering centrifuges have a finely perforated wall or a perforated wall covered with filter cloth. Centrifugal force causes solids to deposit as a cake on the wall or cloth and then forces liquid through the cake. Wash liquor may also be forced through the cake.

Such centrifuges are used to recover crystals (e.g., sugar and citric acid) from concentrated mother liquors, to separate freeze-concentrated vinegar and beverage extracts from ice, and to recover juice from milled fruit. Centrifugal force generated by rotational flow is produced by tangentially injecting solids-laden liquid or gas into hydroclones or gas cyclones. These units have a cylindrical shell attached to a conical bottom, a central outflow port at their top, and a bottom port through which solids or solids-rich underflows leave. The injected liquid or gas first spirals downward in an outer vortex and then spirals upward in an inner vortex. The vortex flow tends to cause particles to move outward toward the wall. In gas cyclones virtually all the gas leaves through the top port. Gas cyclones are sized to provide enough flow residence time to permit wall impact by all particles greater than a desired size.

Gas cyclones are used to separate fine spray-dried material from discharged drying air, pneumatically elutriated materials from carrier gas, chaff from coffee roaster discharge gases, and pneumatically conveyed material from conveying air. In hydroclones, most entering liquid leaves through the upper port as overflow, but enough leaves through the lower port to provide a flowable slurry, the underflow. The centrifugal force generated by vortex flow is partly counterbalanced by drag force generated by radial flow of liquid from the outer to the inner vortex. Consequently, small particles or low-density particles may migrate inward and be carried away with the overflow. Larger or denser particles leave with the overflow. Large banks of miniature hydroclones used in parallel are employed to separate starches from protein- and solute-rich solutions during the wet milling of corn.

Expression

Pressing of fluid-rich biological matter is used to recover fruit and sugar juice, vegetable and fish oil, pectin released by cooking apple pomace, and protein-rich juice from leaves. It is also used to expel whey and to fuse curds together during cheese manufacture and to dewater spent food processing wastes, by-products, and processing intermediates. Expression is carried out in a wide variety of equipment. In screw presses, feed is transported down the length of a perforated barrel by a rotating single screw or by twin screws. The feed compacts, and fluid is expelled through barrel perforations because the screw's root diameter increases, its pitch or rotary speed decreases, or the barrel diameter decreases, or because constricted solids outflow from the barrel causes a progressive buildup of pressure. In roll mills, a grooved, rotating roll is hydraulically pressed against two similar counter rotating rolls that lie beneath it. As the rolls rotate, they drag feed (usually shredded sugarcane) through nips between the top and bottom rolls. This compacts the feed and expels juice. Some of the juice flows back through the advancing feed; some flows laterally into the grooves and then backward. The process is analyzed by relating local back-flow and flow pressure drop to the cumulative extent and rate of compaction and the local specific filtration resistance of the compacted feed. In belt presses, feed is deposited on a moving cloth belt and pressed between that belt and an opposing, moving cloth belt driven over a series of solid or perforated rolls. In some cases the belts follow a serpentine path. Expelled fluid flows through the cloths and is collected in troughs beneath the rolls.

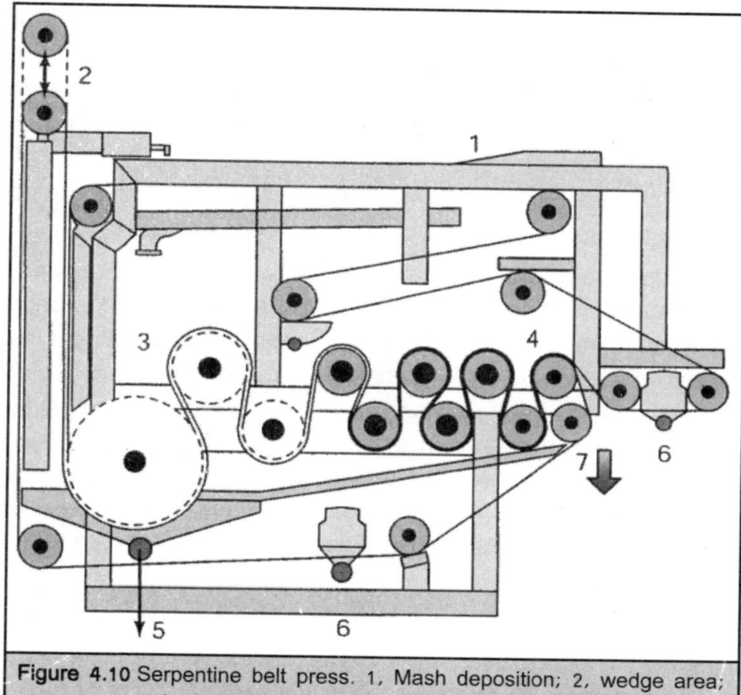

Figure 4.10 Serpentine belt press. 1, Mash deposition; 2, wedge area; 3, low-pressure press zone; 4, high-pressure press zone; 5, juice runoff; 6, belt cleaning; 7, press-cake discharge.

In presses of other types, feeds confined in a perforated cage or in stacked cloth pouches are pressed by a diaphragm or ram. Fluid is also expelled from cakes compacted by centrifugal force in decanter centrifuges. Engineering treatment of expression is based on numerical solution of partial differential equations that describe how flow in press cakes is driven by fluid pressure gradients or related solid stress gradients and how rates of local cake volume change as a function of flow rate gradients. Empirically determined relationships between solid stress, specific filtration resistance, and the solids-based specific volume of the cake are used in solving these equations. Methods have been developed for determining these relationships from constant-pressure or constant-rate pressing tests carried out in suitably instrumented equipment. Some press feeds (e.g., waste treatment sludges and fruit with thin cell walls and weak intercellular bonds) compact excessively as pressure increases. When such feeds are pressed, fluid pressures drop precipitously and solid stress rises sharply over very thin layers near outflow surfaces. Pressing can be improved for such feeds by adding an inert material (e.g., rice hulls) that stiffens the press cake. Other feeds (e.g., spent coffee grounds and oilseeds) have thick cell walls that adhere strongly together. Fluid pressure builds up inside the solid when such feeds are pressed. The extent of in-solids pressure buildup depends on the rate of pressing and the extent of compaction. Pressing of such materials can be improved by treatments that promote cell wall rupture.

REMOVAL OF UNWANTED MATERIALS

Peels, skins, fatty layers, hulls, and shells frequently must be removed from foods. Skins are often manually cut from meat and fish with the assistance of special tools. Wholly mechanical systems are also used. These grab a carcass, carcass part, or fish at a selected spot and drag it past suitably positioned knives and saws. Fat-depth-sensing devices are sometimes used to guide cutting. Moisture contents of shell-encased and hull-encased foods are often adjusted to increase shell and hull friability and separability. Hulls and shells are then cracked by impaction in rapidly rotating pin mills or by passing the food involved through pairs of rollers or plates that turn at different speeds and produce a pressing, rolling action. Thin shell and hull fragments are then pneumatically separated from more compact usable food particles.

Manual peeling of vegetables and fruit is costly; abrasion peeling removes too much peel and is wasteful. Therefore, treatments that break down bonds between cells or decompose cells in or near skins are used. These include short time heating with high pressure steam, short-time immersion in baths of lye, and passing product through a flame. The loosened skin is then brushed off or washed off by water jets, a step that also removes residues of chemical peeling agents. The process can be analyzed in terms of short-term heat or mass transfer in surface layers and the kinetics of protopectin or cell wall breakdown. To sharply limit intercellular bond breakdown, internal temperature gradients near the product surface must be very large and exposure times very short. Even so, undesirable textural and color changes often occur in regions close to the peeled surface.

CLEANING AND DISINFECTION

Washing-based cleaning of food surfaces and cleaning and disinfection of equipment are widely used food processing operations that have been analyzed from an engineering point of view. Efficient cleaning and rinsing can be achieved only when smooth, highly polished, crevice- and pocket-free equipment is used. Equipment conforming to the sanitary design standards of the Technical Committee of Dairy and Food Industry Supply Association and made of type 304 or type 316 stainless steel is frequently used. Parts may also be made of plastics and synthetic materials approved by the U.S. Food and Drug Administration or Department of Agriculture. Equipment cleaning frequently involves an initial water rinse, a caustic detergent wash, a second water rinse, an acid wash (if calcium-rich deposits have formed), and a final water rinse. Conditions that promote corrosion and etching (e.g., use of chlorides) must be avoided.

Disinfection is provided by cleaning or rinsing at high temperatures, by steam heating of equipment, or by use of chemical agents (e.g., hypochlorites). Cleaning formerly involved manual hosing down and brushing of disassembled pipelines and equipment and one-time use of cleaning agents. Manual cleaning procedures are still frequently used in meat processing plants and in food processing pilot plants. Spray wands operating at pressures up to 3.5 MPa (500 psig) are used to ensure removal of adherent material. Clean-in-place (CIP) systems that accomplish thorough cleaning without disassembling piping and permit some reuse of cleaning agents have been developed for many plants. Because cleaning agents and rinse water run through pipelines and narrow-bore equipment in CIP systems, they must be laid out to permit hang-up–free drainage. Product contact surfaces in tanks and large-bore equipment are cleaned with the aid of sprays issuing from permanently installed spraying devices.

Globe-like spray heads or reciprocating sprays are used to provide complete washing and rinsing of internal surfaces. Removal of cleaning agents during rinsing has been successfully modeled so that rinsing can be accomplished without excessive use of water. Soil deposition and fouling interfere with process heat transfer and greatly reduce permeation rates in membrane-based processes. Therefore fouling and soil deposition mechanisms and kinetics for such processes have been extensively studied. The results of these studies have been used to optimize frequencies of cleaning and to select operating conditions and surface treatments that retard fouling.

HOMOGENIZATION AND EMULSIFICATION

Fat in freshly drawn cow's milk is dispersed as butterfat globules that individually contain a fatty core coated with a lipoprotein layer roughly 10 nm thick. Depending on the type of cow involved and how long the animal has been lactating, mean volume/surface diameters of butterfat globules range between 2.5 and 4.5 μm. Unless treated, the globules rise, producing a layer of cream on top of milk. To prevent this, and to minimize butterfat clumping, the globules are broken into droplets whose mean volume/surface diameter ranges between 0.22 and 0.5 μm. This is usually done by passing milk through spring-loaded, narrow-clearance valves in high-pressure piston pumps, called homogenizers. Cavitation produced by rapid increases in velocity in the pump valve and flow-induced shear break up the globules. The mean size of the resulting droplets is inversely proportional to the homogenization pressure used raised to the 0.6 power.

Brownian motion would very rapidly cause collision, clumping, and coalescence of incompletely coated droplets. Therefore the lipoprotein coating must redistribute itself rapidly enough to completely coat freshly created drops before they collide with each other and stick together. Dispersions of small drops of oil in a continuous aqueous phase are often created by intensive agitation of a mixture of an oil and an aqueous solution. The agitation produces turbulent velocity fluctuations, which in turn cause local shearing intense enough to overcome the interfacial tension forces that tend to hold drops together. Mean drop sizes produced by such agitation are frequently proportional to the interfacial tension raised to the 0.6 power and inversely proportional to agitation power expenditure per unit volume raised to the 0.4 power.

Simple dispersions rapidly coalesce and separate when mixing stops. To avoid this, surfactants or agents that coat drops and prevent or greatly retard drop coalescence are added. Dispersions stabilized in this way are called emulsions. Stiff emulsions that act like soft solids (e.g., mayonnaise) contain so much dispersed material that drops butt against each other. A great deal of work has been done to develop gelled low-fat food systems with textural properties resembling those of stiff, oil-rich emulsions.

SOLIDIFICATION

Gelling and Coagulation

Fluid foods often are converted into fluid-rich solids with desirable textures by means of gelling and coagulation. Acid produced by bacterial action and bond site production due to the action of the enzyme rennet serve to coagulate milk casein proteins and produce cheese. Acidification is used to precipitate and gel food proteins with low isoelectric pH values (e.g., casein, milk whey proteins, and soy proteins solubilized by base addition). Calcium salts precipitate soy proteins and cause them to gel during the production of tofu. Heat-induced gelation occurs during the cooking of many protein-rich solutions and slurries. Other protein solutions (e.g., gelatin) gel when cooled and melt when heated. High-methoxy pectin causes acidic sugar-rich fruit juices and slurries to gel when jams and jellies are produced. Fully hydrated (gelatinized) starches are used to produce cold-set soft puddings. Liquid foods gel because reactions or physical chemical forces bond together sites on adjacent, long-chain food molecules or chains of globular proteins.

Low extents of bonding produce increases in viscosity. Gelling occurs when bond junction densities become large enough to form coherent three-dimensional polymeric networks. Gelling occurs only if the gelling agent concentration C exceeds a critical value C_o. Gelling times are often inversely proportional to $(C - C_o)^n$, where n is a constant. Simple kinetic considerations indicate that n should equal 1, but it is frequently larger than 1. Gelling and coagulation rates may be controlled by the rate of bond site production or by rates of bond forming reactions or events. Cross-linking continues after gels form, and the gels progressively stiffen. This reduces the mobility of network sections, slowing and eventually stopping the bonding process. At gelling agent concentrations substantially greater than C_o, gel strength tends to increase roughly linearly as $(C/C_o)^2$ increases.

Cooling-Induced Solidification

Hard sugar candies are made by evaporating water from a concentrated solution of mixed sugars until the solution is essentially converted into a hot melt containing less than 1 % moisture. Then the melt is cooled. So viscous that crystal growth does not occur, the melt ultimately becomes a vitreous glass. Flavoring and coloring agents are blended into the melt. When the hardening melt becomes stiff enough, it is processed in devices that convert it into pieces of appropriate size and shape. Chewy candies are made by cooling, flavoring, coloring, and shaping evaporatively concentrated mixtures of sugars, corn syrup, fat, and milk solids containing 12-15% moisture. In both cases, solidification is effectively caused by large cooling-induced increases in viscosity and flow yield strength.

Cooling is controlled to prevent excess stress from developing and to permit shaping (e.g., molding, rolling, and extrusion) before the completion of hardening. Chocolate is made by molding and cooling finely milled, thoroughly worked mixtures of molten cocoa butter, cocoa, sugar, and in some cases milk solids. Cocoa butter can exist in four polymorphic forms that solidify or change form at different temperatures. Only one, the β form, is stable at room temperature. Therefore carefully controlled cooling and seeding with fine β-form crystals is used in the solidification of chocolate.

DESIRABLE SHAPING

Rolling and Flaking

Food processing intermediates used to produce bread, noodles, cookies, and chewing gum are often rolled into sheets before being further shaped by cutting and other forming operations. Cooked cereal grits are rolled to form flakes that are dried and toasted to make ready-to-eat breakfast cereals. Oilseed grits are rolled into flakes to prepare them for extraction. Ropes of partially hardened candy melts are pulled through sequences of pairs of rollers turning at progressively faster speeds to produce narrow-diameter ropes, which are then cut into pieces that are pressed into shape by other devices. Engineering aspects of these operations can be dealt with by methods developed to analyze the calendaring of plastics.

These methods involve the use of partial differential equations to describe flow fields and pressure gradients that develop in plastic materials passing between rolls. These equations can be solved to determine the pressing force and torque acting on the rolls and the power required to drive them. Dimensional analysis has been applied to circumvent complications due to the complex rheological characteristics of food doughs.

Extrusion

Foods are frequently shaped, mixed, cooked, and texturized in screw-driven extruders. Flow driven by shear between the rotating screw and the stationary barrel conveys, mixes, and kneads the material being processed and forces it through dies that shape the product. Rotating or oscillating knives cut the extruded material into pieces of suitable size. Action of the screws on the processed material can generate a great deal of frictional heat, particularly when screws with shallow flights are used and the feed's moisture content is low. Supplemental heating is often provided by heated jackets. Cooling jackets are sometimes used. Initially, single-screw extruders were used; now, extruders containing co-rotating twin screws that contain changeable sequences of conveying, kneading, and pressurizing elements are frequently used. Flow, mixing, and power expenditure in

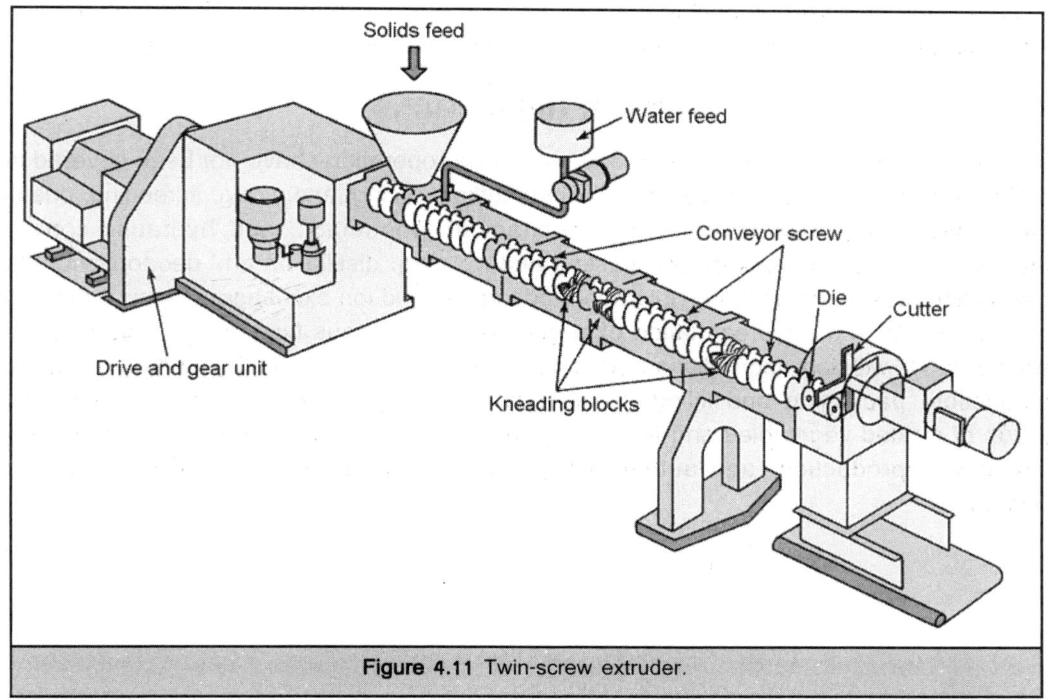

Figure 4.11 Twin-screw extruder.

extruders of both types have been treated by methods developed for extrusion processing of polymers. Foods processed in extruders usually contain 15–31% water so that they can be made plastic when worked on by the screw. Extruders operating close to room temperature and containing screws with deep flights are used to form pasta by forcing semolina-based dough through large arrays of parallel holes. Die hole inserts that provide nonuniform clearance are used in the production of elbow macaroni.

High-temperature extrusion, based on use of screws with shallow flights or elements that provide intense kneading, is used to produce confectionery doughs and puffed and textured foods. Temperatures between 130 and 180°C are frequently used in making these products. Extrusion is used to convert soy protein into products that resemble meat in structure and texture. Such texturization appears to involve the formation of links between ε acids and ε amines on amino acids in neighboring protein chains. Kinetic models for the reaction indicate that its rate depends on shear rates in the extruder and die as well as the concentrations of the reacting species. Concentrated purees or slurries containing components capable of entering into setting reactions are pumped or extruded through dies into baths or vapor-rich atmospheres containing setting agents.

Examples include the use of extruded, slurried, partially dissolved collagen or viscose solutions to form sausage casings and use of pumped slurries containing alginates and pureed fruits or vegetables to form artificially shaped fruit and vegetable pieces. Pump- or piston-based extrusion is used to form products that are subsequently heat set (e.g., doughnuts, masa-based puffed products, and certain types of cookie dough). Extrusion is also used to stuffmeat emulsions into sausage casings. Coextrusion is used to form fruit-filled bars. Moist feeds used to produce shaped cereal pieces and pelletized animal feeds and pet food are pushed by rollers through die holes in thick, rotating, perforated rings. Extrudate emerging from the holes is cut into short pieces by knives.

OTHER OPERATIONS

Because of space limitations, many food processing operations have not been covered in this article. These include mixing; grinding, and milling; cutting, slicing, and dicing; screening; pneumatic separation and classification; pumping; mechanical, pneumatic, and hydraulic conveying; slaughtering and carcass disassembly; casting and molding; distillation and deodorization; liquid-liquid extraction; gas absorption; crystallization; adsorption and ion exchange; chemical modification of food ingredients (e.g., hydrogenation of vegetable oils); various types of cooking and baking; treatment of foods with ionizing radiation; pore generation and puffing; coating and enrobing; sorting and inspection; packaging and filling; fermentation, including brewing, wine making, and the production of pickled vegetables and fermented dairy products; aquaculture; shellfish depuration; automated egg production, and automated growing of plants and plant tissue in controlled environments.

5

CONTAMINATED FREE ANTIBODIES

This chapter deals with the purification of antibodies to make the products for human use. This application requires the use of multistep procedures, with steps of high resolution. Many uses of antibodies do not require such stringent purity. Our objective is to present methods that when combined are capable of achieving parts-per-million levels of contaminants, with the understanding that such criteria are not needed for many uses of antibodies. In such cases, single steps of many of the methods presented will prove to be adequate. Our focus is also primarily on *monoclonal antibodies,* which now dominate the antibody field, but we include a special section on polyclonal antibodies. Similarly, we describe purification processes for the IgG and IgM classes only, because these account for almost all uses of antibodies. Occasional reference is made to purification of the IgA subclass, whose use is rare. Our discussion is limited to methods of purification. The choice of equipment needed to carry out these methods is a very broad topic. It depends on whether process development or production is being performed. It also depends on the scale, on whether the method is good manufacturing process (GMP) or not, and on the nature of the antibody being purified. Thus, details of this selection process are beyond the scope of this article.

CHARACTERISTICS OF ANTIBODIES

Typical feedstock contains serum proteins such as albumin and transferrin as well as cell-degradation products such as DNA and cytosol proteins. In addition, if the antibody is in ascites fluid, the host animal's (typically mouse) antibodies would be present as well. By a combination of chromatographic or precipitation methods, one seeks to purify the desired antibody away from these contaminants. More recently, cell-culture media have become available that are described as "protein-free" and usually contain only low molecular weight digests of proteins or amino acids. If the antibody-secreting cells can grow and produce well in such media, the purification process becomes much easier, because the IgG molecules at a molecular weight (MW) of 150 kDa are readily separated by size from the components of such media. As described in other articles, IgG antibody molecules are multichain glycoproteins that have been divided into various subclasses based on structure and properties. Their carbohydrate content is only 2% and is invariably located in the CH_2 domain of the Fc portion.

The carbohydrate is sequestered, in that the carbohydrate chains are directed inward and are the actual contact residues between the heavy chains in this region, with the two heavy chains bowed outward to accommodate the carbohydrate. Thus, lectins generally do not bind to IgG, and purification must be directed toward the polypeptide portion. It should be noted that there are occasional reports of a carbohydrate site in the variable region of the IgG, in which case a purification method could be directed toward the carbohydrate moiety. IgM molecules, on the other hand, contain about 9 to 12% carbohydrate, attached at five locations throughout the constant domains. Therefore, carbohydrate-directed methods have been developed for IgM and will be discussed later. Specific affinity interactions for IgG and to some extent for IgM have been exploited in their purification.

These affinity methods fall into three categories. First are those that rely on the key functional activity of the antibody, that is, its binding to antigen. Thus, a classic method of purification of antibody is binding on a solid-phase absorbent to which antigen has been immobilized. The second category relies on certain proteins that specifically bind to IgG. In this category are the well-known bacterial proteins A and G as well as second antibodies raised to bind against the desired antibody. Third, less-specific but often considered affinity chromatographic techniques include those relying on the interaction of the antibody with immobilized metals (IMAC), dye-ligands, hydroxyapatite, and thiophilic adsorbents. The isoelectric points vary widely among different antibodies, mostly between about pH 6–9, but within different subclasses the range is generally much tighter.

The isoelectric point directly affects the ability to use ion exchange, hydroxyapatite, chromatofocusing, and preparative electrofocusing techniques to purify antibodies. For example, the isoelectric point of serum albumin at 4.5 to 5.0 indicates that its removal by ion exchange should be feasible for most antibody-purification schemes. Transferrin, at pI = 5.5-6.0, can be more challenging. The hydrophobicity and solubility of antibodies are important in their purification by such methods as hydrophobic interaction chromatography and precipitation methods. In general, antibodies are more hydrophobic and less soluble in the presence of precipitating salts than feedstock contaminants. Finally, the size of antibodies, 150 kDa for IgG and 900 kDa for IgM, makes this a key property to exploit in their purification. Size-exclusion chromatography is widely used, as are ultrafiltration methods. Most contaminants are of lower molecular weight than antibodies. The nucleic acids DNA and RNA, which vary widely in size, can be digested using nucleases to convert them to small oligonucleotides that are readily removed by sizing methods.

TECHNIQUE FOR PURIFICATION

The choice of a technique for purification is based on a number of different factors: the antibody, the feedstock, scale, economics, timing, and desired purity. As mentioned above, this article will concentrate on the most demanding type of antibody purification, that of multigram lots of monoclonal antibodies produced under GMP and for use in parenteral formulations. The purification process can be divided into four parts. The purpose of the first part, *sample preparation*, is to ensure the readiness of the sample for the first purification step. Sample preparation is sometimes not necessary, depending on the nature of the crude feedstock and on the choice of the first purification step. Sample preparation is followed by the first purification step, often known as *initial product capture*. Most typically, the methodology for this step is chosen to give a high purification factor. Indeed, for antibody products with less stringent purification requirements (e.g., for in vitro use), this may be the only purification step required, if properly chosen. The third part of the procedure is any *additional purification step(s)* necessary to reach the purity goal. These steps are sometimes referred to as polishing steps. The final part of the procedure is to put the product into its *final formulation*, that is, the desired final concentration and physical state. This may sometimes be accomplished by the final purification step itself, eliminating the need for separate additional steps.

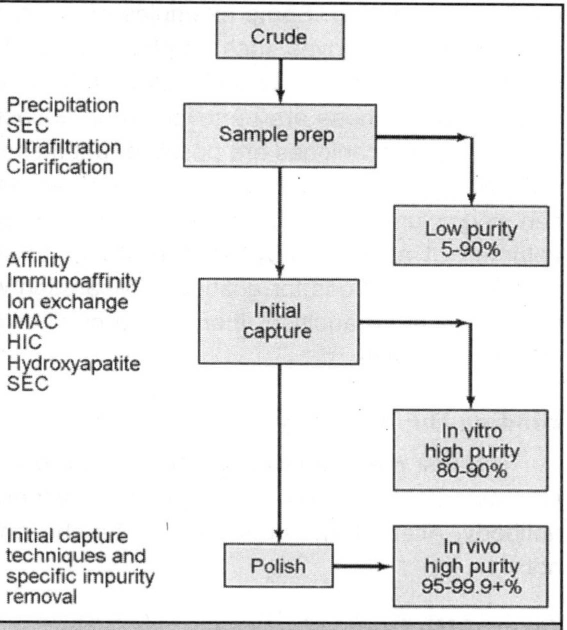

Figure 5.1 A general flow path for antibody purification, showing (on left) techniques typically used at each step.

Sampling Technique

The purpose of sample preparation is to modify the chemical and physical state of the crude product into a state compatible with the initial capture step. From a chemical standpoint, this could involve changing the ionic strength, pH, or type of ion in the product solution. In the simplest cases, this may involve dilution or the addition of salt as a solid or concentrated solution. In a current GMP (cGMP) manufacturing setting, such methods are not particularly desirable. Addition of volume means increased costs. Solids that may be added will generally not be sterile, thus requiring an additional sterile filtration of the modified sample. A more typical process at manufacturing scale would be a buffer exchange by size-exclusion chromatography (SEC) or by ultrafiltration or diafiltration. Because of its time- and volume-intensive nature, the related technique—dialysis—is typically used only on a laboratory scale.

From a physical standpoint, there are two typical processes carried out. One is concentration by ultrafiltration. The second is removal of particulate matter, by microfiltration or centrifugation. A very dirty feedstock (e.g., a batch fermentation) would clog a filter rapidly, making centrifugation the choice. However, for a relatively clean feedstock, the simplicity of filtration makes it the economic choice. On a small scale, precipitative techniques such as ammonium sulfate, capyrilic acid, and polyethylene glycol (PEG) are also in common use. These techniques actually have at least modest purification power and generally have a concentrating effect as well. These techniques are less common at large scale, especially sterile operations, due to engineering considerations. Also, recovery from these operations can be low. In addition to the above very common sample preparations, there are a number of other possibilities.

It is sometimes desirable to remove certain contaminants early in the process, even though their concentration is low. Their removal is needed because their presence could affect subsequent purifications step(s). Some examples of these contaminants are DNA and RNA, endotoxins, and cell culture media dyes such as phenol red, which binds tightly to certain chromatography resins and could diminish their capacity. Bound dyes may also change the selectivity of a chromatographic resin. Many of these are removed when antibody is prepared by the precipitative methods noted earlier. Other techniques are possible, but these are generally used as secondary purification steps and will be described as such later. Another example of a troublesome contaminant is lipid, which can clog columns, affecting their performance and lifetime. Bulk lipids, especially as found in small volumes of ascites, may be partially removed by high-speed centrifugation. As before, the precipitative methods for antibody preparation will generally leave lipids in solution, thus effecting their removal. In addition, there are a number of extractive and chromatographic techniques that can remove lipids.

Binding the Antibody

The first major purification step in a process typically involves binding the antibody to some chromatography matrix, while impurities either flow through or are differentially eluted from the antibody. Alternately, conditions can be chosen where contaminants bind, and the antibody does not.

Immobilizing technique

The most common matrix used to bind antibodies is immobilized protein A. This *Staphylococcus aureus* cell surface protein binds to many, although not all, antibodies and is widely available. Another bacterial cell surface protein, known as protein G, has been introduced more recently. Protein G binds more types of antibodies than does protein A, and it binds them more tightly. Although more versatile than protein A due to its wider specificity, the tighter binding of protein G can cause problems, as will be described later. Both of these are widely used because of their broad specificity, wide availability, and the high degree of purification. Indeed, purities of more than 95% are readily obtained, and antibodies of this purity are usable in many applications. Historically, there have been three problems related to the use of protein A. First, it is expensive. However, given proper care, it is possible to reuse protein A resins many times, decreasing the long-term cost. Second, early protein A resins tended to leach protein A into the product. This is no longer a common problem, although leaching should be tested for, especially for products intended for in vivo use.

Protein A assays are now readily available for such testing. The third problem is that many antibodies require a fairly low (pH 3-4) pH to elute from protein A. There are some antibodies that cannot withstand exposure to these conditions, even if the condition is neutralized as soon as possible. Needless to say, this is an even greater problem with the tighter binding of protein G. In those instances where an antibody is acid sensitive, but it is still desirable to use protein A or G, a number of potential alternative elution conditions have been reported. These include the use of basic pH conditions, chaotropic ions or basic pH followed by ethylene glycol, and peptides that mimic the protein A or G binding site. In addition to purifying bulk antibodies, proteins A and G have other applications in the field. For instance, they can be used to remove residual whole antibody and Fc fragments in preparations of Fab and $F(ab')_2$ fragments, since both proteins A and G bind to the Fc region of antibodies.

Another use is subclass separation from a polyclonal antibody feedstock (e.g., serum). Although this is not always the case, often different subclasses of antibodies have different affinities for proteins A and G, which allows chromatographic fractionation of the subclasses. The differential binding of protein A for different subclasses is not only a blessing, but also a curse and a need for caution. One subclass that does not generally bind well to protein A under physiological conditions is murine IgG1. This is currently the most common monoclonal antibody type seen. However, this problem has been overcome, in that a number of different regimens, in particular the use of slightly alkaline binding buffers of high ionic strength, have been developed to efficiently purify murine IgG1 on protein A. The need for caution lies in that not even antibodies of the same subclass will behave exactly the same. Thus, although general guidelines for the binding and elution of different classes of antibodies are helpful, the protein A or G purification of each antibody must be individually optimized. Similar natural affinity ligands exist with other specificities. Some of these are based on the fact that antibodies are glycoproteins. Examples are mannan binding protein, which targets the mannan sugars of IgM, and the lectin jacalin, which binds human IgA1.

Forms of Antibodies

The most selective technique is that of immunoaffinity, which can take a number of forms. In the most specific, the antigen for the desired antibody, or an anti-antibody against it, is used to make a chromatographic resin. In a somewhat less-specific fashion, antibodies against specific types of light or heavy chains can be used. However, immunoaffinity suffers from a number of problems. First, the antigen or anti-antibody must be produced and purified. Second, binding is often so strong that very vigorous and potentially harmful elution conditions must be used. On a production scale, the economics of producing the antigen or anti-antibody often make this technique undesirable.

Finally, for those antibodies intended for in vivo use, assay techniques must be available to show that the product is free of antigen or anti- antibody. Furthermore, unless process steps downstream of the immunoaffinity column provide adequate viral clearance, as shown by validation studies, the ligand used to make the immunoaffinity support will need to have been purified from a process validated to remove viruses. Also, it must be demonstrated that either the product is free of any leached ligand (as noted above) or the ligand does not affect product safety. These additional requirements would add significantly to the cost of immunoaffinity chromatography.

A renaissance of this technique may come from the advent of chemically and genetically engineered pseudoantigens. Developed for specificity and reasonable binding affinity and produced by recombinant or chemical methods, these compounds (often small peptides) could overcome most, if not all, of the earlier objections. A similar case is the development of mimics of protein A and of generic anti-antibodies. Generic anti-antibodies are those targeted against constant regions of the antibody, making their usage more general. A number of these are available commercially.

Procedure for chromatography

Another very widely used technique is ion exchange chromatography (IEC). Although less specific than proteins A and G, it can often provide purities of more than 80% in one step, which is adequate for many purposes. Both cationic and anionic IEC can be used, in either binding or flow through modes. The choice of method and its utility depend on the relative ionic characteristics of the antibody and its impurities. Care must be taken in selecting the resin and conditions to be used, not only from the standpoint of resolution, but also of capacity. If conditions are chosen where both the antibody and impurities bind, the capacity of the resin for antibody could be dramatically reduced. The pl of a protein is not an exact predictor of a protein's behavior on IEC. The charge distribution on the surface of the protein and a number of other minor effects also play a role. Nonetheless, the pl is a good first approximation.

In one study of 15 murine monoclonal antibodies, pls of 4.9 to 8.3 were reported. Bovine albumin has been reported to have a pl of about 4.5 to 5.0. The pl of another common impurity, transferrin, ranges from 5.5 to 6.0, depending on the species and degree of iron saturation. Thus, it is quite possible to have an antibody that is acidic enough to be inseparable from transferrin and even albumin by IEC. One of the first steps in developing an antibody purification procedure is to determine the pl of the antibody and the contaminants. This may be done quickly and easily and will give a good indication of how useful IEC might be.

Initial purification

In addition to the widely used techniques described in the previous section, a number of others maybe used for the initial purification of antibody. One technique that is growing in use is immobilized metal affinity chromatography (IMAC). Metal ions, usually Ni^{2+}, are held to a chromatography resin by ligands. In turn, these metal ions can interact with proteins, primarily with histidine groups. Many engineered antibodies and fragments are being produced with a polyhistidine tail, which gives this technique a high specificity. IMAC can sometimes also be used for nonengineered antibodies that have naturally occurring surface histidines, although histidine is an uncommon amino acid.

Hydrophobic interaction chromatography

Hydrophobic interaction chromatography (HIC) is another technique that can be used for initial capture. Antibodies tend to be among the most hydrophobic of the proteins in the crude feedstock. Thus, this technique can be quite powerful. An example is given in Figure 5.2, where bovine serum proteins are separated from antibody. On the down side, experience shows that the loading buffer must be carefully chosen. As with IEC, conditions where most of the contaminating proteins bind along with the antibody should be avoided for the initial capture. In such a situation, capacity is decreased and selectivity diminished. Also, selection of an appropriate media can be somewhat more tedious than for IEC, because the choice of potential binding strengths is very broad. Not only is there supplier-to-supplier variation in ligand density and surface chemistry, variabilities also seen in IEC, but there are a large number of binding groups, such as ethyl, propyl, butyl, phenyl, ether, phenylether, and others. Even reversed-phase media, such as octyl, can be used in the aqueous HIC mode with weakly hydrophobic proteins. Because of the variation in manufacturing, very little can be said in a general fashion about the strengths of the various resins. Within an individual manufacturer's line, usually the longer the alkyl chain length, the stronger the binding.

In practice, however, tests should be made. It is important to make these tests, because the selection of a resin is an important way to manipulate selectivity. Another means of altering selectivity is through the choice of buffer. In general, the stronger the lyotropic effect, the stronger the induction of hydrophobic binding by a salt. However, strong salt solutions can have a detrimental effect on the integrity of proteins, so the choice should be limited, if possible, to those with strong salting out effects. Ammonium sulfate is the most commonly used salt, but it has the disadvantage of outgassing ammonia above pH 7.5. Sodium sulfate and potassium phosphate are also good choices, but are somewhat limited in solubility. Because of the limited choice of buffers, the proper choice of resin becomes important. In addition to salt, the selection of pH can have an effect on retention.

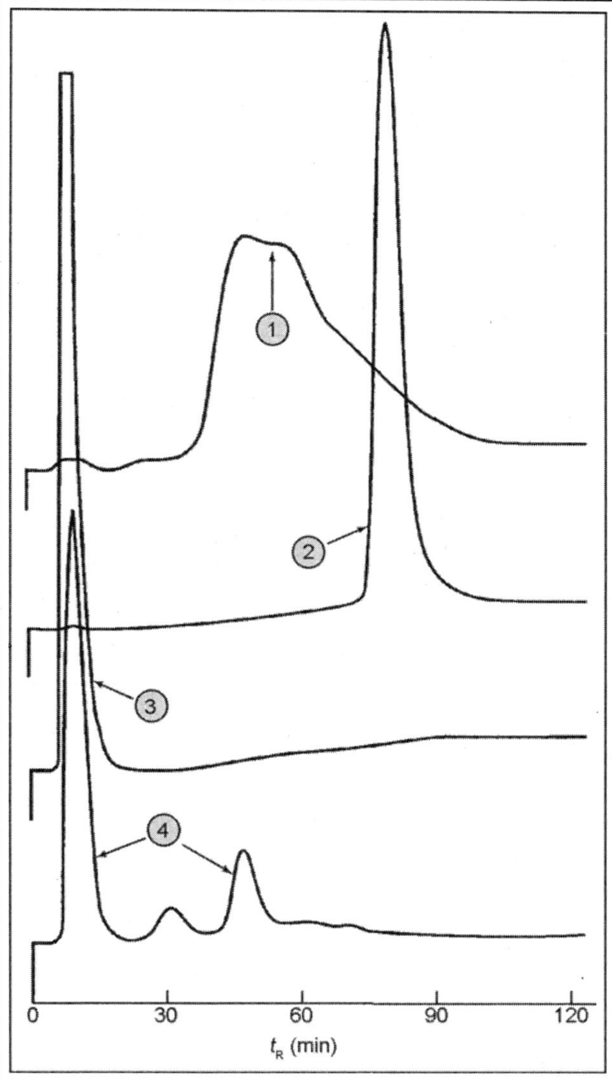

Figure 5.2 HIC separation of mouse polyclonal IgG from fetal bovine serum proteins and bovine polyclonal IgG. Chromatogram 1, bovine polyclonal IgG; 2, mouse polyclonal IgG; 3, partially purified fetal bovine serum; 4, neat fetal bovine serum.

In general, above pH 9 and below pH 4, retention increases, probably a result of denaturation of the protein. Between these pHs some proteins show an effect, but others do not. It is not possible to make generalizations; rather, the effect must be tested. Solution modifiers such as sugars, alcohols, PEG, and urea can also affect both binding and elution and can play a role in special situations. HIC is perhaps deserving of more attention than it has received. Not only is it potentially a powerful technique itself, but for multistep procedures, it can be an ideal complement to IEC. IEC uses a low-ionic-strength load and a high-ionic-strength elution. HIC used a high-ionic-strength load and a low-ionic-strength elution. Thus, it is quite possible that conditions can be developed to allow one column to flow directly to the next, without intermediate sample handling.

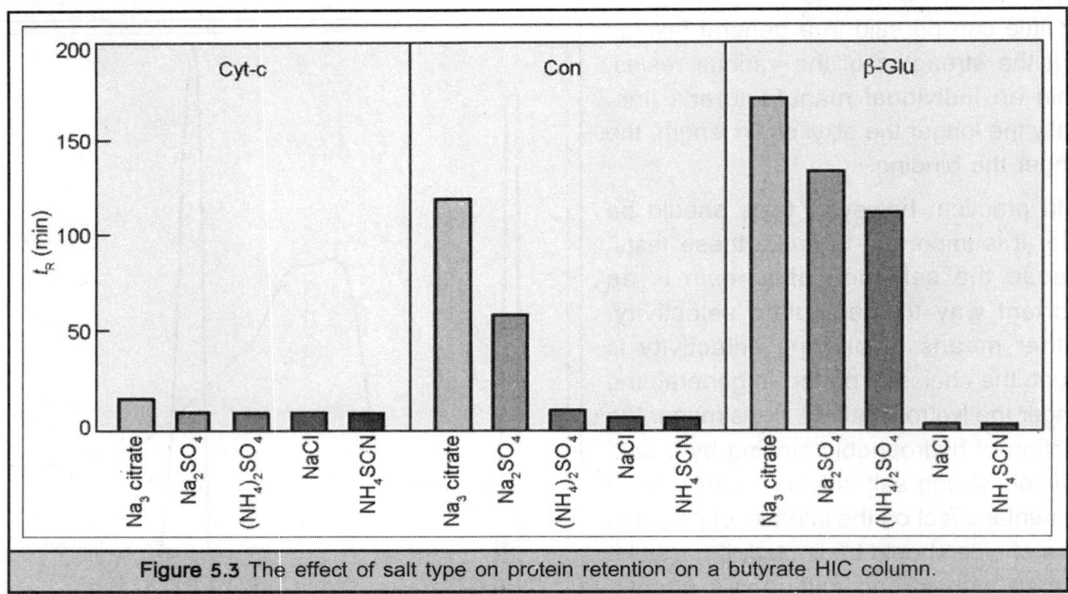

Figure 5.3 The effect of salt type on protein retention on a butyrate HIC column.

Hydroxyapatite chromatography (HAP)

Hydroxyapatite chromatography (HAP) has seen considerable use on a small basis, but little use on the large scale. The main reason for this is that until very recently, all versions of this material were mechanically fragile. Low flow rates were typical, and reuse of the columns was limited. The mechanism of separation is not fully understood, but interaction of the protein with the calcium and phosphate in the resin matrix plays an important role, because a phosphate buffer gradient is most typically used for elution. As with other nonspecific techniques such as HIC and IEC, antibodies exhibit a wide variety of behavior on HAP, and generalizations are somewhat unsafe. However, HAP has been used not only for purifying antibodies, but also for separating different antibody subclasses. With the recent development by several companies of more robust versions of this resin, its use for large-scale production of antibodies is bound to expand.

Size-exclusion chromatography (SEC)

SEC is typically a low-resolution, low-throughput technique. Thus, in general, it is more often used for buffer exchange or polishing steps. However, there are instances when it can be used for initial purification. This is the case when dealing with very large antibodies such as IgM and IgA. IgM is typically a pentamer, with a weight nearing 1 million Da. IgA is usually found as a dimer, with a weight of about 300,000 Da. Thus, these are much larger than most contaminating proteins and can be purified by SEC.

Other methods

Another method that has seen some use is thiophilic chromatography. In this technique, the sulfur ligand of the resin is thought to interact with the aromatic groups of the proteins. It has been shown that this technique can be used to effectively purify proteins. Like HIC, this technique generally loads proteins in high-salt buffers. However, unlike HIC, high salt concentrations also promote elution. Needless to say, different salts have different effects. In addition to chromatographic techniques, there are a number of other techniques that can substantially purify antibodies. The precipitative techniques were mentioned earlier in "Sample Preparation." Ultrafiltration and diafiltration may also be used. Certainly, low molecular weight impurities may be removed by this technique. However, in addition, by careful selection of the membrane and buffer used, some contaminant proteins may be removed as well, such as albumin from IgG. The technique of preparative electrophoresis has gradually been advancing during the past few years.

Currently it is can be considered useful for small-scale (milligram) bench preparations, especially non-GMP. Techniques for true production-scale electrophoresis that can be performed aseptically are being developed. Given the resolving power of analytical electrophoresis, success at developing production-scale techniques could add a valuable new tool to the development of purification methods.

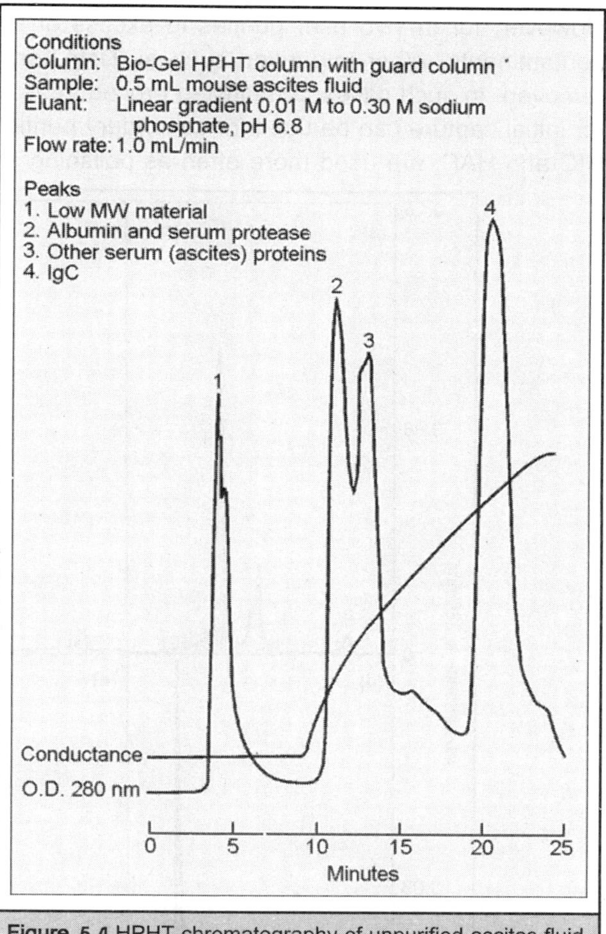

Figure 5.4 HPHT chromatography of unpurified ascites fluid containing monoclonal antibody.

In purification of antibodies on a large scale, process time is very important for economic reasons. One way to increase the efficiency of a process is to combine steps. In the past few years, a number of approaches to combining the concentration and clarification aspects of the sample prep step and the initial capture step have been commercialized. These approaches are generally either a fluidized bed, large-diameter bead resins, or wide-channel tangential flow filters. They are all designed so that cells and cell debris will flow through without clogging the setup, thus achieving clarification. The antibody is to bind to the resin or membrane, and after clarification is complete, it can then be eluted, providing both concentration and purification.

Advanced Method of Purification

Depending on the feedstock, the technique selected, and the optimization, the initial capture step can often give purities in the range of 80 to 95%. This is often adequate for in vitro usages.

However, for in vivo use, purities in excess of 99% are usually required. In addition to protein contaminants, other impurities such as DNA, endotoxins, viruses, and aggregates need to be removed. In such cases, a multistep procedure is almost inevitable. All the same techniques used for initial capture can be used for secondary purification. Indeed, many of the techniques, such as HIC and HAP, are used more often as polishing steps than as initial capture steps.

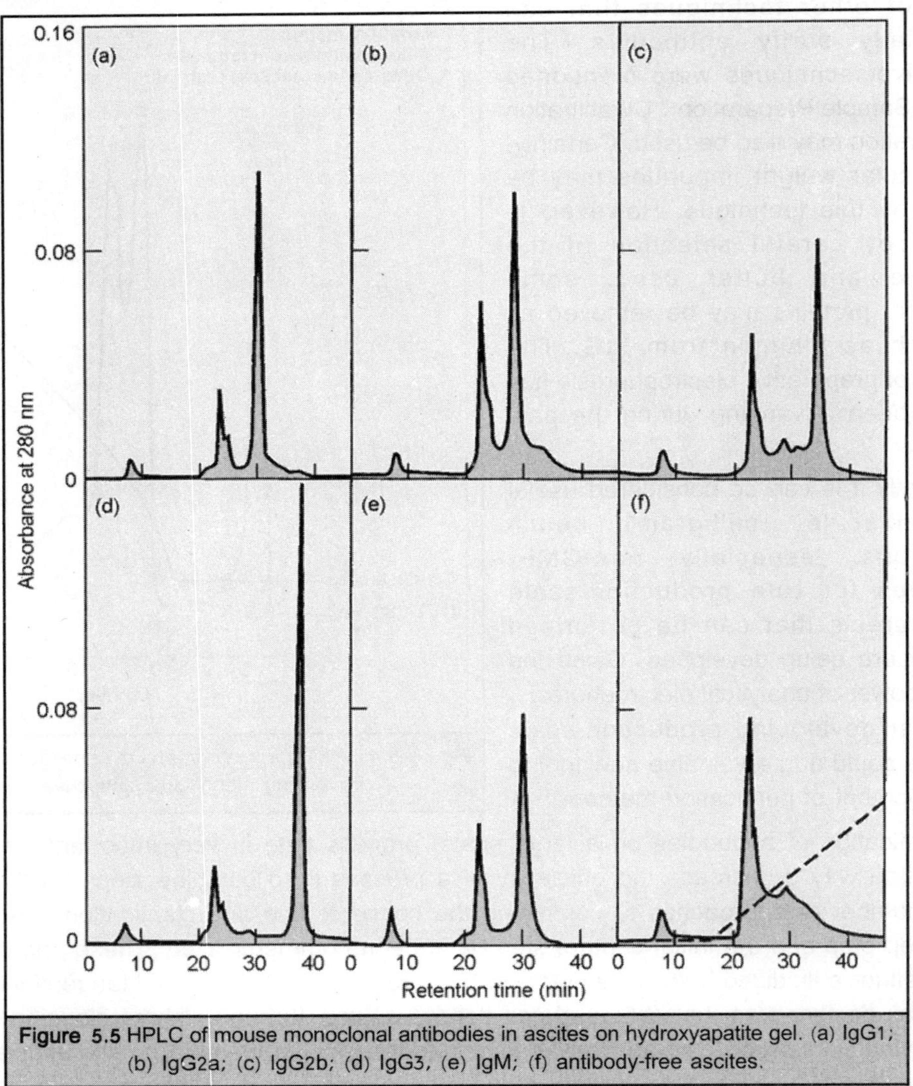

Figure 5.5 HPLC of mouse monoclonal antibodies in ascites on hydroxyapatite gel. (a) IgG1; (b) IgG2a; (c) IgG2b; (d) IgG3, (e) IgM; (f) antibody-free ascites.

The orientation of these secondary steps tends to be different from the initial capture step. With initial capture, the goal is to remove as many and as much contamination as possible. The secondary purification steps tend to be planned to remove individual contaminants. For instance, DNA and RNA tend to be highly negatively charged at just about any physiological pH. Thus, anion exchange resins are often used to remove these nucleic acids from antibodies, with conditions usually selected so that the nucleic acids bind while the antibody flows through. In addition to the techniques mentioned in the initial capture step, there are a number of other techniques aimed at removing specific contaminants. For example, materials are available commercially that are intended specifically for the removal of impurities such as endotoxins, viruses, DNA and RNA, and albumin. In one way or another, these products take advantage of specific attributes of the impurity to remove them.

Some general examples of secondary purifications are discussed in this section. These include both techniques covered in the initial capture section and special polishing techniques. The removal or inactivation of viruses must be validated for any animal-sourced or animal cell-sourced products, including antibodies, intended for in vivo use. All the techniques mentioned to date have the potential for removing virus. A specific attribute that is applicable to all viruses is their size. Being significantly larger than antibodies, they can be removed by a size basis. Although this can be done by SEC, a higher degree of virus removal is typically seen by ultrafiltration, with a membrane specifically selected for a molecular weight cutoff appropriate to retain virus and pass antibody.

In addition to virus clearance, virus inactivation is often performed. The most common techniques used are detergent treatment, low pH, or high temperature. Needless to say, the antibody must be tested to be resistant to these conditions before their routine use. As mentioned earlier, one characteristic of DNA and RNA is the high negative charge. As well as anion IEC, HAP is well known to be effective at binding nucleic acids because of this high charge. HIC also may remove nucleic acids, because nucleic acids do not bind, whereas antibodies do bind to HIC resins. Other properties of nucleic acids are the long, thin nature of the polymer, and, of course, the nucleotide sequence itself. Various precipitation techniques take advantage of this to remove DNA and RNA. Also, complementary nucleotides bound to chromatography supports can be used as an affinity adsorbent. Finally, as alluded to in the "introduction," DNA and RNA can be enzymatically digested using nuclease enzymes (e.g., Benzonase) to generate very small fragments that can be readily removed from antibodies by sizing techniques.

Endotoxins are lipopolysaccharides. Many can be removed by anion exchange chromatography. As with the nucleic acids, HIC has been reported to be able to remove endotoxins for antibodies. In addition, both precipitative and chromatography techniques focusing in on the lipid portion of the endotoxins have been developed. Albumin is probably the most typical and highest concentration contaminant in most feedstocks, whether serum, ascites, or cell culture media. As with the other contaminants, the techniques already mentioned have the potential for removing albumin. However, if additional specific removal is required, antialbumin affinity chromatography is possible. Also, many serum proteins, including albumin, can be removed using dye-ligand chromatography. A well-known technique in laboratory chromatography, it has not been much used in preparative-scale work. The dyes used were originally taken from the textile industry and were not particularly pure. In addition, there was concern about ligand leakage. However, modern dye-ligand resins are made with high-purity dyes prepared especially for this purpose; coupling procedures have improved, minimizing leakage; and assays are often available to track what leakage there may be.

Perhaps the greatest challenge for the polishing steps is the removal of form variants. The easiest form variant to remove is aggregate, which can be removed by SEC, a step that may be needed for final formulation (see next section) anyway. In addition, there may be light or heavy chain variants if a cell-line producing antibody is not stable and specific. There is the possibility of host antibody from serum, ascites, or serum-supplemented cell culture. And there can be intramolecule variants caused by variations in glycosylation, deamidation, proteolytic digestion, and a number of other possible causes.

It is difficult to make a blanket statement in regard to how or even if such variants need be removed. If these variants prove to be extremely difficult to remove, and the product can be proved to be safe and effective with their presence, the economics of the situation may dictate that they not be removed. Nonetheless, variants may need to be removed, and some examples of possible techniques follow. For chain variants and other source antibodies, if the physical characteristics are significantly different compared to the target antibody, they probably can be removed by IEC, HIC, etc. But, if not, one possibility is HAP. This technique has long been known to sometimes be able to separate different idiotypes. Dye-ligand chromatography has also been shown to be capable of separating subclasses. In addition, the different affinity of different antibodies for protein A and protein G allows these resins to be used for variant removal at times.

Immunoaffinity directed against the undesirable variant heavy or light chain(s) is a powerful technique. Glycosylation variants may sometimes be removed by lectin or immunoaffinity chromatography directed against specific sugars. If physical changes such as deamidation do not change the antibody's charge enough for separation by IEC, preparative electrophoresis or chromatofocusing may sometimes be successful. But, basically, this is a struggle that must be developed on a case-by-case basis and often requires a unique solution. In general, the difference between initial capture and additional purification steps may be summarized in one word: resolution.

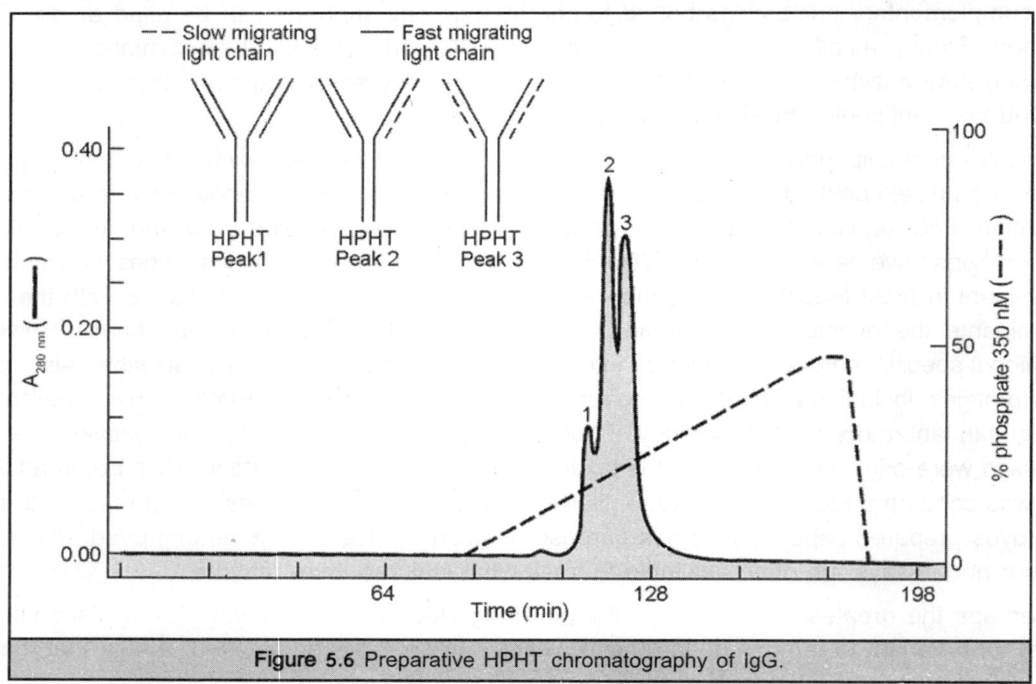

Figure 5.6 Preparative HPHT chromatography of IgG.

In initial capture, the feedstock is a moderately to highly complex mixture. The feedstock may be dirty, that is, contain particulates, if sample preparation is not performed. On the production scale, large volumes are the norm. The goals in initial capture are to remove most of the impurities, with good yield, and to concentrate the product. With secondary purification, the feedstock is relatively clean, most of the protein is the target antibody, and some residual impurities may be physically very similar to the target antibody. Thus, the goal in secondary purification is to achieve high-resolution separation.

Final Formulation

Final formulation can be considered as a part of purification in that it removes conditions that would impair the stability or utility of the antibody in its intended use. One goal of a purification strategy should be to limit the additional processing required to put the antibody into the final formulation. For example, ideally the antibody should be recovered from the final purification step in the same buffer needed for its final formulation. Often this can be done if this goal is kept in mind when the purification process is being developed. Final formulation may be as simple as a straightforward sterilization by membrane filtration through a sterile filter with pores 0.2 μm or less. Another relatively simple formulation step is adjusting the antibody concentration, either by dilution with buffer or by concentration on ultra- filtration. In other circumstances, the buffer composition may need to be changed to achieve optimal stability of the antibody. For this purpose, SEC or diafiltration is widely used. A more complex formulation step would be the addition of excipients to confer stability. Finally, the antibody solution may need to be lyophilized to confer stability, and the liquid formulation may be changed to be compatible with the lyophilization process. Such complex formulation strategies are beyond the scope of this article, so the reader is referred to reviews on this subject.

OBTAINING PURE POLYCLONAL ANTIBODIES

Purification of polyclonal antibodies is actually more of a challenge than the purification of monoclonal antibodies. This is the case for two reasons. The first is related to the method of production. Polyclonal antibodies are produced naturally, by the immune response of an immunized animal. The antibody is purified from serum, with the necessity of removing all the serum impurities, including host antibodies. Monoclonal antibodies, however, are produced in incubators. Historically, for small batches, the incubator has been a mouse, with the resulting ascites fluid containing the desired antibody. This ascites fluid is related to serum and can be an equally complicated mixture. For large lots, and increasingly for small lots, monoclonal antibodies are produced in an artificial reactor, using at least a semidefined culture media.

Again, historically, culture media could be considered only semidefined because it was supplemented with serum. However, even in this case, the concentration and number of macromolecular contaminants to be removed are lower than those in serum. Also, the current trend for monoclonal antibody production is toward use of serum-free or even protein-free media. The second reason that polyclonal antibodies may be more challenging to purify is intrinsic to the antibody itself. Monoclonal antibodies are not necessarily absolutely homogenous. Often, there are differences in glycosylation. And there will be aging differences, such as in degree of deamidation. They will, however, all be of the same isotype, directed against the same epitope. This is not the case for polyclonal antibodies.

Polyclonal antibodies tend to have a broader pI range than monoclonal antibodies, making IEC and related techniques such as HAP more complicated. They also tend to give broader peaks on hydrophobic interaction. The same techniques used for purification of monoclonal antibodies are used for the purification of polyclonal antibodies. Indeed, most of the techniques were developed before the advent of monoclonal antibodies. As with monoclonal antibodies, the key to selecting the right purification techniques lies in knowing the desired yield, the desired purity, and the nature of the contaminants. Success at producing high-purity polyclonal antibodies centers around two points: extensive preparation and the use of affinity techniques. In regard to preparation, characterizing all the individual impurities in a crude polyclonal mixture would be out-landishly time consuming. However, a valuable approach is to run standard IEC, HIC, and HAP conditions, where most if not all the proteins initially bind.

Analysis of the elution fractions by an analytical technique such as two-dimensional electrophoresis will indicate which column technique can remove which impurities. The limitation to this approach is the well-known fact that the binding conditions used in chromatography affect the resolution of elution. The affinity and pseudoaffinity techniques such as dye-ligand, IMAC, and thiophilic chromatography can be important in several roles. These techniques can be selected not only to bind to the target antibody, but also to remove specific impurities. For instance, one of the most vexing problems in antibody purification is removal of host antibody. If a specific isotype is desired, affinity techniques that bind to the other isotypes can be used to remove these unwanted host antibodies.

Generic affinity techniques such as protein A and G are still very powerful tools. And, if practical to use, immunoaffinity has excellent selectivity. In summary, there is nothing unique in purifying poly- clonal antibodies compared to monoclonal antibodies. To reach a given degree of purity, the polyclonal antibody will probably take more planning; more preparation; and for high purity, use of more powerful techniques, especially affinity techniques.

Unwanted Characters

As has been seen, antibodies can be remarkably diverse. One problem seen is when an antibody does not behave in a fashion similar to others of the same isotype. The solution of this type of problem is just good old-fashioned process optimization. A much more serious problem is the purification of an antibody with limited solubility. A wide variety of additives can potentially be used for solubilization of antibodies, including urea, guanidine HCl, detergents, and sugars. Needless to say, there is the potential for these agents to interfere with purification steps as well. Thus, solution of this problem tends to require rather extensive development.

A similar problem is lack of stability. This may be intrinsic to the antibody, that is, it occurs even under physiological conditions, or it may be caused by a harsh purification condition, such as a low pH protein A elution or a virus-killing step. In the second case, generally alternate conditions are possible. Examples of alternates for both the examples are given earlier in this article. The solution of an intrinsic stability problem is similar to that of a solubility problem. These problems can be quite complicated and are beyond the scope of this article, but good references are available.

ANTIBODIES OF PHARMACEUTICAL PURPOSE

Although purities of 95% or greater are often suitable for antibodies intended for nonpharmacological use, such as in vitro diagnostic reagents, much greater levels of purity are required for those intended for use as pharmaceuticals. In general, there are three types of contaminants of concern to regulatory agencies: microbes, viruses, and their products, such as endotoxins, proteins, and DNA. These contaminants must be removed to trace levels. The need for sterility of a biological product intended for use in humans is a given. This is primarily assured by strict adherence to both aseptic technique and cGMP (100). Less apparent is the need for a validated process to remove any *potential* viruses that may be present in the crude material, including murine retroviruses, which are present in almost all hybridoma cell lines. These issues have been addressed in some detail by the regulatory authorities. These regulatory authorities, such as the Center for Biologics Evaluation and Research division of the FDA, have relied on a three-tiered approach to controlling any infectious agents that may be present in the crude product or acquired during the downstream processing.

1. Careful selection of raw materials, including media components, for the absence of biological contaminants
2. Validating that the purification process can remove or inactivate viruses
3. Full testing of the product at appropriate stages in the process for the presence of viruses and infectious agents

Each of these three tiers of control must be adhered to by those engaged in the purification of antibodies for human use. A full treatment of these regulatory considerations is beyond the scope of this article. The reader is encouraged to refer to the cited references and keep abreast of current regulations issued by the various governmental agencies. The antibody molecule is one of the most versatile biological molecules created. Advances in immunology, biochemistry, and molecular biology have yielded a wealth of information about the structure and function of this class of proteins. The huge diversity of their binding sites allows one to acquire exquisitely specific binding reagents for almost any molecular target. Yet, the conserved nature of their constant domains as well as their characteristic three-dimensional structure allows the purification specialist to approach their recovery with much understanding of the behavior of these biomolecules. In spite of this, the adage that each monoclonal antibody is unique must be remembered, because even monoclonal antibodies from the same species, class, and subclass may behave very differently in purification procedures. With this caveat in mind, it is our hope that this review will provide a helpful guide to the purification of these fascinating biomolecules.

6

ANAEROBIC MICROORGANISMS

There are both aerobic and anaerobic microorganisms that are important to the advancement of biotechnology. Anaerobic microorganisms include obligate anaerobes that require oxidation–reduction potentials of –150 to –420 mV and facultative anaerobes that can grow at oxidation-reduction potentials between +300 and –420 mV. Oxygen must be excluded in working with obligate anaerobes. Many important fermentation products are produced under anaerobic conditions. These include food products such as bread, yogurt, cheeses, wine, beer, sufu, and sauerkraut. Acetone, butanol, and ethanol are examples of industrial chemicals produced by anaerobes. Specialty chemicals produced by anaerobes include vitamins and pharmaceutical products. Major references describing the microbial products and the microorganisms that produce them include those by Erickson and Fung, Reed, Steinkraus, and Zeikus and Johnson.

Anaerobic digestion and biodegradation processes are applied widely to treat process waste products and in environmental restoration. Environmental microbiology is significant because of the beneficial impact of natural and nurtured processes and because of the isolation and identification of useful microorganisms that have evolved in natural environments. Anaerobic processes occur in production agriculture; cattle and sheep are still the greatest commercial success in harvesting and utilizing cellulosic plant products because of the anaerobic fermentations that occur in the rumen of these animals. The taxonomy of anaerobes has evolved as new information has become available. Based on 16/18 S rRNA sequence comparisons, a universal phylogenetic tree has been described that includes bacteria, archaea, and eukarya. Anaerobes are found in all three branches of this tree.

CULTIVATION TECHNIQUE

The microbial world consists of aerobic and anaerobic organisms, defined by the ability to tolerate molecular oxygen. Sonnewirth traced the evolution of anaerobic methodology and recorded that Leeuwenhoek demonstrated that some life forms could exist in the presence of gases other than oxygen and that Pasteur discovered anaerobiosis in 1861, coining the terms *aerobes* and *anaerobes*. The separation between these groups is not clearly defined. A convenient way to categorize these organisms is by observing their ability to grow at different oxidation– reduction potentials (E_h). E_h can be measured and expressed + or – millivolts (mV). Thus, aerobes can grow between +300 and –50 mV; facultative anaerobes, +300 to –420 mV; aerotolerant anaerobes, +180 to –350 mV; and obligate anaerobes, –150 to –420 mV. The microbiological procedures for studying obligate anaerobes are similar to those used for the more familiar aerobic organisms, with the exception of the need to exclude oxygen from the working environment during manipulation, inoculation, and incubation of cultures and samples. As a class of organisms, anaerobes are very diversified in their behavior in terms of oxygen tolerance and nutritional requirements for growth.

There is no single set of procedures for isolation of all anaerobes. Clinically important anaerobes constitute the most studied class. Specific methods were developed for isolating anaerobes from various sites in patients. A detailed treatment of clinically important anaerobic organisms is recorded in the *Manual for the Determination of the Clinical Role of Anaerobic Microbiology* by Gall and Riely. This manual contains a definition of clinically important anaerobic bacteria methods for the collection and transportation of anaerobic specimens, culturing of anaerobic specimens, and identification of anaerobes. The book *Isolation of Anaerobes*, by Shapton and Board, contains chapters describing the isolation of clostridia from animal tissues, feces, soil, and foods. Some chapters deal with the isolation of anaerobes from rumen as well as with sulfate-reducing bacteria and photosynthetic bacteria. Zeikus describes the biology of methanogenic bacteria and methods of studying these organisms. In the area of food microbiology, very little attention has been given to the anaerobic microbiology. The *Compendium of Methods for the Microbiological Examination of Foods* contains some chapters on anaerobes, mainly on the spore formers and the detection of toxins in foods.

Anderson and Fung reviewed the status of anaerobic methods, techniques, and principles for food bacteriology in detail. The history of cultivation of anaerobes was detailed by Hall and more recently summarized by Fung. The general conclusion on isolation procedures is that samples must be free from contact with air during collection time. The best method of collecting samples is by aspiration with needle and syringe. Sample should be placed in media that are designed to handle anaerobes. A variety of liquid and solid media are marketed for anaerobic cultivation by such companies as DIFCO, BBL, UNIPATH/OXOID, and others. Currently, anaerobic cultivation systems include anaerobic jars, anaerobic glove boxes, roll tubes, anaerobic tube systems, and biological membrane systems.

Anaerobic Jars

Anaerobic jars that use both H_2 and CO_2 were developed by Brewer and Allgeier. The chemical generator consisted of a sodium borohydride tablet and a citric acid– sodium bicarbonate tablet in a convenient package. These two tablets, when activated by the addition of water to the package, produce H_2 and CO_2 gas, respectively. The system required no vacuum or electric current because the $H_2 + O_2$ reaction was catalyzed by a cold catalyst (alumina pellets coated with 0.5% palladium). CO_2 is used in the system for the promotion of growth of some anaerobic bacteria. Some companies are marketing these anaerobic jars (e.g., BBL, B.T.L., and UNIPATH/OXOID). The advantages of the anaerobic jar include: (i) ease of operation, especially for the small laboratory; (ii) economy of space; (iii) convenience of prepackaged materials for the generation of gases; and (iv) safety of the units. The disadvantages include (i) the culture plates are inoculated under aerobic conditions; (ii) the jar remains aerobic for 15 to 20 minutes even while the Gas Paks are activated, which may affect some strict anaerobes; and (iii) occasionally the jar fails to achieve anaerobiosis.

Anaerobic Chamber

To eliminate the chance of exposure of anaerobes to the natural environment during bacteriological manipulation, the anaerobic chamber, or glove box, was developed. The basic principles for achieving anaerobiosis in the glove box and for promoting binding of residual O_2 and H_2 in the presence of a catalyst (Laidlaw principle) are to remove air first and then flush the chamber with inert gas. Reducing agents in the medium also help to reduce the E_h to a desired low level. Companies marketing anaerobic glove boxes include Format Scientific, Don Whitney Scientific, and Microflow MDH Ltd. The advantages of the glove box include: (i) the bacteriological procedure can be operated under anaerobic conditions from the sampling stage to the incubation stage; (ii) the area for manipulation is large; and (iii) in combination with temperature control, the culture can be incubated in the chamber without further disturbance of the cultural environment. The disadvantages are: (i) high initial cost; (ii) high operating expenses (gas and electricity); (iii) high maintenance cost; (iv) requirement for operator training; (v) oxygen contamination through leakage, operator error, or mechanical malfunction; (vi) occasional high labor requirements; and (vii) moisture buildup in the box.

Common Method

One of the most widely used anaerobic cultivation systems is the Hungate roll tube, developed by Hungate. Sterile tubes were flushed with CO_2 while the medium was aseptically added to the tubes to prevent air contamination. The tubes were then inoculated, sealed with rubber stoppers, cooled (48°C), and hand spun so that the agar with cultures forms a thin film in the inner wall of the test tube. After incubation the colonies could be examined with ease. Isolation of colonies was also feasible with minimum effort by flushing CO_2 into the cavity to allow maintenance of reduced condition while performing microbiological operations. A popular version of the roll tube system is the VPI Anaerobic Culture System, developed by Moore. This apparatus has a platform so that the tubes can stand upright with cannulas positioned above each tube. The system also allows for streaking cultures into the roll tube anaerobically. The system is continuously flushed by CO_2 to exclude O_2 from the system. The advantages of the Hungate roll tube system include: (i) good anaerobiosis with minimum investment, (ii) unit operation for convenience of small sample numbers; and (iii) ease of observation of colonies and isolation of cultures for further studies. Disadvantages include: (i) time requirement for large number of samples; (ii) need for manual dexterity; and (iii) occasional failure of some tubes to achieve anaerobiosis.

Autoclaving Method

Ogg et al. designed a tube with double wells so that agar with cultures can be introduced into the tube and form a thin film. Anaerobiosis is initially achieved by boiling and autoclaving of the medium. By the action of reducing agent in the medium and minimum reabsorption of O_2 due to the design of the tube (commercially called the Lee tube), good anaerobiosis is maintained in the system. The glass cylinder forming the inner wall (the inverted smaller tube) makes it easy for enumeration and observation of colony morphologies. The Lee tube has the advantage of ease of operation and low operating cost (no gas or special equipment needed). The disadvantages include: (i) difficulty in picking isolates from the agar; (ii) difficulty in cleaning the tubes; and (iii) fragility of the system.

Fung and Lee developed a double-tube system for anaerobic cultivation of bacteria from foods. It consists of a glass tube (15 x 1.0 cm o.d.) placed into a larger glass tube (15 x 1.5 cm o.d.) with anaerobic agar (designed for specific organisms) and culture sandwiched between the two tubes. The screw cap of the larger tube when closed seals the system and makes it anaerobic. Anderson and Fung were successful in cultivating strict anaerobic rumen microbes, such as *Megasphaera elsdenii, Butyrivibro fibrisolvens, Clostridium perfringens, Eubacterium limosum,* and anaerobes from beef and poultry.

Ali et al. successfully used this system to isolate *C. perfringens* from meat. Aramouni et al. used the double-tube system to study *C. sporogenes* in home-style canned quick bread. The advantages of the system include: (i) ease of operation; (ii) inexpensive and ease for cleanup; (iii) ease of enumeration and observation of cultures; (iv) ease of obtaining colonies (by removing the inner tube and pick colonies); and (v) flexibility and ease of adaptation by other laboratories. Disadvantages include: (i) operation of samples in the open environment and (ii) occasional breakage of agar by gas-producing organisms.

Membrane Fragments of *E. coli*

A novel method using biological materials to achieve anaerobiosis was developed by Adler and Crow. Sterile suspensions of membrane fragments derived from *Escherichia coli* were able to remove oxygen rapidly and efficiently as a result of the presence of a cytochrome-based electron transport system that transfers hydrogen from suitable donors in the bacteriological media to oxygen and produces water as an end product. The membranes are stable, nontoxic, and active in both liquid and solid media. The history of the development of this concept and manufacturing of membrane fragments were described recently by Adler and Spady. The commercial product is named Oxyrase and marketed by Oxyrase Inc. Fung et al. described the use of Oxyrase to stimulate the growth of *Listeria monocytogenes, Campylobacter jejuni, E. coli* O157:H7, etc. to high numbers (10^6 CFU/mL) so that secondary detection methods such as polymerase chain reaction (PCR), enzyme-linked immunosorbent assay, and DNA/RNA probes can detect the presence of these facultative anaerobic pathogens much faster for food safety concerns.

Fung et al. also detailed the use of Oxyrase and membranes derived from *Acetobacter* and *Gluconobacter* to stimulate the growth of starter cultures to produce fermented food faster. Claerbout described the applications for bacterial membrane fragments in sterile pharmaceutical quality control. Thurston and Gannon made a detailed comparison between Oxyrase anaerobic agar plates and conventional anaerobic chambers for the isolation and identification of anaerobic bacteria from clinical infections and concluded that Oxyrase is a valuable compound for anaerobic cultivation of bacteria.

IDENTIFICATION OF ANAEROBES

The identification of anaerobes is a concern for applied microbiologists. A variety of rapid and automated methods in microbiology have been developed for the isolation, enumeration, and identification of microorganisms, and many of these new methods can be adapted to anaerobic bacteriology. Identification of anaerobes can be achieved by the following methods and procedures: conventional methods, miniaturized methods, commercial diagnostic kits, physical and biochemical methods, immunological methods, and PCR and related methods.

Conventional Methods

Identification of microorganisms using conventional procedures is summarized in Table 6.1. For anaerobes, the most important aspect is to have all the liquid and solid media incubated in anaerobic environments, as described in the previous section. After the reactions are obtained, microbiologists will determine the identities of the cultures by a variety of schemes, charts, and flow diagrams. The conventional method is time consuming and highly dependent on the knowledge and skill of the microbiologist involved.

Table 6.1 Needed information for identification of microorganisms

Needed information
Morphology under magnification and on agar plates
Gram reaction and special staining properties
Biochemical activity profile and special enzyme systems Pigment production, bioluminescence, chemi-luminescence, and fluorescent compound production
Nutritional and growth-factor requirements
Temperature and pH requirements and tolerance
Fermentation products, metabolites, and toxin production Antibiotic sensitivity pattern (antibiogram)
Gas requirements and tolerance
Genetic profile; DNA/RNA sequences and fingerprinting Pathogenicity to animals and humans
Serology and phage typing
Cell wall, cell membrane, and cellular components
Growth rate and generation time
Ecological niche and survival ability
Motility and spore formation
Extracellular and intracellular products
Response to electromagnetic fields, light, sound, and radiation Resistance to organic dyes and special compounds
Impedence, conductance, and capacitance characteristics

The morphology is one of the important considerations in identifying microorganisms by conventional methods. Under the microscope, the morphology of anaerobic microorganisms are similar to the aerobic counterparts. The morphology of bacteria are rod-shaped, coccal-shaped, and spiral-shaped organisms. There are also other bacteria that are pleomorphic (variable shapes depending on growth conditions). The morphology of yeast and mold are similar to the aerobic counterparts.

Miniaturized Methods

Fung and Hartman developed a variety of miniaturized microbiological methods for the identification and enumeration of aerobic bacteria. The entire concept and system were adapted for miniaturized anaerobic bacteriology simply by operating the procedures in an anaerobic chamber. Guy Miller and T.G. Nagaraja at Kansas State University found that miniaturized techniques could be used for studying rumen bacteria effectively. The advantages of miniaturized techniques are savings of material, space, time of incubation and operation, and labor. Another advantage is the flexibility of the system to be able to adapt to many different laboratories, especially for environmental and industrial settings where commercial diagnostic kits are not available for specialized organisms.

Advanced Techniques

Around the beginning of 1970s with the advancement of miniaturized microbiological techniques and concepts developed by Fung and colleagues and other scientists, many diagnostic kits appeared on the market mainly for clinical microbiology. These methods gradually found their ways into food microbiology laboratories and sometime into industrial and environmental laboratories. Currently commercially successful diagnostic kits are marketed toward clinical settings. This may change if the need arises in the near and far future. There are two major types of diagnostic kits. One type depends on *growth* of the pure culture and biochemical reactions obtained in the media. This type of test is typically slow, requiring overnight incubation and anaerobic incubation of cultures. Another type is by monitoring *enzyme systems* of the pure culture. In this system, anaerobiosis is not necessary and the results can be obtained in about 4 hours.

Kelly and Schipp made a detailed analysis of the approaches and designs of anaerobic diagnostic kits, including discussions on substrate concentration, organism concentration, pH of reaction, buffering capacity, additives, incubation time, and growth medium for the identification of anaerobes. In that article, important variables affecting system performance and issues concerning comparative analysis of data reported in journal articles were also studied. To compare data reported in the literature, one needs to address the issues such as source of strains, database version, the "gold standard," number of species/strains tested, need of supplementary tests, and accuracy. In general, a diagnostic kit providing 90 to 95% accuracy compared with the conventional method is considered good. Any system with less than 85% agreement with the conventional method is considered unacceptable. The major growth-dependent commercial kits for anaerobes are API 20A, marketed by bioMerieux Vitek Inc., Hazelwood, Mo., and Minitek Anaerobe system, marketed by BBL Microbiology Systems, Cockeysville, Md. The API 20A kit consists of 20 microtubes with dehydrated substrates.

A liquid suspension of a pure culture is made and then introduced into each microtube. The kit is then incubated anaerobically, and after 1 to 2 days the reactions are read and data recorded in a convenient chart. The results are converted into a number system. From a code book the identity of the culture can be ascertained. These types of results provide a statistical probability of the unknown culture by comparing the attributes in the computerized database. Thus, an organism may be identified as *C. perfringens* at 0.9500 level. Sometimes the printout may include requests for additional tests, and in some cases the organism may be given two identifications with probability of each listed. This general format of identification is used by all commercial diagnostic kits. Minitek Anaerobe II uses paper disks saturated with substrates. Thirty-five disks are available. After a suspension of the pure culture is made, the liquid is dispensed into wells of the units containing the appropriate disks. The assemblage is incubated anaerobically for 2 days. Reactions are read by color changes and recorded. The identity of the unknown can be obtained by following the procedure outlined for the API system.

Use of enzyme test systems include Rapid ID 32A, AN-Ident, RapID ANA II, Vitek ANI Card, and MicroScan Rapid Anaerobe Panel. All these methods are incubated aerobically because enzyme activities will not be influenced by the presence of air. Rapid ID 32A and RapID ANA II are similar in that the cultures are made into a liquid suspension, and the liquid is inoculated into microwells. After 4 hours, the color reactions are read and the identities are ascertained from a code book. Vitek ANI Card involves injection of liquid sample into a plastic card that is the size of a credit card with 30 wells. After 4 hours of incubation, the results are read and data are entered into the ANI program on the VITEK computer. Identification is made by the computer. In the MicroScan Rapid Anaerobe Panel system, the culture is inoculated into 24 substrates and two color interpretation controls. After 4 hours of aerobic incubation and addition of reagents, the panels are read either manually or with the aid of a Touchscan light box. Data are then interpreted by computer for identification. It should be emphasized that the aforementioned systems are designed for clinical bacteriology. Identifications of unknowns from environmental, food, and industrial sources may not be as accurate because the databases of these systems are from clinical isolates.

Biological Technique

Biological technique deals with the physical and biochemical analysis of components of cells and metabolites of cells have also been used for identification of anaerobes. Fingerprinting techniques have been explored to identify anaerobes by matching profiles from unknown cultures with known profiles in the data base. One such approach is the gas–liquid chromatography method popularized by the VPI Anaerobic Laboratory. Mitruka provided a detailed discussion on this method. The Hewlett-Packard company developed a procedure to fingerprint the fatty acid profiles of microorganisms, including anaerobes, using a high-performance liquid chromatography technique. The AMBIS system involves computer-integrated fingerprinting of electrophoretically separated labeled proteins of bacteria to match profiles of unknowns versus knowns. In all these systems, these cultures must be pure and grown anaerobically before accurate identification can be made. Again, the database must be from the appropriate sources for the identification to be accurate.

Immunological Technique

Microorganisms have specific antigen properties, and many of these antigens can elicit the production of antibodies in animals. With the appropriate antibodies, one can then use the antibodies to identify the unknown by antigen–antibody reactions. Historically, polyclonal antibodies have been used for detection and characterization of bacteria. More recently, monoclonal antibodies have been used for the reactions. Both types of antibodies are useful for anaerobic microbiology. The reactions can be agglutination of cells directly with antibodies or indirect agglutination of antigens reacting with antibodies fixed on a variety of support systems, including blood cells, latex beads, plastic steel balls, etc. The most popular format of antigen–antibody reactions used is ELISA. In this test, the antibody is fixed on a support surface and the antigen is captured. A second antibody then reacts with another site of the antigen.

The second antibody is conjugated with an enzyme. A substrate is then applied, and if the enzyme is present a color reaction will occur. The intensity of the color reaction will indicate the presence of the antigen. A cutoff color intensity is established to indicate a positive or negative test of the presence of the antigen. At first, this test was done manually, but recently, it has been completely automated. The analyst presents the instrument with a suspect sample, and the instrument will automatically complete the test with a computer printout. An exciting new development in immunological analysis is the field of immunomagnetic capture technology. In this technology, beads are magnetized and then antibodies or other capture particles are attached on the surface of the beads.

Antibodies against whole cells, antigens, and other target particles are fixed on the beads. The charged beads are then introduced to a broth or food sample to interact with target bacteria or other molecules. In this phase, the sample is mixed thoroughly with the charged beads tumbling in the liquid food to interact with target organisms. After the interaction, a powerful magnet is applied to the side of the reaction tube. The magnet holds all the beads with or without captured organisms. The rest of the debris is discarded, and the beads are released and washed. The captured organisms can be subject to growth in a special broth or agar, or these may be used for further immunological reaction, PCR, DNA/RNA reactions, or other tests.

Nucleic Acid Hybridization

DNA and RNA hybridization tests have been developed to detect organisms whose specific gene sequence of interest is known. At first, radioactive compounds were used to report the hybridization, but recently, the reaction has been reported by enzyme and color reactions. The truly revolutionary development in genetic type of identification of microorganisms is in the field of PCR technology. In this technology, the gene sequence of the organism of interest (an anaerobe, for example) is known at the DNA level. A piece of the DNA can be unfolded by heat, and after cooling of the unfolded DNA, two primers will interact with specific regions of the single DNA strands. A heat-stable polymerase will complete the complementary strand in the presence of nucleotides from the 3′ end and the 5′ end. Thus, in one cycle, one DNA molecule will become two, and after another cycle, two will become four, etc. In about 2 hours, one molecule of DNA can be amplified to 10^6. To use this technology, the genetic sequence of the target organism must be known, and the appropriate primers must be developed for successful use of this powerful tool.

There are many new variations of this technology, such as the search for a cold amplification method to bypass the heating cycle and the development of an ELISA-type detection scheme to bypass the need to do electrophoresis of PCR products. Another technology is called ribotyping.

In this technology, the DNA of a target organism is extracted from the cells, an appropriate enzyme or a collection of enzymes is then used to cut the DNA. The DNA fragments are then separated by electrophoresis, and the pattern of the bands are photographed and quantified, resulting in special fingerprints.

Table 6.2 Anaerobic fermentation products and microorganisms that can produce significant quantities of the product

Product	*Microorganism*
Methane	*Methanobacterium*
	Methanococcus
	Methanosarcina
	Methanobrevibacter
	Methanospirillum
Acetate	*Clostridium thermoaceticum*
Ethanol	*Saccharomyces cerevisiae*
	Saccharomyces uvarum
	Saccharomyces carlsbergensis
	Zymomonas mobilis
	Clostridium thermocellum
Lactic acid	*Lactobacillus*
	Streptococcus (*Lactococcus*)
	Leuconostoc
	Pediococcus
Propionate	*Clostridium propionicum*
	Clostridium tetanomorphum
Acetone	*Clostridium acetobutylicum*
Isopropanol	*Clostridium beijerisckii*
Acetoin	*Clostridium acetobu tylicum*
Butanediol	*Enterobacter aerogenes*
	Bacillus
	Serratia
	Erwinia
	Aerobacter
Butyrate	*Clostridium butyricum*
	Clostridium tyrobutyricum
Butanol	*Clostridium acetobutylicum*
	Clostridium beijerinskii
	Clostridium tetranomorphum

The fingerprints can be matched with fingerprints of known cultures for identification of unknowns. One added feature is that for certain organisms (e.g., *E. coli*) there are different fingerprints, even for the same organism. This becomes very important when one wants to trace the occurrence of an organism in the case of an outbreak of food- borne disease. For example, finding *E. coli* O157:H7 in several foods and the environment in an outbreak is not good enough to trace the etiology of the outbreak. With ribotyping, the culprit *E. coli* O157:H7 can be traced to a particular food by matching the ribotyping pattern of the strain that caused the outbreak with the origin of the strain in certain foods or in the environment. This will help pinpoint the source of the infection. Dupont's Qualicon Division is leading the way for the development and commercialization of this technology.

There are, of course, many other methods for identification of unknowns, such as impedance, conductance, capacitance, pyrolysis pattern, microcalorimetry, flow cytometry, etc. *Automated Microbial Identification and Quantitation: Technologies for the 2000s*, an excellent recent reference book edited by Olson, highlights many of these developments. There will be an explosion of technologies in applied microbiology in the near future that will directly and indirectly influence the enumeration, characterization, isolation, and identification of anaerobes.

COMMERCIAL PRODUCTION

Many products are or have been produced by fermentation in commercial quantities. Food products, industrial bulk chemicals, solvents, enzymes, and amino acids are produced with the help of anaerobic microorganisms. In this work, the emphasis is on those processes carried out by anaerobes, which generally means that important pathways yield useful products under conditions where oxygen is not required. However, oxygen is present in modest quantities to enhance microbial growth in some cases. Alternative manufacturing processes have been developed for some products; they can be produced by fermentation and by chemical processes that do not require microorganisms. Often these products are made by a microbial process in some countries and a chemical process in other countries; both processes may be used in the same country as well. The choice may depend on local raw materials, taxes, and scale of operation; often local economics determines which process is used. Table 6.2 lists several compounds produced by anaerobes and some of the microorganisms that can produce significant quantities of the product.

The pathways by which these products are produced have been investigated; many of the well-known industrially important pathways such as the Embden-Meyerhof-Parnas pathway are described in detail elsewhere. Ethanol is produced in large quantities for use as a motor fuel in several countries. Methane is often produced by mixed cultures as a product of anaerobic digestion of waste materials. Lactic acid is widely produced in food fermentations where milk, meat, or vegetables are fermented. Table 6.3 lists some of the many food fermentations and microorganisms that are present and active in these fermentations. Detailed descriptions of food fermentations are given by Reed and Steinkraus. Flavors are introduced into fermented foods as a result of the fermentation process; however, preservation is often as important as flavor in food fermentations. Knowledge of the biochemical pathways and the enzymes that facilitate the biochemical transformations is needed to commercialize and optimize many anaerobic fermentations. The control systems are often altered to overproduce desired products. In the case of biodegradation, it maybe necessary to add plasmids that provide genetic material necessary for a biodegradation pathway to function. As new products such as gasoline oxygenates enter the soil and water environment, microorganisms capable of biodegradation of methyl-*tert*-butyl ether (MTBE) have evolved.

Table 6.3 Fermented food products and associated microorganisms

Product	*Microorganisms*
Yogurt	*Streptococcus thermophilus*
	Lactobacillus bulgaricus
Sauerkraut	*Enterobacter cloacae*
	Erwinia herbicola
	Leuconostoc mesenteroides
	Lactobacillus brevis
	Lactobacillus plantarum
	Pediococcus cerevisiae
Cucumber fermentation (pickles)	*Lactobacillus brevis*
	Lactobacillus plantarum
	Pediococcus cerevisiae
	Leuconostoc mesenteroides
Sausage	*Lactobacillus*
	Leuconostoc
	Pediococcus
Bread and sourdough bread	*Saccharomyces cerevisiae*
	Lactobacillus
	Candida milleri
	Leuconostoc mesenteroides
	Saccharomyces cerevisiae
Cheddar cheese	*Streptococcus* (*Lactococcus*) *cremoris*
	Streptococcus (Lactococcus) *lactis*
Swiss cheese	*Lactobacillus bulgaricus*
	Streptococcus thermophilus
	Propionibacterium shermanii
Blue cheese	*Streptococcus* (*Lactococcus*) *lactis*
	Streptococcus (*Lactococcus*) *cremoris*
	Penicillium roquiforti
Camembert cheese	*Streptococcus* (*Lactococcus*) *lactis*
	Streptococcus (*Lactococcus*) *cremoris*
	Penicillium camemberti

Product	*Microorganisms*
Cottage cheese	*Streptococcus (Lactococcus) lactis*
	Streptococcus (*Lactococcus*) *cremoris*
Acidophilus milk	*Lactobacillus acidophilus*
Kefir	*Streptococcus* (*Lactococcus*) *lactic*
	Streptococcus (*Lactococcus*) *cremoris*
	Torula or *Candida yeast*
Buttermilk and sour cream	*Streptococcus (Lactococcus) lactis*, (including sub-species diacetylactis)
	Streptococcus (*Lactococcus*) *cremoris*
	Leuconostoc cremoris
Wine	*Saccharomyces cerevisiae*
	Saccharomyces uvarum
	Other *Saccharomyces*
Beer	*Saccharomyces cerevisiae*
	Succharomyces uvarum
Soy sauce	*Asperigillus oryzae*
	Saccharomyces rouxii
	Torulopsis versatilis
	Pediococcus soyae
Miso	*Pediococcus halophilus*
	Saccaromyces rouxii
	Torulopsis
	Streptococcus faecalis
Tempeh	*Rhizopus oligosporus*

ANAEROBIC FORMATION OF PRODUCTS

Anaerobic growth and product formation are often investigated together when the product is growth associated. In anaerobic fermentations, adenosine triphosphate (ATP) needed for growth is often formed as a result of substrate phosphorylation. Products such as ethanol are produced because the organism is growing and needs ATP for synthesis of microbial cell mass. Mass balances and yield parameters that describe growth and product formation can be written in terms of the mass of substrate, biomass, and product or in terms of the equivalents of available electrons of each of these.

Consider the anaerobic production of a simple product, such as ethanol, by anaerobic fermentation. The chemical balance equation is as follows:

$$CH_mO_l + aNH_3 = y_cCH_pO_nN_q + zCH_rO_sN_t + cH_2O + dCO_2 \qquad \ldots(1)$$

where CH_mO_l, $CH_pO_nN_q$, and $CH_rO_sN_t$ give the elemental compositions of the carbon, hydrogen, oxygen, and nitrogen in the substrate, biomass, and extracellular product, respectively. For glucose as substrate, $m = 2$ and $l = 1$; for ethanol as product, $r = 3$, $s = 0.5$, and $t = 0$. For the valences C = 4, H = 1, O = −2, and N = −3, the available electron balance is

$$y_c \frac{\gamma_b}{\gamma_s} + z \frac{\gamma_p}{\gamma_s} = 1.0 \qquad \text{...(2)}$$

or

$$\eta + \zeta_p = 1.0 \qquad \text{...(3)}$$

where γ is the reductance degree and the subscripts s, b, and p refer to substrate, biomass, and product, respectively.

Luedeking and Piret reported that product formation kinetics and growth kinetics were related; product formation was modeled using a growth-associated term and a maintenance-associated term; that is,

$$\frac{dP}{dt} = \alpha \frac{dX}{dt} + \beta X \qquad \text{...(4)}$$

where X is biomass concentration and P is product concentration. Growth is dependent upon the availability of substrate according to the Monod model

$$\frac{1}{X}\frac{dX}{dt} = \mu = \frac{\mu_{max} S}{K_s + S} \qquad \text{...(5)}$$

Here S is substrate concentration, μ is specific growth rate, μ_{max} is maximum specific growth rate, and K_s is the saturation constant. The organism uses substrate for growth and maintenance as follows:

$$-\frac{dS}{dt} = \frac{1}{Y_{x/s}^{max}} \frac{dX}{dt} + m_s X \qquad \text{...(6)}$$

where $Y_{x/s}^{max}$ is the true growth yield corrected for maintenance and m_s is the maintenance coefficient. In this process, the product formation kinetic parameters and the bioenergetic parameters are related. Equation 6 may be written as (2)

$$\frac{\mu}{\eta} = \frac{\mu}{\eta_{max}} + m_e \qquad \text{...(7)}$$

and equation 4 becomes

$$\frac{\mu}{\eta}\zeta_p = \frac{\mu}{\eta}(1-\eta) = \alpha_e \mu + \beta_e \qquad \text{...(8)}$$

if the following relationships are used (2)

$$m_e = \beta_e = m_s \frac{\sigma_s \gamma_s}{\sigma_b \gamma_b} = \beta \frac{\sigma_p \gamma_p}{\sigma_b \gamma_b} \qquad \text{...(9)}$$

$$\frac{1}{\eta_{max}} = \alpha_e + 1 = \frac{\sigma_s \gamma_s}{\sigma_b \gamma_b Y_{x/s}^{max}} = \alpha \frac{\sigma_p \gamma_p}{\sigma_b \gamma_b} + 1 \qquad \text{...(10)}$$

The true growth yield in equation 7 must satisfy the relationship (2)

$$\eta_{max} = \frac{(\sigma_b \gamma_b / 12) Y_{ATP}^{max}}{(\sigma_b \gamma_b / 12) Y_{ATP}^{max} + \delta} \qquad ...(11)$$

where Y_{ATP}^{max} is the true growth yield based on ATP, δ is the equivalents of available electrons transferred to products to produce 1 mol of ATP from ADP, and σ_b is the weight fraction carbon in biomass. Information on the theoretical maximum yield with respect to ATP and the value of δ for a particular fermentation allows one to estimate the range of values of the growth-associated kinetic term in the product formation model. For $\delta = 12$ equivalents of available electrons per mole of ATP generated, $Y_{ATP}^{max} = 28.8$ g cells per mole ATP, $\sigma_b = 0.462$, and $\gamma_b = 4.291$; $\eta_{max} = 0.283$ and $\alpha_e \geq 2.52$.

Product formation is directly linked to the specific growth rate, true growth yield corrected for maintenance, and the maintenance coefficient as shown in equations 1 to 11. Equation 3 shows that the product yield is related to the biomass yield; the largest product yields are expected when the cells are using carbon and energy primarily for maintenance and nearly all the available electrons are being converted to product. When oxygen is present, some of the available electrons in the carbon substrate are transferred to oxygen, and the product yield is reduced accordingly. Growth kinetics, product formation rates, and product yields are affected by temperature, pressure, water activity, osmolality, oxidation–reduction potential, bioenergetics, and the concentrations of organic substrate, inorganic nutrients, biomass, hydrogen ions (pH), carbon dioxide, products, and electron acceptors such as oxygen, nitrate, and sulfate.

FERMENTATION PATHWAYS

The product yields under anaerobic conditions depend on the operating conditions and the fermentation pathway for the selected strain of microorganisms. Equations 2 and 7 can be rearranged as follows:

$$\zeta_p = 1 - \eta \qquad ...(12)$$

$$\eta = \frac{\mu}{\mu / \eta_{max} + m_e} \qquad ...(13)$$

If equation 13 is substituted into equation 12, one obtains

$$\zeta_p = 1 - \frac{\mu}{\mu / \eta_{max} + m_e} \qquad ...(14)$$

Equation 14 indicates that the product yield is a function of the biomass true growth yield, which depends on the growth yield with respect to ATP and the efficiency with which ATP is formed through substrate phosphorylation. The product yield also depends on the values of the maintenance coefficient and specific growth rate. Table 6.4 shows how the true growth yield varies with the growth yield with respect to ATP and the number of moles of available electrons of substrate required to produce 1 mol of ATP according to equation 11. It also illustrates how the product yield varies as a function of specific growth rate and maintenance coefficient.

The product yield decreases as the specific growth rate increases. As values of the maintenance coefficient increase, the product yield increases also. Estimated values of the true growth yield and maintenance coefficient are presented in the book by Erickson and Fung for several sets of experimental data reported by a variety of investigators. These values are within the expected range of values based on theoretical yields and known pathways; however, not all of the estimated values reported by Erickson and Fung are below the estimated theoretical yield. It is common to have experimental error because of the difficulty of making measurements.

Table 6.4 Effect of ATP yield, true growth yield, maintenance coefficient, and specific growth rate on biomass yield and product yield

Y_{ATP}^{max}	δ	η_{max}	m_e	μ	η	ζ_p
28.8	12	0.283	0.1	0	0	1.0
28.8	12	0.283	0.1	0.1	0.221	0.779
28.8	12	0.283	0.1	0.2	0.248	0.752
28.8	12	0.283	1.0	0.1	0.0739	0.926
10.5	12	0.126	0.1	0.1	0.112	0.888
10.5	12	0.126	1.0	0.1	0.056	0.944
10.5	24	0.0674	0.1	0.1	0.063	0.937
10.5	24	0.0674	1.0	0.1	0.04	0.960

When data consistency is checked by making carbon and available electron balances and yields are compared with theoretical yields, it is possible to identify experiments where errors are present; however, it is not always possible to determine which particular measurements have the largest errors associated with them. When errors are present in the data, the covariate adjustment method should be used in data analysis. The results in Tables 6.4 may be related to metabolic pathways. The Embden-Meyerhof-Parnas pathway commonly produces 2 mol of ATP per mole of glucose ($\delta = 12$ equivalents of available electrons per mole of ATP generated), whereas the Entner-Doudoroff pathway produces 1 mol of ATP per mole of glucose ($\delta = 24$). *Zymomonas mobilis* has the Entner-Doudoroff pathway; thus, the lower values of true growth yield are expected. As shown in Table 6.4, the product yield is predicted to be larger for *Z. mobilis*.

BIODEGRADATION

Anaerobic digestion is widely used in wastewater treatment to degrade the solids from primary and secondary treatment. Anaerobic processes are also applied in the treatment of industrial effluents and in bioremediation of contaminated soil and ground water. Although environmental microorganisms have been functioning and serving the needs of humans for a very long time, the commercialization and engineering of this aspect of bioprocess technology was started after the discovery of microorganisms. Wastewater treatment is the second oldest bioprocess technology after food processing. Anaerobic degradation processes depend on the electron acceptors that are present. Stouthamer points out that when several electron acceptors are present, the microorganism selects the electron acceptor that yields the largest amount of energy by repressing the formation of reductase enzymes for the other electron acceptors.

When oxygen is present, aerobic processes are observed and anaerobic processes are generally absent unless there are spatial regions or time periods where oxygen is absent. Nitrate is the electron acceptor of choice when oxygen is absent, and when nitrate and oxygen are both absent, sulfate is the favored electron acceptor. Methane production, which occurs when water is the electron acceptor, is generally inhibited if sulfate is present. Stouthamer points out that other electron acceptors may participate, including other oxidized forms of nitrogen and fumarate. Many bacteria have been found that can use nitrate as the terminal electron acceptor. The first product of nitrate respiration or nitrate reduction is formation of nitrite, which can be further used as an electron acceptor by many microorganisms. In denitrification, the end product of further reduction is nitrogen gas; however, in other cases the end product is ammonia. Stouthamer reviewed the enzymology and bioenergetics of nitrate reduction.

Sulfate-reducing bacteria include *Desulfovibrio, Desulfobacterium, Desulfotomaculum, Desulfococcus, Desulfosarcina, Desulfomonile, Desulfonema, Desulfoarculus, Desulfobulbus, Desulfobotulus,* and *Desulfobacula*. For these genera, many species have been described in the literature. These organisms have great diversity in biochemical properties, and they can completely oxidize a variety of organic substrates. Methanogenic bacteria are widely found under anaerobic conditions in nature. Methane is produced in soil, landfills, anaerobic digesters, rumens of animals, and many other places. In anaerobic digestion, methanogenic bacteria are found together with hydrolytic fermentative microorganisms and syntrophic acetogenic bacteria. The hydrolytic fermentative organisms, which include eubacteria, fungi, and often protozoa, produce fermentation products such as ethanol, lactate, butyrate, succinate, and acetate. The acetogenic bacteria (*Pellobacter, Desulfovibrio, Syntrophobacter, Clostridium, Syntrophomonas,* and others) produce acetate, formate, bicarbonate and hydrogen from the ethanol, lactate, butyrate, and other organic acids and alcohols with more than two carbons.

The methanogenic bacteria use the acetate, formate, and hydrogen as substrates. In anaerobic digesters, about 75% of the available electrons in the complex organics are first transformed to fermentation products other than acetate, formate, and hydrogen, which make up the remaining 25%. More than 60% and as much as 72% of the available electrons is converted to acetate, which is then used by methanogens to form methane. More than 25% of the available electrons are converted to hydrogen before being incorporated into methane. Thiele and Zeikus and Thiele reviewed mixed culture interactions in methanogenesis and provided information on selected anaerobic microorganisms that are active in anaerobic digestion. Many microorganisms that are active in environmental microbiology have been studied and described; however, each year many new organisms are described for the first time. New anaerobic treatment processes for specific applications, such as nitroaromatic munitions compounds, are being developed. Although anaerobic processes are often slower than aerobic processes, they have been found to be economically attractive in a wide variety of applications.

MIXED CULTURES

Mixed cultures are used extensively in food fermentations and in biodegradation of complex substances. Mutualism, commensalism, amensalism, competition, and prey–predator relationships are the main interactions observed in mixed cultures. Mutualism occurs when both species benefit from the interaction, whereas commensalism describes processes where only one species benefits and the other is not affected. Amensalism occurs when one species restricts another, whereas competition affects both species because of nutrient limitations. The predator consumes the prey in prey–predator interactions. In actual situations, more than two species are often present, and several types of interaction may be taking place. In the recent book on mixed cultures by Zeikus and Johnson, most of the contributions are related to either food fermentations or environmental microbiology. There are food fermentation chapters on bread, milk, oriental foods, wine, and vegetables and environmental microbiclogy chapters related to degradation of polysaccharides, methanogenesis, detoxification of hazardous waste, corrosion, and leaching processes in mineral biotechnology. In San Francisco sourdough bread, *Lactobacillus sanfrancisco* ferments only the maltose, whereas *Candida milleri* uses all the other free sugars but not maltose. Lactic acid, acetic acid, and carbon dioxide are the main products of the fermentation; however, small amounts of propionic, isobutyric, butyric, α-methyl-*n*-butyric, isovaleric, and valeric acids have been found in commercial San Francisco sourdough breads.

Anaerobic cellulose degradation is accomplished by a mixed culture of Clostridium cellulolytic bacteria that produce cellobiose, glucose, and cellodextrins; fermentative bacteria that produce propionate, butyrate and other fermentation products from glucose and other intermediates; syntrophic acetogenic bacteria that convert the fermentation products to acetate, hydrogen, and carbon dioxide; homoacetogens that convert hydrogen and carbon dioxide to acetate; and methanogens that produce methane from acetate, formate, and hydrogen. Leschine reports that *Clostridium papyrosolvens* grow mutualistically in co-culture with a noncellulytic *Klebsiella*; *C. papyrosolvens* hydrolyzes cellulose for the *Klebsiella,* whereas *Klebsiella* excretes vitamins required by the *C. papyrosolvens*. The soluble sugar products of cellulose hydrolysis provide substrates for many noncelluloytic commensal organisms that depend on the cellulolytic bacteria, but do not provide known compounds in return. In the natural environment, the human environment, and even in the industrial environment, mixed cultures of bacteria, yeast, mold, viruses, protozoa, nematodes, and higher organisms live in antagonistic, cooperative, or inert coexistence.

Pure culture systems are the result of understanding of the fundamental role of microbes in a particular process and the isolation and cultivation of such cultures for specific reactions. This section deals with single-culture processes mixed pure-culture processes, and mixed natural-culture processes, using food systems as examples. This information certainly can be applied to other industrial and fermentation processes. Other publications deal adequately with the subject of single-cell process. The key success of single-culture process is to provide the culture with a sterile substrate and environment with no contamination during the process. Single-cell process is a man-made situation classified as a controlled process because the substrate is prepared and processed in such a way as to minimize contamination. Examples of this type of process are wine making, beer making, bread making, single-culture dairy product fermentation, and vinegar production. The kinetics of growth and product formation are easier to control and monitor.

Process of Mixed Pure-Culture

For some processes, it is desirable to have more than one pure culture for proper product development. The mixed pure cultures can be a controlled mixture of bacterium with bacterium or with a combination of yeast or mold, or both. The relationships among these pure cultures during growth become very complex.

Natural Process

In many situations, such as in nature and noncontrolled fermentation, the flora in the process are mixed natural cultures. The interactions of these microbes are even more complex than those of mixed pure-culture process. The microbial successions in these processes follow a special sequence. If the products are treated properly, the desirable cultures provide the characteristic end results. On the other hand, if the conditions are not treated properly, the process will fail.

Mixed Pure Bacteria Culture

An example of mixed pure bacterial culture fermentation is yogurt fermentation, which typically uses *Streptococcus thermophilus* and *Lactobacillus bulgaricus*. The interactions are commensal. Figure 6.1 illustrates the development of acidity of pure and mixed cultures of *S. thermophilus* and *L. bulgaricus* in liquid. In mixed cultures, *S. thermophilus* grows faster at the beginning, reduces the pH by lactic acid production, and also produces formic acid, which is a stimulatory compound for *L. bulgaricus*. *L. bulgaricus* produces amino acids that stimulate the growth of *S. thermophilus*. Higher acid production with mixed cultures was obtained and compared to pure cultures. This is an example of commensalism in mixed cultures. Details of the kinetics and mathematical relationships of interactions between *S. thermophilus* and *L. bulgaricus* are presented in an article by Fung et al.

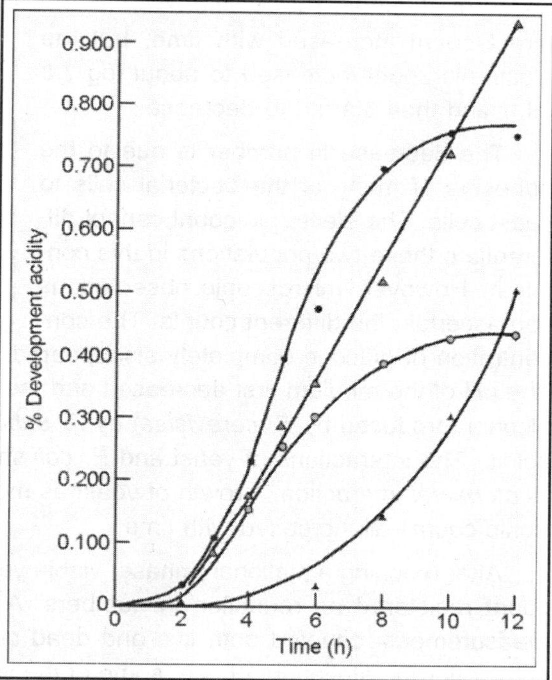

Figure 6.1 Development activity of pure and mixed cultures of *L. bulgaricus* and *S. thermophilus*.

Interaction between Bacterium and Yeast in a Mixed Culture

Yeast is one of the most important organisms for industrial use in alcoholic fermentation. Bacterial contamination may occasionally occur during yeast fermentation, which may pose serious problems for the fermentation industry. This section deals with the interaction of *Saccharomyces cerevisiae* with *Acetobacter suboxydans* or *E. coli* in mixed- culture growth. *S. cerevisiae* was grown in mixed cultures with either *A. suboxydans* or *E. coli* in a culture medium and glucose at 28°C. Growth was monitored by viable cell counts of yeast and bacteria using selective agars, direct count by microscopic reading of cells in a Petroff-Hauser counting chamber, and electronic counts by the Coulter counter. By selecting the appropriate threshold windows, the electronic instrument can differentiate yeast count and bacterial count in the same liquid. During the growth period, pH values were monitored by a conventional pH meter and glucose was determined by the Glucostat, a commercial kit designed to detect glucose in solution. A vinometer was used to measure alcohol level of the solution.

Figure 6.2 illustrates the interaction between *S. cerevisiae* and *A. suboxydans* as reported by Fung et al. For the yeast, both direct count and electronic count increased with time. The direct count registered about 10 times more cells than the electronic count. Because *A. suboxydans* does not grow well in solid agar medium, viable cell count data were not obtained. For *A. suboxydans,* the direct count increased with time, but the electronic count increased to about log 7.8 mL^{-1} and then started to decrease.

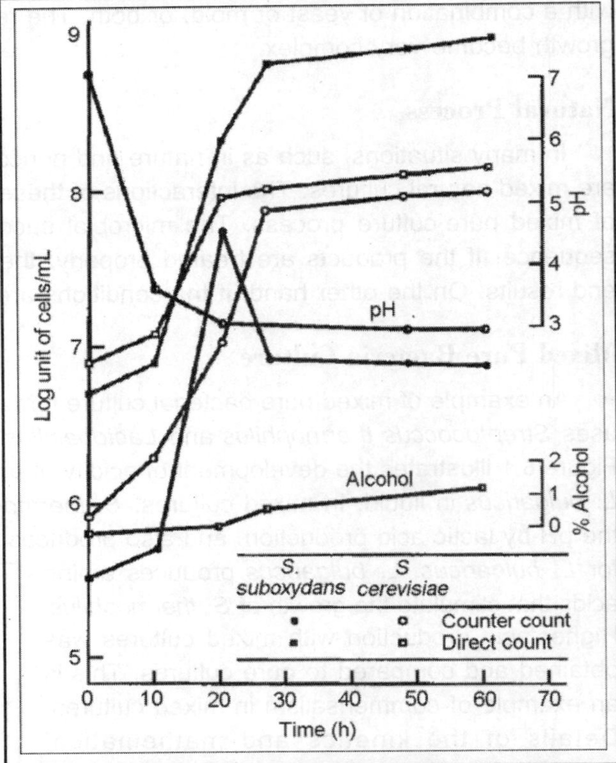

Figure 6.2 Interaction of mixed cultures of *S. cerevisiae* and A suboxydants in terms of cell numbers monitored by direct count and electronic count as well as product formation.

The decrease in number is due to the adhesion of many of the bacterial cells to yeast cells. The electronic count cannot differentiate these two populations in this condition. However, microscopic observations can ascertain the different counts. The concentration of glucose completely disappeared in the first 10 h of yeast and bacterial interactions. The pH of the medium first decreased and then stabilized at pH 3. This is because of oxidation of alcohol (produced by *S. cerevisiae*) by *A. suboxydans* to acetic acid, making the reaction mixture acidic. The interactions of yeast and *E. coli* show interesting contrasts compared with the yeast–*Acetobacter* interaction. Growth of yeast as monitored by viable cell count, direct count, and electronic counts all increased with time.

After reaching a stationary phase, viable yeast count decreased, but direct count and electronic count registered no reduction in numbers. A likely explanation is that the former two biomass measurements counted both live and dead cells, but the viable cell count only registered living cells. After the stationary phase, some of the yeast cells went through the death cycle. The growth curves of *E. coli* showed an increase in the viable cell count and direct cell count. The electronic count reached log 8 mL^{-1} and then declined, exhibiting a trend similar to that observed in the yeast–*Acetobacter* interactions.

Alcohol contents increased to about 2%, and the pH first dropped to around 4 and then rose to 6. This pattern differs from that obtained from the interaction for yeast and *Acetobacter*. *E. coli* cannot oxidize alcohol and thus did not create large amounts of acid to counteract the basic metabolites in the reaction vessel. This resulted in a medium that reverted to a more alkaline state.

Interaction between Bacteria and Mold in a Mixed Culture

Another example of mixed culture interaction is ensilage of crops. Dalamacio and Fung and Dalmacio et al. studied the influence of ammonia on bacteria and mold during ensilage of high-moisture corn. Bacteria were not affected by ammonia treatment as much as mold at the onset of the treatment. Bacterial populations remained detectable after treatment of 1.0, 1.5, and 2% ammonia. However, no mold was detected for corn treated with 1.0, 1.5, 2.0% ammonia. Corn treated with 0.5% ammonia had high bacterial and mold counts. After 2 months of storage, the bacterial population increased to more than 10^7 g^{-1}. Mold count increased to around 10^4 g^{-1} for the corn treated with 0.5% ammonia. After 4 months of storage, the bacterial count reached saturation level (higher than 10^{10} g^{-1}) but the mold population remained at around 10^4 or 10^5 g^{-1}. Not only did the bacteria and mold counts change in the fermentation, but the genera of bacteria and mold also changed concomitant to the fermentation stages. At the beginning, most of the bacteria isolated were *Bacillus,* but as the fermentation progressed most of the bacterial isolates belonged to *Lactobacillus.* At the beginning of the storage of ammonia-treated corn (in the fall season), the predominant mold was *Mucor,* which is considered a field mold. As the storage continued into the cold winter, *Penicillium* isolates increased, and at the end of the storage in spring, *Scopulariopsis* predominated. These successions occurred for a variety of reasons. For example, the winter months favored the development of *Penicillium,* which is cold tolerant. As temperature increased in the spring, *Penicillium* population was overtaken by *Scopulariopsis,* which can use complex nitrogenous foods better than *Penicillium.* The aforementioned examples of microbial interactions are just some of the typical interaction patterns in food and in nature. Countless other interactions can occur. The main point to emphasize is that mixed-culture interactions are very complex, but scientists can use this knowledge to optimize process control in obtaining desirable products in food and industrial microbiological processes.

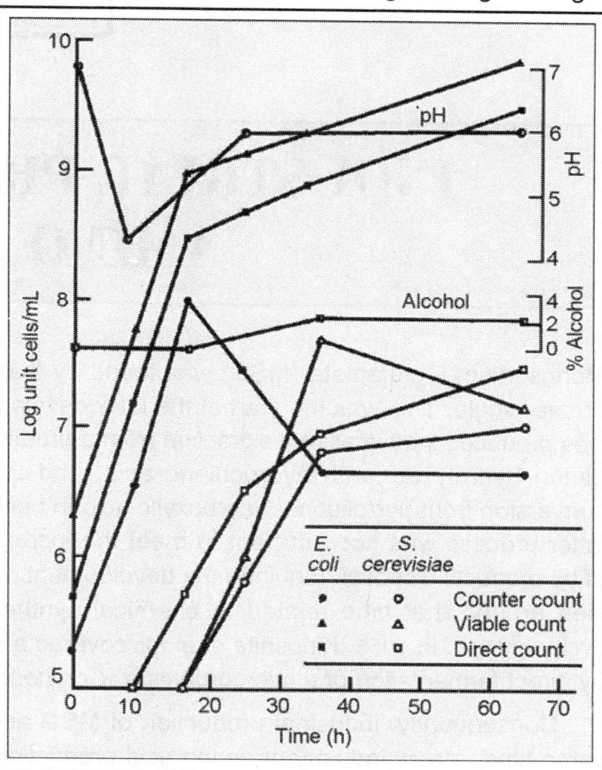

Figure 6.3 Interaction of mixed cultures of *S. cerevesiae* and *E. coli* in terms of cell numbers monitored by direct count, electronic count, and viable cell count as well as product formation.

7

INDUSTRIAL PRODUCTION OF AMINO ACIDS

Monosodium L-glutamate (MSG) was found by Ikeda in 1908 as a flavoring component of *konbu*, or sea tangle. This was the start of the later development of the amino acid industry. Initially, MSG was produced industrially by extraction from a protein hydrolysate, more specifically, first from wheat gluten hydrolysate with hydrochloric acid, and then from defatted soybean hydrolysate or by conversion from pyrrolidone-5-carboxylic acid in beet molasses, for use as a seasoning agent. This latter process was not sufficient to meet the increasing demand because it involved the problem of by-products disposal, requiring the development of new production processes. Patent applications filed around that time related to chemical synthesis of MSG from succinate semialdehyde or cyclopentane. In 1956, Kinoshita et al. discovered a microbial process for L-glutamic acid production by direct fermentation of a microorganism (*Corynebacterium glutamicum*) from sugar and ammonia.

Consequently, industrial production of MSG as a seasoning was rapidly enlarged, and at the same time, a new industry of amino acid production by fermentation of microorganisms emerged. In 1958, Kinoshita, Nakayama, and Kitada found that an auxotrophic mutant of *C. glutamicum* requiring homoserine accumulated L-lysine in a medium, which enabled industrial production of L-lysine by fermentation. Establishing the basis for the development of the amino acid fermentation industry, this technology suggested the possibility of producing various other amino acids by auxotrophic mutants and also the importance of research in fermentative production of biological components with regulatory mutants. Since then, direct fermentation of various amino acids has been broadly studied.

Reduction of the cost of amino acids, owing to the development of fermentation techniques together with the establishment of their production processes, facilitated extended application of amino acids to uses other than as a seasoning. Particularly, L-lysine, an essential amino acid that is lacking in several food proteins, has been enjoying remarkably increased demand as a feed additive. Table 7.1 shows the production of the major amino acids. Most of the amino acids produced in large quantities are still manufactured by fermentation, except for glycine, which does not have optical isomers, and methionine, which has a similar effect as a feed additive in both L- and DL-forms.

Table 7.1 Estimated worldwide production of amino acids

Amino acids	*Estimated production (ton/year)*	*Process*
Glycine	22,000	Chemical synthesis
L-Aspartic acid	7,000	Enzymaic method
L-Arginine	1,200	Fermentation
L-Cysteine	1,500	Enzymaic method
Monosodium L-Glutamate	1,000,000	Fermentation
L-Glutamine	1,300	Fermentation
L-Histidine	400	Fermentation
L-Isoleucine	400	Fermentation
L-Leucine	500	Ex, Fermentation
L-Lysine HCl salt	250,000	Fermentation
DL-Methionine	350,000	Chemical synthesis
L-Phenylalanine	8,000	Fermentation, Chemical synthesis
L-Proline	350	Fermentation
L-Threonine	4,000	Fermentation
L-Tryptophan	500	Fermentation, Enzymaic method
L-Tyrosine	120	Extraction

In addition to being used as a seasoning agent and a feed additive, amino acids are currently used widely as a raw material for the sweetener Aspartame (*N*-L-α-aspartyl-L-phenylalanine 1-methyl ester), for pharmaceuticals and agrichemicals, for infusion and oral nutrition, for surfactants, and so forth, based on their reactivities and nutritional, pharmacological, and flavoring effects.

In connection with amino acid fermentation, many reports have been published on the physiological properties of amino acid–producing microorganisms. The following discussions stress problems from the standpoint of practically manufacturing amino acids on an industrial scale.

AMINO ACID FERMENTATION

Amino acid fermentation may be defined in two ways. One is a definition in the narrow sense and refers to direct accumulation of amino acids in a medium containing a sugar, ammonia, and other nutrients, and the other is a broad definition that covers the fermentation process involving the addition of specific precursors and amino acid production by reaction using enzymatic functions of microorganisms. Here, mainly direct fermentation of the former type is discussed.

Fermentation processes generally have a number of advantages and benefits that accrue to the user of the process. On the other hand, they also have drawbacks. Both are summarized as follows:

Uses

1. Mild conditions are used both in fermentation and in product recovery; hence, little product degradation takes place.
2. Fermentation requires relatively less complex operation.
3. Once the plant is built and operation has begun, there are relatively low maintenance costs.
4. Only L-form amino acids are obtained.

Obstruction

1. Operations provide low product concentrations compared with chemical synthesis processes and require large volumes of water, large fermenter capacity, and comparatively high capital investment.
2. The requirements for strict sterility add to capital costs and operation costs.
3. Large amounts of energy for oxygen transfer and mixing are required.
4. Product recovery may be complex, difficult, and expensive.
5. The process time necessary to reach maximum concentrations of the desired product is usually comparatively long.

The aforementioned narrow-sense definition of direct fermentation of amino acids includes three types of fermentation: batch-type, fed-batch-type, and continuous.

Batch Fermentation in Industrial Process

Industrial fermentation is mostly performed using batch processes. In a batch process, a large volume of a medium containing nutrients and substrate material is inoculated with a viable culture of one or more appropriate microorganisms. Microbial growth and biochemical synthesis are allowed to proceed until an optimum yield of metabolite or a desired biochemical transformation has been obtained. Although this process is the basic form of fermentation, it is not frequently practiced on an industrial scale because productivity is limited by the amount and nature of the nutrients present at the time of inoculation. In most amino acid fermentation, enhanced product concentration is important to improve production efficiency and requires more advanced fermentation processes.

Process of Fed-Batch Fermentation

This fermentation method is aimed at efficiently carrying out fermentation and is characterized by a low concentration of components in the initial medium to minimize metabolic regulation. At the time of inoculation the medium promotes initial growth of microbes; subsequent supplies of more raw materials drive the desirable increase in metabolite biosynthesis. Feeding is effected either intermittently or continuously. Industrial fermentation of most amino acids is accomplished with this method.

Fed-batch fermentation requires feeding equipment in addition to the equipment required for batch fermentation and therefore leads to higher fixed costs. However, the process can provide improved productivity as a whole because of the enhanced yield and reduced fermentation time.

Actual product cost depends on the cost of the carbon source or specific precursors that are used for feeding.

The nitrogen source is supplied generally through pH control with ammonia. This process is of particular importance in industrial operations where the reaction of the microbial catalyst does not last long; hence, continuous fermentation cannot be practiced, unlike in the case of L-glutamic acid fermentation from cane molasses using penicillin, described next.

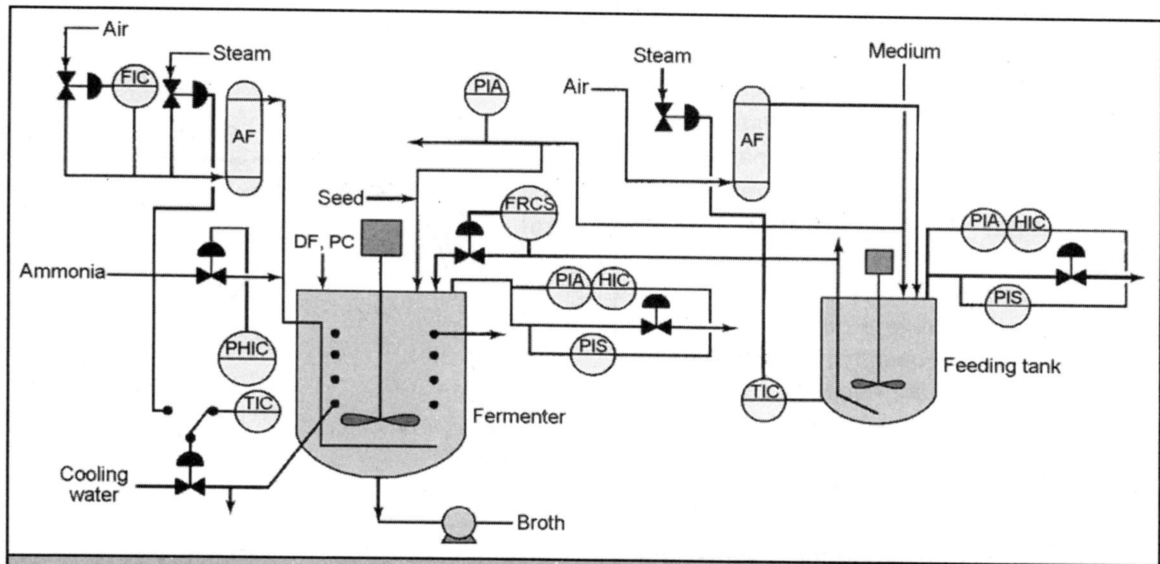

Figure 7.1 Typical system for fed-batch fermentation. AF, air filter; DF, defoamer; PC, penicillin; FIC, flow indication control; PHIC, pH indication control; TIC, temperature indication control; PIS, pressure indication sum; PIA, pressure indication alarm; HIC, highest indication control; FRCS, flow record control sum.

Table 7.2 Classification of fed-batch microbial reaction processes

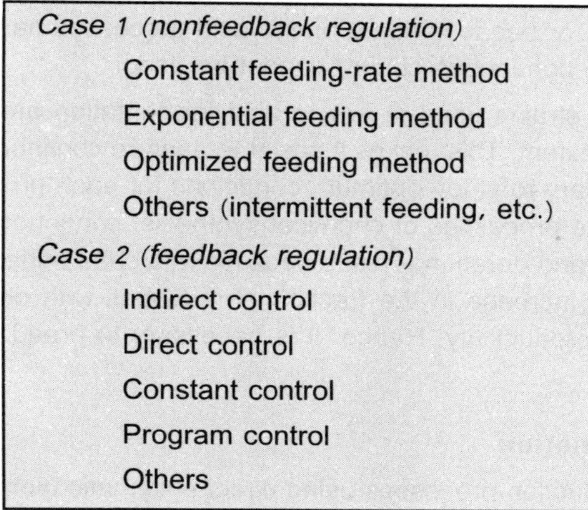

Case 1 (nonfeedback regulation)
Constant feeding-rate method
Exponential feeding method
Optimized feeding method
Others (intermittent feeding, etc.)
Case 2 (feedback regulation)
Indirect control
Direct control
Constant control
Program control
Others

Continuous Fermentation

In *continuous fermentation,* a complete medium is fed to a fermenter after an appropriate period of batch fermentation, and the same quantity of broth is continuously taken from the fermenter to maintain the fermentation broth at a fixed volume. This may be performed either by the chemostat method using a substrate or limiting substance, or by the turbidostat method in which the cell level is adjusted to maintain constant cell mass. Because the continuous fermentation process allows improvement of productivity compared with the ordinary fermenter, the initial investment in equipment is small relative to the production volume, and operation cost is low. However, one drawback is that it is not suitable for small-scale production, and the challenges of sterile operation and equipment maintenance are more necessary than they are for batch and fed-batch fermentation. This process shifts from batch fermentation to continuous fermentation when productivity per unit time in the former is relatively high. There are many reports analyzing the steady-state condition in continuous fermentation. Most of them relate to cell culture, but a few reports are available specifically on amino acid fermentation. One of the reasons may be that studies on amino acid fermentation have been mainly directed to the influence of metabolic regulation, as in the case of the penicillin addition method in glutamic acid fermentation, or because some amino acid processes have a distinct growth phase and production phase, and continuous culture cannot be used.

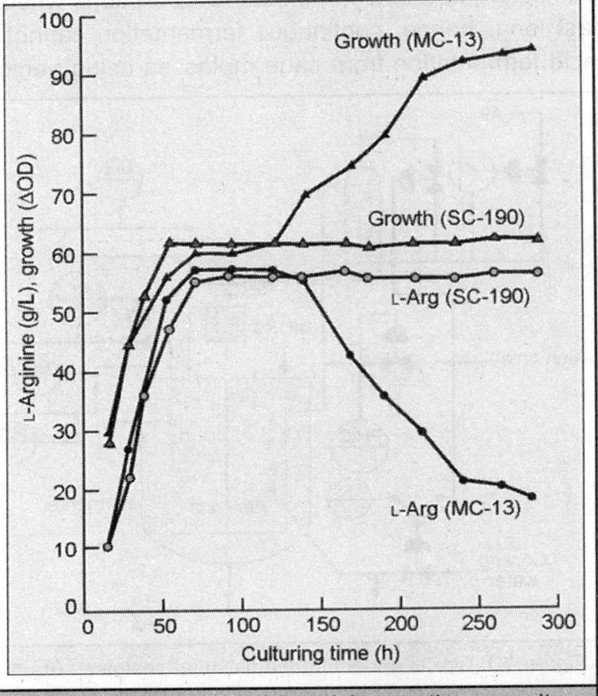

Figure 7.2 Comparison of L-arginine continuous culture profile of MC-13 and SC-190.

Many of the microbial strains used in amino acid fermentation are released from metabolic regulation to a remarkable extent. This makes it easier to analyze continuous processes and thereby optimize them. It is necessary to study optimum conditions for each process and to optimize their industrial application. Unlike processes of chemical synthesis, continuous fermentation processes have their own restrictions and duration. This is because microbes undergo spontaneous mutation within the system, and an increase in the fraction of microbes with decreased productivity may lead to rapid reduction in productivity. Hence, it is necessary to breed a strain with high genetic stability.

Enzymatic Biotransformation

Of the amino acid production processes using direct enzymatic biotransformation, those for L-alanine, L-aspartic acid, L-lysine, and L-tryptophan have been the most extensively studied, and some of the results have led to standard industrial processes. Although the enzymatic production process of L-lysine from DL-aminocaprolactam did not result in practical application, this technology is interesting because of its use of petrochemical products for fermentation raw materials.

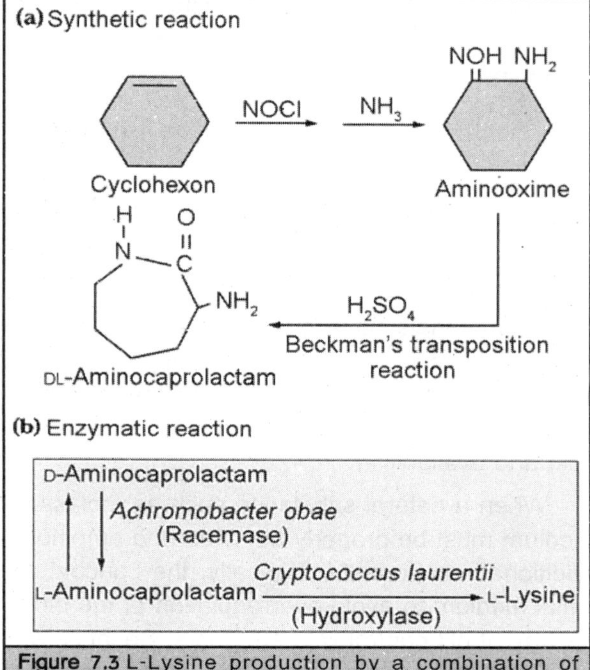

Figure 7.3 L-Lysine production by a combination of synthetic reactions.

MICROORGANISMS

In amino acid fermentation, screening/ breeding of microorganisms to be used for the process is the most important step. These microorganisms are classified into (1) wild-type strains isolated from nature to meet a specific purpose, (2) mutants that provide the desired properties that are developed by spontaneous or artificial mutation of wild-type strains, and (3) recombinant strains bred or constructed through genetic recombination technologies.

Wild-Type Strains

Microorganisms suitable for the purpose are selected from nature. They are cultured under adjusted fermentation conditions so as to produce excess amounts of the desired products. Typical examples are *C. glutamicum* and *Brevibacterium flavum* used for L-glutamic acid fermentation.

Mutant Strains

Auxotrophic strains and amino acid analog–resistant strains fall into the mutant strain category. Most amino acids can be produced by these types of microorganisms. Currently, most fermentative production of amino acids is accomplished in this fashion.

Table 7.3 L-Lysine-producing microorganisms

	L-Lysine HCl productivities	
Microorganisms	*Conc. (g/L)*	*Yield (%)*
C. glutamicum (Hse^-)	13	13
B. flavum (Thr^-, Met^-)	34	34
C. glutemicum (Hse^-, Leu^-, $Pant^-$, ACE^r)	42	42

Recombinant Strains

There are many reports on the strains prepared by recombination of the various genes required for amino acid biosynthetic enzymes, as well as their possible applications to amino acid fermentation. Self-cloning recombinants are often reported in the literature because they are highly practical from the viewpoint of regulation. They will be put to use in the future as a primary technology for amino acid fermentation. Whichever types of strains are constructed, conditions such as raw materials, temperature, process, and so on, as well as the by-products to be produced must be determined.

CULTURE MEDIA

Components of a medium are decided upon with regard to the microorganism, cost, purification process, waste treatment, and so forth. Except for particular cases, the cost of the carbon source, a main raw material, accounts for most of the cost of raw materials. Hence, its selection is of primary importance. Cane molasses, beet molasses, glucose, methanol, *n*-paraffin, and other carbon source materials have been studied. In enzymatic processes and fermentation processes with addition of precursors, petrochemical products may be used as a precursor for the desired product. Carbon sources currently used for the industrial production of amino acids are mostly sugars, including cane molasses, beet molasses, and corn syrup (glucose), because of their relatively low cost and availability.

When a natural substance such as molasses is used as a carbon source, composition of the medium must be properly adjusted and empirically optimized because natural substances contain additional components. Typically, the concentration of the nitrogen source is relatively low in an initial medium to avoid overregulation of the biosynthetic pathways, and nitrogen is supplemented by way of pH control with ammonia. Because amino acid fermentation is a process of converting ammonia into amino groups, the importance of the nitrogen source is second only to that of the

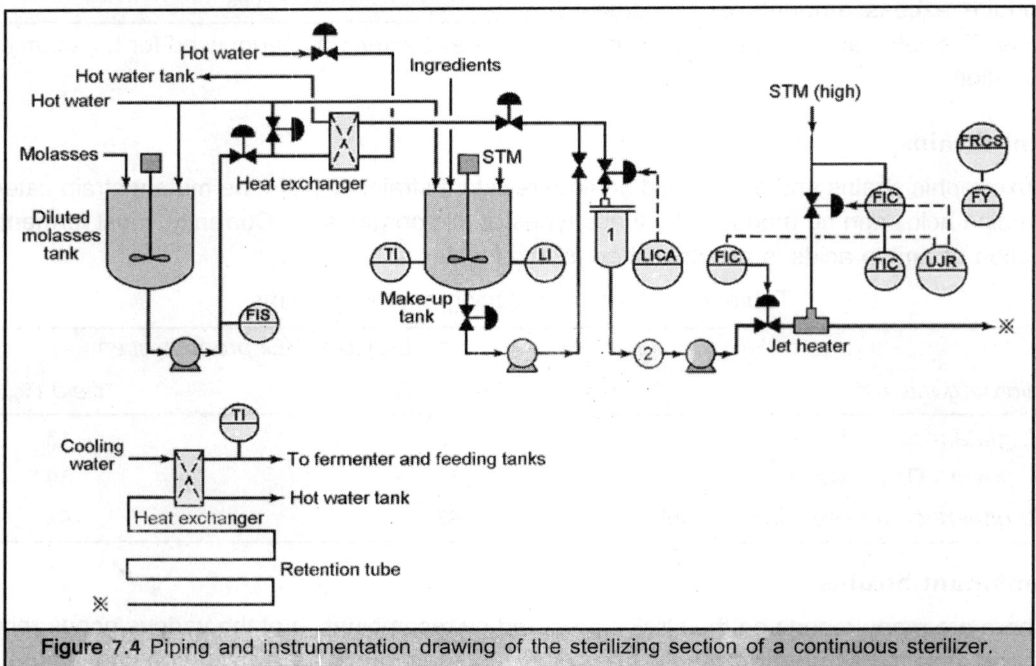

Figure 7.4 Piping and instrumentation drawing of the sterilizing section of a continuous sterilizer.

carbon source. In addition, phosphate and other minerals are added in the required amounts. When an auxotrophic strain is used for fermentation, the required substance is added at an optimum concentration. Furthermore, an antifoam agent must be added to prevent foaming of the broth at the sterilization and fermentation stages.

Sterilization

When the fermenter has a capacity of approximately 10,000 L or less, the preparation of raw materials and steam sterilization are often both carried out within the fermenter. On the other hand, when a large fermenter (40,000-L capacity or more) is used, as in the case of amino acid fermentation, each piece of equipment is separately sterilized with steam and cooled with sterile air, and raw materials other than sugar are prepared in a make-up tank and subjected to steam sterilization and cooling. Sugar is sterilized separately and sent to the fermenter with other nutrients. Similar procedures apply to preparation of the feed tank and the feeding solution in fed-batch fermentation. Feeding is carried out from the feed tank through a sterilized line while controlling the flow.

INDUSTRIAL OPERATION

Figures 7.1 and 7.5, respectively, show a process flow diagram of amino acid fermentation. The process comprises the steps of laboratory seed culture, factory seed culture, and fermentation, and the size of culturing is scaled up in this order.

Seed Culture

The most important feature in the steps of laboratory and factory seed culture is to ensure active microbes. This promotes initial growth of microbes at the fermentation step and stabilizes

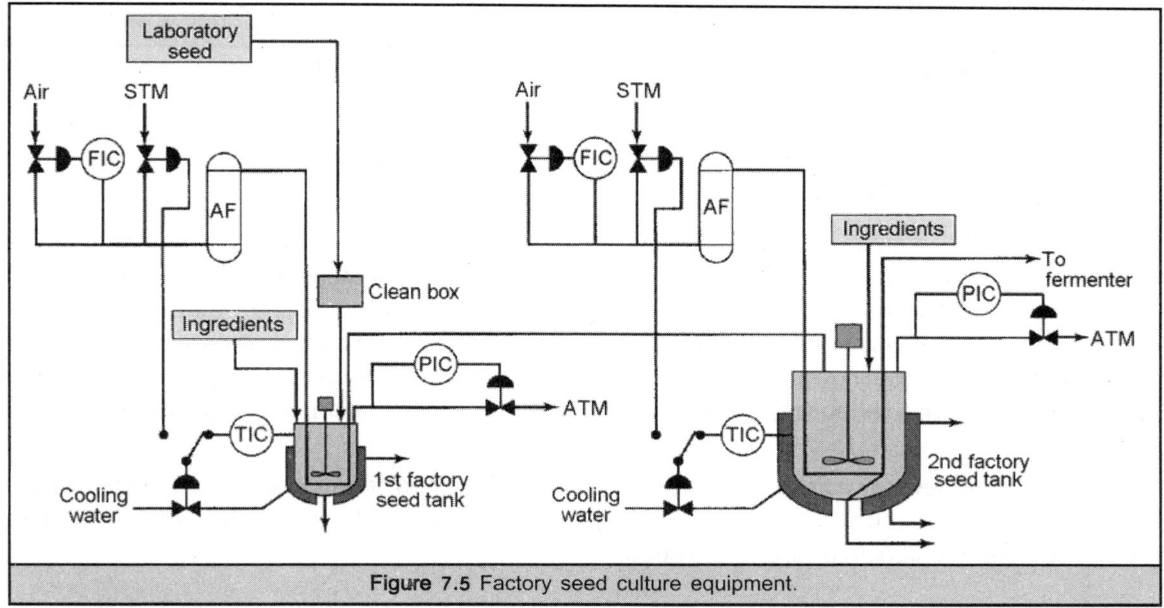

Figure 7.5 Factory seed culture equipment.

fermentation. Although the cost required for these steps accounts for a relatively large portion of that of the whole process, the steps are of much importance to properly operate the process; it is thus necessary to carry out the steps only under optimum conditions.

Laboratory seed culture requires an exceedingly high level of sterility because all subsequent steps for all of the lots may be lost as a result of contamination at this step. Likewise, transfer from laboratory seed culture to factory seed culture must be performed under highly sterile conditions. Typically, all transfer steps after laboratory seed culture are closed, and the whole fermentation train of equipment must be sterilized carefully before the initiation of transfer. Because amino acid fermentation is generally carried out under aerobic conditions at a neutral pH, the process will easily support the growth of a contaminant.

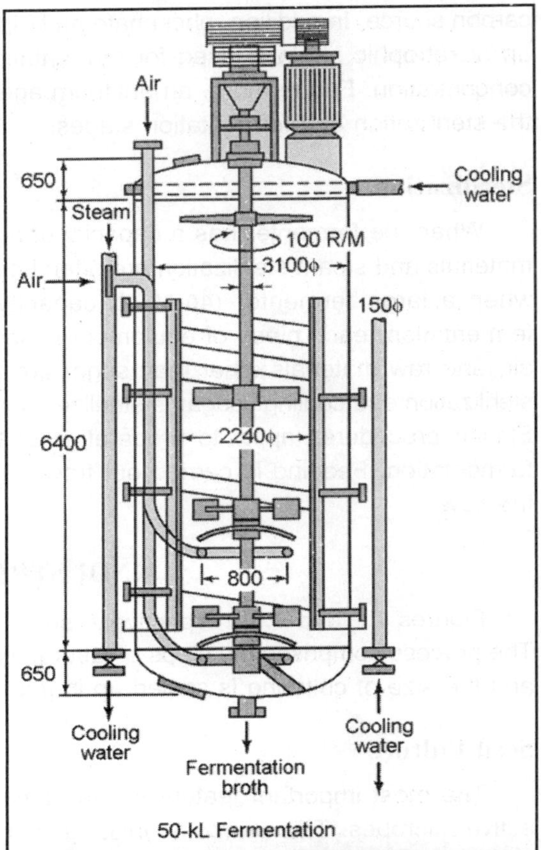

Figure 7.6 Example of an industrial fermentation.

Equipments Required

Amino acid fermentation is performed aerobically, so an efficient oxygen supply is required. The effect of oxygen supply on amino acid fermentation indicates that the sufficiency rate of oxygen required (r_{ab}/KrM) by microbes is controlled by the oxygen transfer efficiency of the fermenter. Oxygen transfer efficiency correlates with aeration and agitation conditions. It strongly affects the results of fermentation and at the same time constitutes the main factor of the cost of utilities.

Measurement and Regulation

In a fermentation process, aeration, pH, feed rate, dissolved oxygen, temperature, and foaming level must be measured and controlled. Parameters automatically measured during fermentation generally include fermenter temperature, volume and temperature of cooling water, volume of air supplied into the fermenter, concentrations of O_2 and CO_2 in the exhaust gas, pressure within the tank, feed volume, agitation rate, agitation energy, pH, dissolved oxygen, and degree of foaming. Sampled broth is analyzed mainly for optical density, sugar concentration, amino acids, and organic acids.

Fermentation temperature, pH, aeration, feed volume, and foaming level are automatically controlled. In continuous fermentation, equipment capable of controlling the volume of both discharged and the liquid surface is provided.

Downstream Processing

Amino acids are typically recovered and purified either by chromatographic methods or by concentration–direct crystallization methods. Although most amino acids may be isolated and purified using an ion exchange resin, this tends to increase the cost of waste-liquor treatment because of the large volume of diluted waste liquor. On the other hand, with the concentration–direct crystallization method, the discharged mother liquors have a high biological oxygen demand and contain a large amount of plant-growth-promoting factors and may thus be used as fertilizer. This makes the cost of waste-liquor treatment less burdensome; however, this method cannot be applied to all amino acids because the chemical composition of fermentation broth strongly affects the purification steps.

Wastewater Treatment

In recent years, treatment of fermentation waste liquor has become more and more important in the planning process. When chromatographic purification is employed, the low concentration and large volume of waste liquor is burdensome, and the cost of treatment by the active sludge method is high, exerting much influence on the cost of the final product. In this respect, the direct crystallization method is advantageous, though it may not be applicable to many products due to characteristics peculiar to those products. It is necessary to construct a process that takes waste treatment into consideration as soon as the raw materials have been selected.

OBJECTIVES

As already mentioned, economy of the fermentation processes must consider the cost of waste-liquor treatment. It is difficult to accurately estimate the economic contribution of waste treatment because this factor is affected by many factors other than fermentation, such as environmental regulation around the location of the plant and so forth. In this discussion, process economics is limited to the steps that begin with the selection of raw materials and end with the desired products. A primary part of the cost of fermentation processes is the cost of the carbon source, which is decided by the (1) unit price of the carbon source, (2) fermentation yield, and (3) purification yield. In L-glutamic acid fermentation, because the carbon source to be used directly affects these three points, selection of the carbon source is particularly important. On the other hand, in the case of fermentation of other amino acids in which auxotrophic strains are used, the cost of the required nutrient also is a major factor. Thus, it is necessary to use a strain requiring a low concentration of such a nutrient.

Production of pharmaceutical-grade neutral amino acids requires difficult separation of analogous amino acids by- produced to the level at which they contribute to increased cost. To overcome this problem, it is necessary to minimize by-produced amino acids to a level at which purification of the desired product is not affected, and to this end, breeding of suitable strains and improvement of fermentation processes that result in few by-products are important. The cost of utilities supporting fermentation is generally high because although the fermentation is carried out mostly at normal temperature and atmospheric pressure, the following factors create additional demands:

1. Sterility must be secured.
2. Aerobic fermentation is involved.
3. It is necessary to more intensely concentrate the fermentation broth because of low product concentration compared to other processes.

PRODUCTION OF AMINO ACIDS

Corynebacterium glutamicum, which produces L-glutamate, requires biotin for growth but excessively forms L-glutamic acid at a biotin concentration suboptimal for growth. Under the condition of excess biotin, L-glutamic acid can be produced by adding penicillin or a surface-active agent, or culturing an oleic acid–requiring strain at growth-limiting oleic acid concentrations. Cane molasses, beet molasses, and corn syrup are mainly used as the carbon source because of their low cost. By way of example, L-glutamic acid fermentation by the fed-batch method using cane molasses is described here. Although cane molasses contains most of the carbon, nitrogen, phosphate, and minerals for microbial growth, its use as a carbon source is basic, and the growth medium is optimized by supplementing with other nutrient sources. Such medium contains an excess concentration of biotin. Potassium salt of penicillin G (8 units/L) is added to the culture broth late in the logarithmic growth phase. Upon addition of penicillin, the microorganism produces L-glutamic acid—mainly due to a change of the permeating system and/or the enzyme activities in TCA cycle—and L-glutamic acid is excreted into the broth.

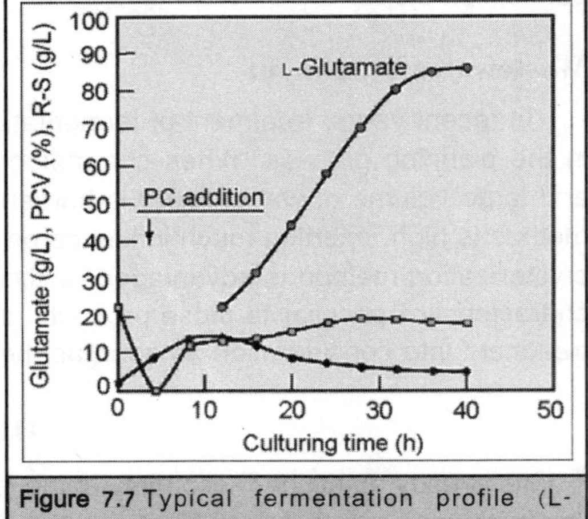

Figure 7.7 Typical fermentation profile (L-glutamate)

Fermentation is continued while feeding molasses and controlling the pH to neutrality with liquid ammonia. Molasses is fed until L-glutamic acid is no longer produced in the broth. The final L-glutamic acid concentration is around 95 g/L. The method of penicillin addition provides remarkably high productivity of this amino acid compared with fermentation under suboptimal biotin conditions. For purification of L-glutamic acid, two methods are available: the direct crystallization method and the chromatographic ion exchange resin method. Because it is well known that the resin method can be used to purify all amino acids produced by fermentation, only the direct crystallization method is discussed here. The purification process may comprise an L-glutamic acid (GA) extraction process and an MSG process to obtain MSG from L-glutamic acid.

The GA extraction process discharges impurities out of the system as a GA mother liquor (containing microbial cells) to recover high-purity GA from the fermentation broth and an MSG mother liquor produced from the MSG process. High purity can be secured by efficiently washing crystals with water at a pH range near the isoelectric point. To this end, the crystal type (there are α- and β-types) must be in the α-form, which can be easily separated by centrifugation. In the MSG process it is essential to maintain the purity of the product while securing a high yield. These may be achieved by enhancing the purity of L-glutamic acid in the GA process and efficiently using the carbon source to achieve minimum levels of residues. Because the MSG mother liquor contains a high concentration of dissolved MSG, it is sent back into the GA process to augment the total yield of L-glutamic acid.

The GA mother liquor (containing spent cells) and a decolorized active carbon cake are the main fractions discharged out of the system. Therefore, it is clear that the properties of the fermentation broth strongly affect the total yield of MSG. How the mother liquor of the GA process is used is also important. Hence, with the direct crystallization method of MSG production, there is unification of the whole process from raw materials to final product. The yield of MSG from the fermentation broth after purification is 65–70 wt %. This is an excellent and effective production system with an attractive process economy if the mother liquor by-product can be used as a fertilizer. As already mentioned, yield is strongly affected by the ratio of L-glutamic acid concentration to total solids in the fermentation broth. This ratio varies, depending upon the purity of raw material sugar and the fermentation efficiency. Whereas the latter depends on the strain of bacteria used, purity of the sugar depends on the quality of the raw material.

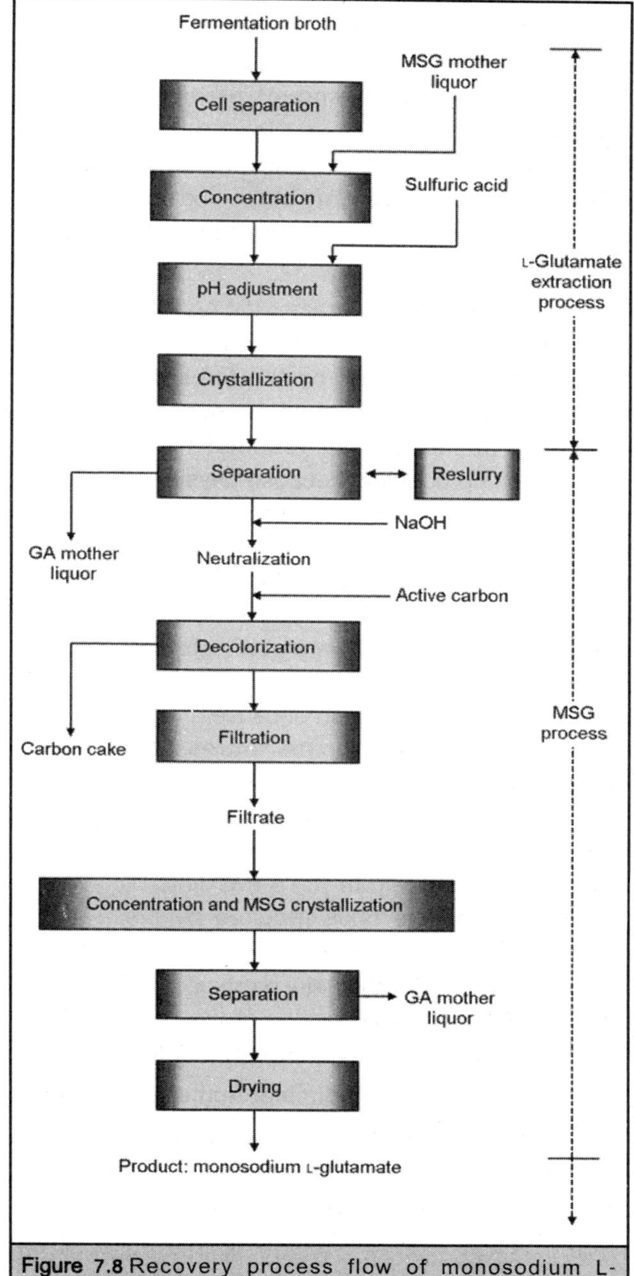

Figure 7.8 Recovery process flow of monosodium L-glutamate.

L-Lysine HCl

L-Lysine HCl is demanded in large quantities for animal feed and must be available at low cost. Reduction of cost through the improvement of productivity is very important, as is the quality of the product. Mutants of *C. glutamicum* (regulatory mutants such as homoserine and/or leucine leaky mutants, which are revertants from auxotrophic mutants and/or lysine analog–resistant mutants) are used. The main raw material for L-lysine fermentation is either molasses or corn syrup. Addition of a biotin source may be required, depending on the raw materials. Molasses contains a sufficient amount of biotin. Because the strains used for L-lysine fermentation are released from metabolic regulation to a remarkable extent, propagation and product formation proceed simultaneously.

This is one of the characteristic features of L-lysine fermentation. In L-lysine fermentation carried out by the fed-batch method, more nitrogen must be supplied compared to the amount required for L-glutamic acid production because L-lysine is a basic amino acid. It is necessary to add ammonium sulfate to the feed solution to supplement the NH_3 supplied for pH control and to neutralize the L-lysine formed. Furthermore, the feeding rate of the solution is controlled to maintain the sugar in the broth at a low concentration because a rapid increase causes catabolite repression. Fermentation is usually complete in about 60 h at a concentration of 95 g/L (as L-lysine HCl). Continuous fermentation of L-lysine is interesting from the viewpoints of productivity and the price required by the market. There has been a report on laboratory-scale steady-state production of L-lysine HCl at a concentration of 80–100 g/L with a productivity of more than 3 g/L per h. To be successful, however, the following conditions must be satisfied:

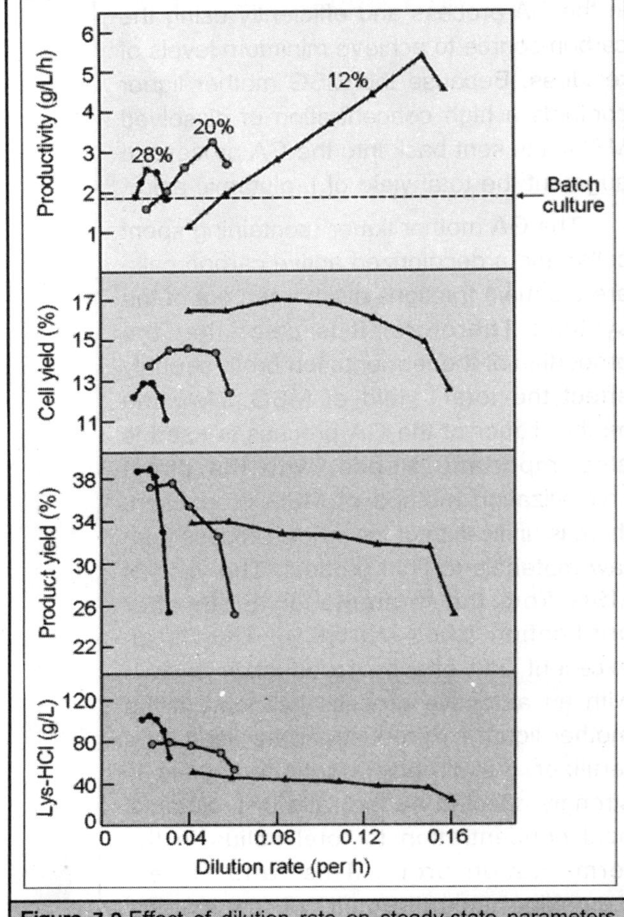

Figure 7.9 Effect of dilution rate on steady-state parameters for *C. glutamicum* B-6.

1. The equipment for industrial fermentation must ensure a pure culture of *C. glutamicum*.
2. The strain's production of L-lysine must not be decreased in the presence of a high concentration of L-lysine.
3. The strain must be genetically stable.

If these conditions are met, high productivity is secured by determining the optimal dilution rate (feeding volume/working volume) for stable operation. Continuous fermentation is superior to batch fermentation in that the productivity per fermenter is higher. However, it is difficult to operate a continuous fermenter for a long period, unlike synthetic chemical processes, due to the three requirements just listed. All factors must be taken into consideration when deciding whether to employ this process. Figure 7.10 shows a flow diagram of the purification process for feed-grade

L-lysine HCl according to the adsorption-desorption method using a cation exchange resin. After separation of the cells, sulfuric acid–acidified broth is passed through the resin for adsorption, followed by washing with water and elution with ammonia. (Lysine is removed from the broth and spent water is sent to the waste treatment step.)

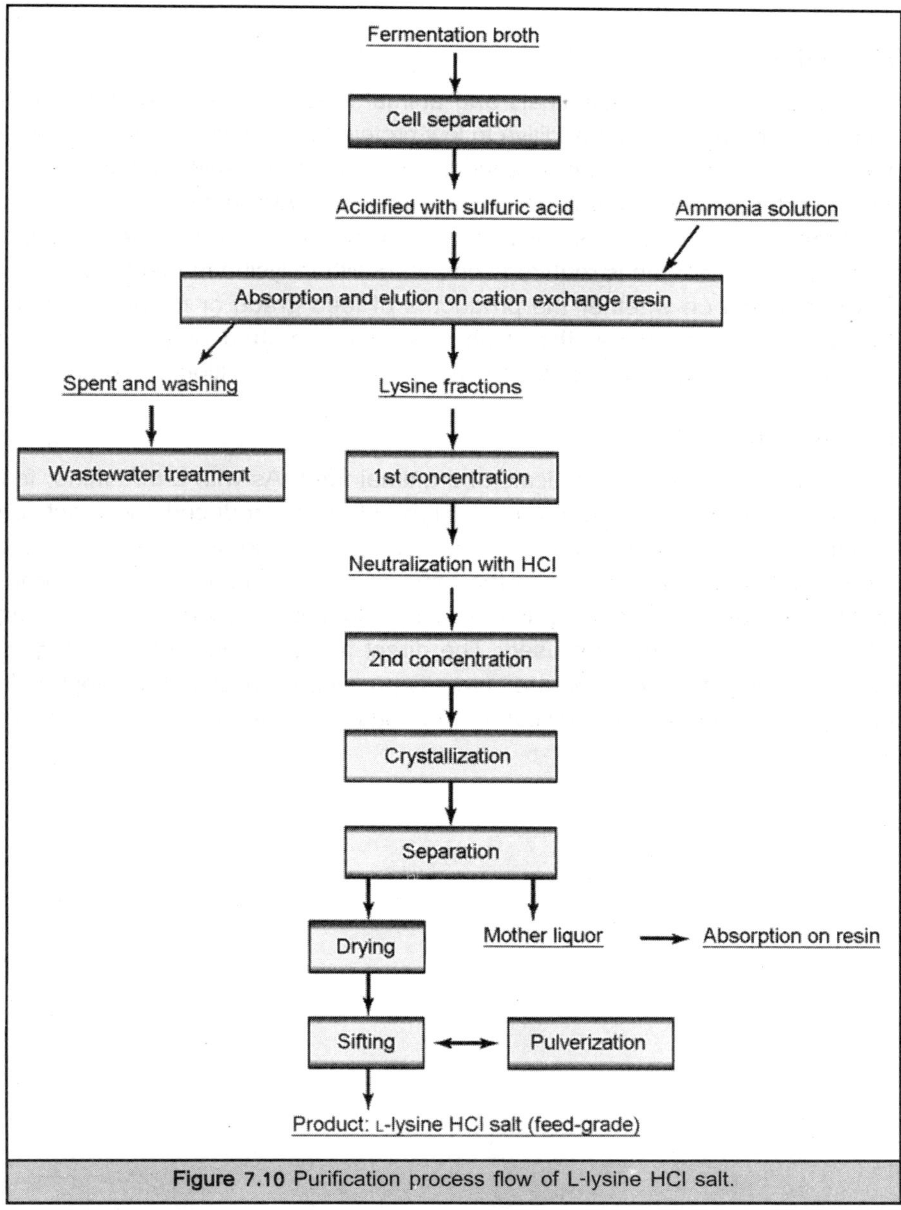

Figure 7.10 Purification process flow of L-lysine HCl salt.

Fractions containing L-lysine are collected and concentrated by removing ammonia (ammonia recovery). The concentrate is neutralized, further concentrated in a vacuum pan, and subjected to crystallization. The resulting mother liquor is spent on the resin-adsorption step and recovered. The yield of the final product from the broth is 80–82 wt %. This process is readily operable and provides stable product quality. However, the relatively high cost of wastewater treatment is a problem because there is a large volume of wastewater with a low-concentration chemical oxygen demand.

Use of L-Threonine

L-Threonine is used in pharmaceuticals and animal feed. In recent years, there has been a growing demand for L-threonine as an additive to low-protein feeds. The strains used include mutants of *C. flavum* and *Escherichia coli* (mutants requiring diaminopimeric acid, methionine, or isoleucine, or their partial revertants) and recombinants of genes involved in the biosynthetic pathway of threonine. The fermentation process is basically similar to that of L-lysine. Because L-threonine is a neutral amino acid, less nitrogen is required compared with L- lysine fermentation. The purification process differs depending on whether the product is of feed grade or for pharmaceutical use. In the case of feed-grade L-threonine, the resin adsorption method or the concentration–direct crystallization methods may be selected depending upon the composition of the fermentation broth.

Use of L-Tryptophan

L-Tryptophan is used in pharmaceuticals and animal feed. As with L-threonine, feed-grade L-tryptophan has been in increasing demand. L-Tryptophan is produced by direct fermentation, fermentation with a precursor, or an enzymatic method using indole and L-serine as raw materials. In the direct fermentation method, strains of the genus *Corynebacterium* requiring aromatic amino acids and having a resistance to aromatic amino acid analogues or recombinants of aromatic amino acids biosynthetic genes in *E. coli* are used. The direct fermentation method using mutants and recombinants and the enzymatic method are currently practiced. Because L-tryptophan is not highly soluble in the aqueous media, the purification methods have been developed to take this into consideration.

8

PRINCIPLES OF FERMENTATION

The term *fermentation* is derived from Latin *fermentum* to ferment, and it has been used to describe the metabolism of sugars by microorganisms since ancient times. Thus, fermentation of fruits is so old that ancient Greeks attributed its discovery to one of their gods, Dionysos. Among the classical fermentation processes are beer brewing, which has documented from as early as 3000 B.C. in Babylonia, soya sauce production in Japan and China, and fermented milk beverages in the Balkans and in Central Asia. Before World War II, however, fermentation processes mainly found their application in the food area, and it was with the introduction of the penicillin production that large-scale fermentation was first used in the production of pharmaceuticals.

Today fermentation processes are used in the production of many different products and these processes can be divided into nine categories, according to the product. Development of fermentation processes can roughly be divided into four phases. First the product is identified. In the case of a pharmaceutical this may be a result of random screening for different therapeutic effects by microbial metabolites (e.g., by high throughput screening of secondary metabolites from *Actinomycetes*) or it may be the result of a targeted identification of a novel product (e.g., a peptide hormone with known function). Outside the pharmaceutical sector the product may also be chosen after a random screening procedure (e.g., screening for a novel enzyme to be used in detergents) or it may be chosen in a more rational fashion. With the rapid progress in the

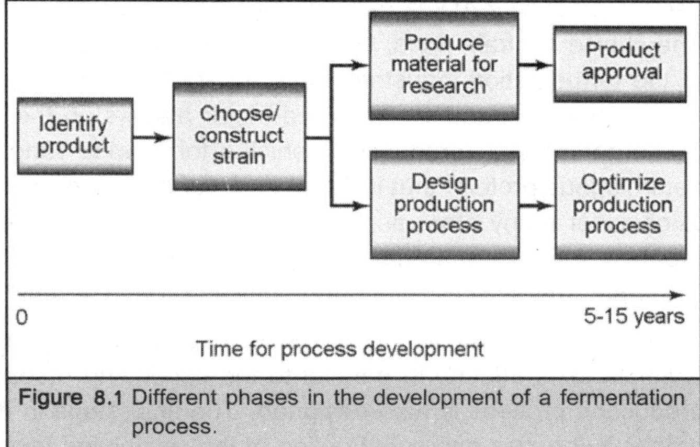

Figure 8.1 Different phases in the development of a fermentation process.

different genomic sequencing programs (the yeast genome was completely sequenced in 1996 and the genome of several bacteria have been completely sequenced), there is much focus on identification of novel products, and today topics such as bioinformatics and gene function are frontline research topics. The next step in the development phase is to choose a strain. In the past this choice was normally obvious after the product had been identified; for example, *Penicillium chrysogenum* was chosen for penicillin production because it was the first organism identified to produce penicillin.

Table 8.1 Categories and typical application of fermentation processes

Category	*Typical processes*
Production of microbial cells	Baker's yeast, single cell protein, lactic acid bacteria
Production of primary metabolites	Ethanol (both in beer and wine production and for technical use), lactic acid, citric acid, acetic acid, amino acids, vitamins
Production of secondary metabolites	Antibiotics (penicillins, cephalosporins, clavulanic acid, tetracycline), skikonin
Production of microbial enzymes	Lipases, proteases, amylases, glucose isomerase
Production of pharmaceutical proteins	tPA, human insulin, erythropoietin, growth hormone, hepatitis B vaccine, monoclonal antibodies
Production of polysaccharides	Xanthan gum, dextran
Production of DNA	Genes for vaccines or gene therapy
Biotransformations	Steroids, L-sorbose (from L-sorbitol)
Production of tissue	Bone marrow, skin

With the introduction of recombinant DNA (rDNA) technology, it is now possible to choose almost any host for the production. Thus, a strain of *Escherichia coli* has been constructed that can produce ethanol at a high yield, and a recombinant strain of *P. chrysogenum* can now be used to produce 7-ADCA (a precursor used for synthesis of cephalosporins) directly by fermentation. The choice of strain does, however, often depend much on tradition within the company, and most of the fermentation industries have a set of favorite organisms that are used in the production of many different products. Thus the major players on the industrial enzyme market have chosen a few organisms as production vehicles for a wide range of enzymes. With the production of a heterologous protein, that is, expression of a foreign gene in a given organism, it is also necessary to consider many other aspects. After the strain has been constructed, one of the first aims is to produce sufficient materials for further research, and this is typically done in pilot-plant facilities. For a pharmaceutical compound sufficient material must be produced for clinical trials.

For other products it may be necessary to carry out tests of the product and examine any possible toxic effects. In parallel to the continuing research in the application of the product, the production process is also designed. The final steps in development are product approval by the proper authorities and construction of the production facility (which in some cases may be through retrofitting of an existing plant). After the process has been developed, there is continuous optimization of the process, and in some cases better strains are introduced. However, for pharmaceutical products new strains are, only introduced when there is a significant improvement in the process because it is costly to obtain approval from the authorities for redesign of the process.

Table 8.2 Advantages and disadvantages of different hosts for production of recombinant

Host	*Advantages*	*Disadvantages*
Bacteria (*E. coli*)	Wide choice of cloning vectors	Posttranslational modifications lacking
	Gene expression easy to control	High endotoxin content
	Large yields possible	Protein aggregation (inclusion bodies)
	Good protein secretion	
Yeast (*S. cerevisiae*)	Generally regarded as safe (GRAS)	Less cloning vectors available
	No pathogens for humans	Glycosylation not identical to mammalian glycosylation
	Large-scale production established	Genetics less understood
	Some posttranslational modifications possible	
Filamentous fungi	Experience with large-scale production	High level ofheterologous protein expression has not been achieved
	Source of many industrial enzymes	Genetics not well characterized
	Excellent protein secreters	
Mammalian cells	Same biological activity as natural protein	Cells difficult to grow in bioreactors
	Expression vectors available	Expensive
		Slow growth
		Low productivity
Cultured insect cells	High level of gene expression possible	Not always 100% active proteins
	Posttranlational modification possible	Mechanisms largely unknown

BASIS OF FERMENTATION PROCESS

The criteria used for design and optimization of fermentation processes depends on the product. Thus, the criteria used for a high-volume–low-value-added product are normally completely different from the criteria used for a low-volume–high-value-added product. For products belonging to the first category (which includes most whole cell products, most primary metabolites, many secondary metabolites, most industrial enzymes, and most polysaccharides) the three most important design parameters are

1. Yield of product on the substrate
2. Productivity
3. Final titer

A high yield of product on the substrate is for these processes very important, since this gives a good utilization of the raw materials, which often accounts for a significant part of the total costs. Thus, in penicillin production the costs of glucose alone may account for up to 15% of the total production costs. Productivity is important because it ensures an efficient utilization of the production capacity (i.e., the bioreactors). In an increasing market it is especially important to increase productivity, because this may prevent new capital investments.

Final titer is of importance for the further treatment of the fermentation medium (e.g., purification of the product). Thus, if the product is present in a very low concentration at the end of the fermentation, it may be very expensive to perform purification of the product. In the production of novel pharmaceuticals, which typically belong to the category of low-volume–high-value-added products, the three previously mentioned design parameters are normally not very important. For these processes quickness to market is generally much more important, and change of the process after implementation is often complicated due to the required U.S. Food and Drug Administration (FDA) approval. In the initial design phase it is, however, still important to keep these three design criteria in mind. The requirement for a high final titer is especially important since the cost for purification often accounts for more than 90% of the total production costs.

BASIC PHILOSOPHY

Our basic philosophy comprises the following simple principles:

1. Each design case has unique requirements and calls for individualized application of the basic design concepts and practices discussed herein. The one-size-fits-all approach usually results in sleeves that drape into your food, arm holes that cut off circulation, pant legs that are made to trip over, or some combination of these features and others. The magic number approach—all problems have simple solutions based on codified numbers such as fixed geometric ratios—leads to the same place.
2. Successful design requires a systems approach. A fermenter is a system that is part of a process system, and the process is part of a plant system. All these systems interact with each other, with real people, with control systems, and with other systems related to regulatory compliance, safety, documentation (including protocols, SOPs, etc.), change control, maintenance, and so forth. Failure to take these interactions into account during fermenter design usually results in considerable pain, not only for those guilty of the omission, but also for innocents who were never asked for planning and/or design input, but who must live with the result. Note that appropriate documentation, protocols, SOPs, change control, and so on should be considered important elements of design and operation of any plant, licensed or not.
3. Compromise is always necessary. Nothing in any project is 100% right, and nothing has to be. He or she who looks for 100% of anything will hold up a project needlessly and will generate a lot of animus.
4. The time to be thinking about all of these points is prior to and during design, not when you're standing on the plant floor trying to validate a continuous mixer that has only one port and no instrumentation.

We discuss next the bases for safety and regulatory considerations, primarily to sensitize readers to these issues early on. They are all too often overlooked during early stages of design, when the focus is on satisfying process requirements (e.g., oxygen transfer). Unfortunately, this frequently leads to important constraints being ignored and hence to designs that require expensive and often cumbersome "fixes."

FACTORS AFFECTING THE PROCESS

All fermenter designs are influenced by safety and regulatory factors, the extent depends on (a) the nature of the product and its intended use, (b) the process, and (c) the nature and location of the facility. For example, even fermentation products not regulated by the U.S. Food and Drug Administration (FDA) must adhere to safety requirements, which in some cases are very strict. Furthermore, no professional design can escape the effects of a long list of code requirements, for example, the ASME pressure vessel code. By the same token, not all products that are FDA regulated are equally affected by safety, regulatory, and code requirements. We must consider each case on the basis of its own unique requirements. Our objective in the rest of this section is to introduce some of the requirements and to provide some rational bases for their implementation, details of which are discussed later. Much of the discussion is directly applicable to fermenters used for production of human or animal health products or for their precursors, but also applies to other cases to varying extents.

Levels of Biosafety

A fermenter must be an integral part of a system designed to insure safety at three levels: product, worker, and community. Many of the methods used to protect the product also serve to protect the plant personnel and the community (e.g., use of closed systems and HEPA filters); however, there are potential points of conflict. For example, some containment practices that call for completely welded hard piping to a contained drain line for any condensate that could be exposed to culture fluid could expose the product to drain line contaminants. One resolution of this problem is to use steam locks in all such lines. All such conflicts encountered to date have been resolved, but not always easily or relatively inexpensively. Insurance of product safety is tied primarily to compliance with FDA (or similar organizations) regulations. This will be considered in the subsection "Product Safety." Here we deal with worker and community safety. Protection must be provided against potential ill effects of fermentation products (e.g., cytotoxins, allergens), organisms (e.g., pathogens, whether recombinant or not), fermentation by-products (e.g., pharmacologically active precursors of the active product), or some combination.

The practice of providing such protection is called *containment*, which is defined as insuring that deleterious fermentation components can't be transported to any area inside or outside the plant before they have been rendered harmless. To accomplish this, containment must be exercised at several levels. We consider in this article only direct or "primary" containment of fermenters; however, one should not lose sight of the fact that the other levels must be considered during fermenter design, if for no other reason than to ensure that the fermenter design and operation will be consistent with the overall containment strategy. Containment levels have been defined by the National Institutes of Health (NIH), based on the potential danger of organisms. These levels, in increasing order of potential danger are GILSP, BL1-LS, BL2-LS, BL3-LS, and BL4-LS. (BL4-LS must be handled on a case-by-case basis in consultation with the Centers for Disease Control and others. It is, thank goodness, well beyond our scope.) There are other classifications that have been developed in the United Kingdom, the European Union, and elsewhere, but these do not differ markedly from the U.S. scheme. It is important to know that (a) all of these guidelines address what must be accomplished but not how they should be accomplished, and (b) there is not complete agreement on what is actually required to implement the guidelines. In addition, the guidelines do not address containment of products.

Although many of the considerations for product containment are the same or very similar to those for organism containment, there are some very important differences. Most existing processes use organisms that require either GILSP or BL1-LS containment; nevertheless, many plants are built to satisfy requirements for BL2-LS (actually, somewhere between BL2-LS and BL3-LS). The reasons for this include satisfying FDA requirements (e.g., safety of the product) and the fact that the additional cost is not large and is relatively cheap insurance. There also is the oft-stated reason that such a route insures greater flexibility for future operation. One might add to this the fact that the rules are not well established and that requirements could change substantially between the time of design and the time production begins. It also is worth noting that there is growing interest in using organisms that do, in fact, require BL2-LS and even BL3-LS containment, which will probably focus greater attention on further development of acceptable implementation practices as well as changes in the guidelines.

Note that equipment requirements and cost for BL2-LS and BL3-LS are currently about the same (this probably will change); BL3-LS facilities requirements and cost are very much greater than for BL2-LS. Finally, it is important to note that containment is nothing new. Highly pathogenic organisms have long been used to produce therapeutics and biological warfare components; hence, there is a considerable body of experience dealing with the subject. There also is considerable guidance to be had from the nuclear industry; nevertheless, the natures, sizes, and large number of new commercial processes, along with new methods, materials, and so on, will force heightened awareness, scrutinity, review, and modernization or change.

Safety at Physical Level

Most physical safety issues are addressed by Occupational Safety and Health Administration (OSHA) requirements and by various national codes such as the ASME Pressure Vessel Code and the National Electrical Code. There also are many local construction codes that must be satisfied. Among these are various earthquake-resistant construction codes that are also applicable to containment considerations. And then there are the requirements of the final arbitors: the insurance companies.

Safety at the Product Level

We focus here on FDA requirements, but the reader is urged to keep in mind that he or she also must consider those of other regulatory agencies and be aware of the unsettling facts that it is not always clear which agency has jurisdiction, and there often are conflicting requirements. The primary purpose of FDA regulations is to insure product safety, purity, and efficacy. The legal bases for the regulations and their enforcement can be found in the *Code of Federal Regulations* (CFR) and a number of other FDA documents such as *Points to Consider, Inspection Guides,* and *Guides to Industry*. Industrial implementation of all the regulations, along with ongoing improvements, is referred to as current good manufacturing practices (cGMP). There is an endless stream of writings in professional journals, short courses, and such concerning the regulations and cGMP practice.

Unfortunately, there is a great deal that is not covered clearly in writing, and there is a constant proliferation of new and/or revised guidelines followed closely by lots of (frequently conflicting) interpretations and opinions. In addition there has been a problem of nonuniform application. For example, regulatory scrutiny has been far stricter for fermenters used to produce active molecules directly than for those used to produce precursors; there has been considerable variation from product to product within a given class; and manufacture of human drugs has been regulated much more rigidly than manufacture of animal drugs. While understandable to some extent, such nonuniformity has caused considerable confusion. These gaps are beginning to be closed, and this may have considerable influence on the design of fermenters—particularly those used to produce precursors and animal drugs. In simplest terms, cGMP requires that the combination of fermentation process equipment and operating protocols must consistently yield product that meets acceptable specifications and is capable of being converted/purified consistently to final product, meeting approved product specifications. From this general sense of intent it can be argued that failure to have control over the fermentation puts the final product at risk. Few would argue with this general concept; however, there is considerable debate as to interpretation and extent. Requirements that flow from the need to have control are based on relatively simple ideas:

1. If an organism is subject to environmental, medium, or other conditions outside the range in which it is known to yield product meeting acceptable specifications and capable of being converted/purified consistently to final product meeting approved product specifications, then we have no guarantee that it does not produce other products with which the recovery system can not cope and that can escape detection by the analytical methods in place.
2. If a fermenter becomes contaminated with other microbes, said microbes could produce toxins that could be carried undetected to the final product. It is possible that this could occur without altering the behavior of the process organism. It also is possible that products of the contaminant could cause the process organism to make toxins that could go undetected into the final product. Given these possibilities, plus the fact that it is not possible to prove that the contaminant will always be the same, evidence of contamination is evidence for lack of control.
3. If a fermenter is not cleaned properly, deleterious microbial or nonmicrobial products could remain to contaminate the next batch in such a way that impurities could be carried undetected to the final product. Similar statements can be made about other contaminants introduced via other routes as a result of poor cleaning practices.

Again, it is unlikely that many would argue with these points, but again there is considerable debate as to interpretation and extent. Requirements 1–3 can be translated to specific requirements for, among other things, controllability and reliability of fermentation conditions, sterilization, and aseptic and cleaning operations. Controllability requires that the fermenter and its subsystems be designed and constructed in ways that make possible control of environmental variables within the operational ranges required. This obviously means that mechanical design must be in harmony with control system design. Sterilization/aseptic operation translates to (i) destroying any microbial contaminants that may be present in any part of the equipment that might contact process fluids and (ii) insuring that no microbes (other than the production organism) can enter after the equipment has been sterilized. The latter embodies the concept of the "sterile barrier." To these ends, the following apply:

1. The fermenter and all its ports and direct attachments must be sterilizable.
2. All piping that will contact process fluids (including additives) and/or provide paths into the system (e.g., the air exhaust line) must be sterilizable initially. Some (e.g., sampling lines, addition lines) must also be sterilizable at any time during a fermentation.
3. Inlet gases (e.g., air, oxygen) and all additives (e.g., medium components, acid, base, antifoam) must be sterilized before they contact any sterile process piping.
4. All penetrations (e.g., drive shaft, probe ports) must be sterilizable.

The preferred sterilization method is automatic sterilization-in-place (SIP) with steam. This is discussed at length later. Clean operation requires, among other things, the following:

1. The system must be designed such that any surface that can contact a process stream can be cleaned consistently to a level that insures that the product will be free of soils resulting from a fermentation.
2. Nothing in the system that comes in contact with process fluid can introduce unacceptable and unidentifiable materials.

Note (for example) that 21 CFR 211 (Part D) (guidelines for design of equipment for licensed facilities) does not have any specific requirements for fermenter sterilization or cleaning; however, the preceding discussion coupled with the intended use and cGMP concepts makes it difficult to argue the point. But again, there remain the questions of rational interpretation and extent of application. Riding along with cGMP is the practice of validation. This is an extensive, expensive, and controversial area for which and about which reams of reams have written. Following is the formal definition of validation: "Establishing documented evidence that provides a high degree of assurance that a specific process will consistently produce a product meeting its predetermined specifications and quality attributes." What this means is that not only does the equipment have to be designed, installed, and operated so as to run consistently within well-defined ranges of process variables to consistently yield product that meets acceptable specifications and is capable of being converted/purified consistently to final product, meeting approved product specifications, but that there must be adequate documentation to prove this before product can be sold.

In addition, one must demonstrate that this goal can be achieved under so called worst-case conditions; there is often considerable debate as to what this really means. It is beyond our scope to consider validation in detail. Suffice it to say that it can translate to checking and testing every component of the equipment and documenting not only its proper function but also its history. The brave of heart are referred to the bulging literature on the subject. All of this means we must take great care to consider *all* the details, such as materials, corrosion issues, machining methods, vendors, welds, and types of steam and water, keeping in mind that all must be considered as part of a system that must interface "seamlessly" with the systems of cGMP and validation. Failure to consider this before and during design is almost always very costly. Design should be done so as to facilitate validation, and subsequently, routine cGMP operation, maintenance, and so forth. "Simple" things such as judicious placement of access valves and piping isolation can go a very long way to ensuring minimum pain and maximum operability. Having said all this, we reiterate that not all fermenters are subject to all these requirements, and even those that are, are subject to them in varying degrees (at least in practice).

EQUIPMENTS

A design basis for fermentation equipment should derive from a facility/process design basis. The latter should include (among other items) the general nature of the facility (e.g., research and development vs. production, single product vs. multiuse); product(s) specifications; regulatory, containment, and other requirements; level of automatic operation; general processing scheme (e.g., batch); nature of individual process steps; productivity, concentrations, and so forth; cleaning requirements; special considerations (e.g., earthquake-proof construction); critical valving and instrumentation; staffing requirements and constraints; architectural and general floor plan constraints; and utilities. Most of these will have some influence on fermenter design—some in more subtle ways than others. It is difficult to overstate the importance of the initial definition that will derive from the design basis.

Obviously, a rational design basis for a fermenter must also be based on fermentation characteristics as well as operating cycle and productivity required (which should derive from the overall process design basis). From these will flow the sizes and number of vessels, and definitions of oxygen transfer, heat transfer, power, and bulk mixing requirements. The design must satisfy these but must also satisfy requirements for regulatory compliance, safety, cleaning, facile operation, and maintenance. All of these factors are highly interactive (e.g., design for oxygen transfer affects design for cleaning); hence, responsible design will almost always require several iterations to ensure the greatest probability of success and to minimize lost time caused by installation, operation, and various problems. The iterative nature of the design (and, unfortunately, construction) process makes most important the existence of a well- crafted, well-implemented, and well-documented change control process. Finally, as with any engineering project, failure to have a solid basis of design, a complete scope, accurate process flow diagrams and piping and instrumentation diagrams (P&IDs), and accurate process timing will almost certainly result in added cost, lost time, and worse.

TRANSPORT PROCESS

One of the first steps in fermenter design is translation of process demands to oxygen transfer, heat transfer, bulk mixing, and power requirements—the so-called transport processes (TPs). Sound translation is not straightforward because the TPs interact not only with each other, but also with vessel geometry and other design factors. The nature of each TP and the relationships among them vary with the process requirements. They also vary significantly with the basic nature of the organism and the culture broth. The major factors here are the rheological nature of the broth and the sensitivity of the organism to fluid mechanical forces. In some cases, sensitivity of the organism to temperature and other environmental factors, such as pH, impose tighter constraints (e.g., wall temperature control).

Single-cell organisms (most bacteria and yeasts) tend to tolerate fluid forces very well (there are a few exceptions) and tend to have low-viscosity, Newtonian broths. Mycelial organisms (fungi and some mycelial bacteria such as streptomycetes) tend to be more prone to damage by fluid mechanical forces than are single-cell organisms and tend to have high-viscosity, non-Newtonian broths. Detailed discussion of all these factors is beyond the scope of this article; however, we now discuss a few points concerning rheological behavior, and point to fluid effects wherever appropriate in the remainder of the article.

Newtonian Viscosity

Newtonian viscosity depends only on composition and temperature; the nature of the fluid motion does not affect it. Non-Newtonian viscosity does depend on the nature of the fluid motion. This dependency is usually expressed in terms of fluid shear rate (a measure of how rapidly fluid velocity changes from one point to another point close by). There are several types of non-Newtonian behavior (e.g., pseudopalstic, Bingham) that can be described quantitatively by a host of mathematical models. One model used frequently to describe non-Newtonian fermentation broths is the so-called power law:

$$\eta = K\gamma^{(n-1)} \quad \text{...(1)}$$

where η is viscosity, γ is shear rate, and n and K are constants. Such relationships can be useful if one has rheological data for the subject broth and if the design correlations used incorporate rheological properties in a meaningful and accurate way. It is seldom that either of these conditions is satisfied, and almost never that both are satisfied. Also, there is a considerable amount of misleading and/or inaccurate information in the literature concerning broth rheology, its measurement, and its use; the reader is cautioned to tread carefully in this area. If accurate viscosity information is available, it can be a valuable qualitative and sometimes semiquantitative guide, if interpreted and used properly.

Mass transfer (oxygen), heat transfer, and bulk fluid motion all depend strongly on rheological characteristics. Generally, all are poorer in non-Newtonian than in Newtonian broths; indeed, the rheological characteristics of mycelial broths can and do impose severe constraints. Two of many examples are the following:

1. Heat and oxygen transfer rates tend to be anywhere from 50 to 5% of what they would be for typical (e.g., *Escherichia coli*) bacterial fermentations.
2. Bulk mixing quality (homogeneity) is much poorer than in typical Newtonian broths; therefore, accurate monitoring and control are much more difficult.

Difficulties associated with viscous, non-Newtonian rheology also extend to other areas such as cleaning.

Oxygen Transfer/Aeration

Requirements

Oxygen transfer rate (OTR) requirements are usually dictated by conditions during the most active part of the growth phase (other phases require oxygen, but supply rate usually is not as high). The requirement is calculated from

$$OTR = (Y_{x/o})^{-1} \, (\mu X)_{max} \quad \ldots(2)$$

where μ is the specific growth rate (h^{-1}), X is the cell mass concentration (g/L dw), and $Y_{x/o}$ is the cell yield coefficient on oxygen (g cells dw/g O_2). Note that the maximum growth rate and maximum cell mass are not always reached simultaneously.

Gas glow and linear velocity

The very first thing we must do in satisfying an OTR requirement is to insure that we have a closed oxygen material balance. For most practical operating conditions, the rate of change of oxygen inventory in a fermenter is very small compared to oxygen flows in the inlet and outlet and to the OTR. It also can be demonstrated easily that the total molar flow rate of gas does not change enough to cause any loss of sleep. Given these practical realities, the oxygen material balance is

$$OTR = (1{,}000 \times 60)F(y_{in} - y_{out})/V_L \quad \ldots(3)$$

where, F is the gas flow rate (mol/min), y is the mole fraction of oxygen in the gas, and V_L is the liquid volume (L). Note that the usual units for OTR are millimoles per liter per hour and that the factors of 1,000 and 60 in equation 3 are conversion factors needed to keep unit consistency. The material balance also can be expressed as

$$F = OTR \times V_L/(1{,}000 \times 60\varepsilon y_{in}) \quad \ldots(4)$$

where ε is the oxygen transfer efficiency and y_{in} is the oxygen mole fraction in the inlet gas. Equation 4 is a more useful form because we know that under practical conditions e will be between 0.15 and 0.35 for typical Newtonian broths, and between 0.05 and 0.15 for typical non-Newtonian broths.

As a general rule, heroic efforts will be required to get OTRs over 300 mmol/(Lh) in large fermenters (>5,000 L). Even if the effort is expended and it is successful, it will usually create heat transfer problems that will require even greater effort to overcome. Our advice is to consider other options very seriously. The material balance provides information about the gas volumetric flow rate, which often is expressed as the standard flow, v_{std}, in units of standard liters per minute (SLPM):

$$v_{std} = 22.4F \quad \ldots(5)$$

This is useful because most flow meters are calibrated in terms of standard flow; however, actual flow is more useful for fermenter design purposes. One problem encountered in calculating the actual flow is that it increases from the bottom to the top of the broth because of pressure change. As a practical matter, however, it is usually adequate to use the average pressure in the tank:

$$v_{act} = v_{std} \times (1/P_{avg})(T_f/298) \quad \ldots(6)$$

where v_{act} is the gas flow at average pressure (L/min), P_{avg} is the average pressure (atm, abs), and T_f is the fermentation temperature (K). One might want to revisit this for very tall vessels.

The standard gas flow rate also is expressed in standard gas volumes per liquid volume per minute (VVM). Among the important information that can be obtained from v_{act} is the gas linear velocity, V_S (cm/min):

$$V_S = 4 \times 1000\ v_{act}/(\pi \times D_t^2) \qquad ...(7)$$

where D_t is the inside diameter of the fermenter. V_S affects mass transfer, deliverable power via the impellers, foaming, gas holdup, and aerosol formation. As a rule of thumb, V_S should be held below 200 cm/min to avoid problems associated with excessive gas holdup, foaming, and aerosol formation. These effects are discussed later.

Mass transfer

The rate at which oxygen can be transferred is dictated by three factors: the driving force, the resistance to transfer, and the contact area between the gas and liquid phases. This is usually expressed as

$$\text{OTR} = K_g a(\Delta P)_{LM} \qquad ...(8)$$

where K_g is the mass transfer coefficient ([mmole M]/[Lh atm]), and a is the interfacial area per unit volume (M^2/M^3). $(\Delta P)_{LM}$ is the log mean pressure driving force, defined as

$$(\Delta P)_{LM} = (P_{O,in} - P_{O,out})/LN[(P_{O,in} - P_O^*)/(P_{O,out} - P_O^*)] \qquad ...(9)$$

where $P_{O,in}$ is the partial pressure of oxygen in the gas inlet (atm, abs), $P_{O,out}$ is the partial pressure of oxygen in the gas outlet (atm, abs), and P_O^* is the partial pressure of oxygen that would be in equilibrium with the dissolved oxygen concentration in the existing liquid (atm, abs). Note that use of the log mean driving force assumes that the liquid is mixed perfectly (homogeneous liquid phase) and that the gas moves in plug flow through the fermenter.

These assumptions are acceptable for most low-viscosity Newtonian broths, but they can be quite poor for high-viscosity, non-Newtonian broths. Unfortunately, no useful alternatives have yet been proposed; therefore, extra caution should be exercised in interpreting calculated results for non-Newtonian broths. Other mass transfer coefficients are used for driving forces other than $(\Delta P)_{LM}$. The most common among these is $k_l a$ (min^{-1}), which is used in conjunction with a dissolved oxygen concentration driving force. The reader is cautioned that $K_g a$ and $k_l a$ are sometimes confused with each other—even in the literature. A given OTR requirement can be satisfied by various combinations of driving force and mass transfer coefficient. It is important to note, however, the following:

1. Factors that influence driving force also can influence $K_g a$.
2. Each combination will have different effects on the ultimate design of the vessel, piping system, and support systems, as well as on operational factors related to regulatory compliance cleaning, maintenance, and so on. Such effects are discussed as we proceed.

Driving force is affected by total pressure, gas inlet oxygen mole fraction, total gas flow rate, and the dissolved oxygen concentration. Total pressure and oxygen mole fraction affect the oxygen partial pressure directly. The effect of total gas flow is a bit more subtle. For a given oxygen transfer rate, increasing the total gas (fixed inlet conditions, pressure, etc.) increases the mole fraction of the outlet gas flow, thereby increasing the average mole fraction of oxygen in the gas; hence, the average driving force is increased. This can be seen quantitatively via an oxygen material balance which gives

$$P_{O,out} = P_{O,in} - \text{OTR} \times P/(1{,}000F) \qquad \text{...(10)}$$

where P is the total absolute pressure (atm) and F is the total molar flow (mmol/h) of gas.

The characteristics of K_g and a are complex and not completely understood. Our first clue to this is that $K_g a$ is expressed as a product in terms of empirical correlations (we don't understand enough about them to get reliable quantitative guidance from first principles). Typical correlations have the form

$$K_g a = \delta \times (P_g/V_L)^{\varphi} (V_S)^{\kappa} \qquad \text{...(11)}$$

where P_g is the power delivered to aerated broth (HP), V_L is the unaerated liquid volume (L), and δ, φ, and κ are usually advertised as constants. There also a several variations on this form. A discussion of all the fine points of such correlations is beyond our scope; however, the reader should take note of the fact that published correlations usually cannot be relied upon to provide accurate predictions. They should be used as qualitative or semiquantitative guides only. Reasons for this caveat include the following:

1. They are usually applicable only under conditions at or near the ones used to develop them. Unfortunately, most are developed at scales and under conditions that are not realistic for production.
2. They can have pronounced dependence on fermenter geometry.
3. They do not scale up well, or at all. The "constants" in equation 11 and its relatives usually aren't constant.

These caveats should be made more emphatically for viscous non-Newtonian broths than for Newtonian broths because non-Newtonian rheological characteristics can have profound effects on $K_g a$ (as well as the other transport properties). Most $K_g a$ correlations do not account for broth rheology. Those that do, usually do so inadequately or in ways valid only for a particular broth. It also is important to note that the broth rheology can vary dramatically during a fermentation and that the rheological characteristics can be affected by the fermentation conditions, ranging from the nature of the inoculum to the history of the agitation speed. As noted earlier, we can get some qualitative guidance from the form of equation 11. We'll do that, but please keep in mind that the relationships are highly nonlinear, the factors that influence $K_g a$ are dependent on each other in practical operation, and it is not always possible to control variables at the levels you would like.

For example, equation 11 predicts that increasing power input per unit liquid volume will increase $K_g a$. This usually turns out to be true in practice if the power can actually be delivered to the fluid and delivered in a manner that will contribute to $K_g a$. Such might not be true in any given case. The fact that a drive's power rating is 100 hp doesn't mean that 100 hp can be delivered to the broth. Deliverability depends on, among other things, geometry (vessel and agitator), rheology, agitator speed, and gas linear velocity. The effects of all are complex and interactive; the interrelations are more complex for non-Newtonian broths than for Newtonian broths. The manner in which power is delivered depends largely on impeller type and broth rheology. For example, oxygen transfer rates to highly viscous, non-Newtonian mycelial broths have been shown to be larger for A-315 (hydrofoil) impellers than for Rushton (disk turbine) impellers delivering the same power. There appears to be no meaningful difference for Newtonian broths. The bottom line is that one should try to obtain experimentally determined correlations for the subject organism and broth at meaningful scales. Failing that (which usually is the case because of time and resource constraints), one should try to use correlations developed for systems as similar as possible to the one at hand. Considerable art is involved here.

Power

Power delivered to the broth is used for micromixing and gas dispersion, which are related to mass transfer, and macromixing, which provides overall homogeneity. Agitated vessel power is delivered via two mechanisms: direct mechanical power from the impellers and gas expansion. The bulk (>90%) of the power comes from the impellers as long as the fluid motion is under their control, a condition that prevails so long as the impellers are not flooded. There are reasonably reliable correlations available to determine flooding conditions for Newtonian broths; however, flooding usually is not a problem under typical conditions used in most Newtonian broth fermentation. Flooding is more likely in highly viscous, pseudoplastic broths typical of mycelial and polysaccharide fermentations because the viscosity near the impeller is much lower than in the rest of the broth. As a result (assuming air is introduced under the impeller), air tends to channel toward the impeller, thereby enshrouding it and decreasing the deliverable power.

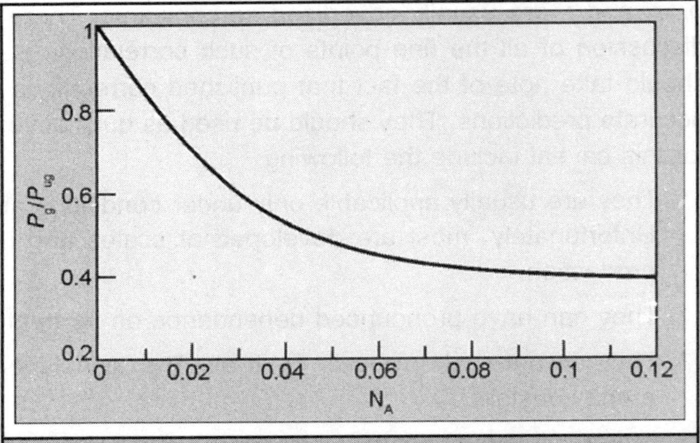

Figure 8.2 Power ratio as a function of aeration number for a Rushton turbine.

This can dramatically decrease overall mass transfer (and heat transfer) rates and quality of bulk mixing quality. Reliable flooding correlations for non-Newtonian broths have not been published, but some companies have accumulated a considerable amount of data for the fermentations they practice. Calculation of power input relies on empirical correlations. As with $K_g a$, many correlations have been published, but they are not particularly reliable for many of the same reasons given for the unreliability of $K_g a$ correlations; therefore, we give the same advice here as we did for $K_g a$. Among the more popular approaches used is the one that relies on the aeration number, *NA* (dimensionless), defined as

$$N_A = v/ND_i^3 \qquad \text{...(12)}$$

where v is the gas volumetric flow rate (m^3/min), N is the agitation speed (min^{-1}), and D_i is the impeller diameter (m). Published correlations give the ratio of gassed power to ungassed power as empirical functions of N_A:

$$P_g/P_{ug} = \text{Func}(N_A) \qquad ...(13)$$

where P_{ug} is the power delivered to the same broth agitated under the same conditions but unaerated. Each such correlation pertains to a specific impeller type, a single fermenter geometry, and a fairly narrow range of operating conditions. In some such correlations, a unique function is presented for each impeller type. In others the relationship is shown not to be unique. In most cases, the correlations are for a single aerated impeller. The situation is more confusing for non-Newtonian broths.

Obviously, to use equation 13 one must be able to calculate the ungassed power. This can be done via other empirical correlations that give ungassed power in terms of the dimensionless power number (N_P), defined as

$$N_P = P_{ug} \times g_c/N^3 \times D_i^5 \times \rho \qquad ...(14)$$

where g_c is Newton's law conversion factor, and ρ is the ungassed broth density (lb/ft^3). There are published correlations that give N_P as a function of Reynolds number for a wide range of single impellers. Each is specific to a particular vessel geometry, liquid height, and so forth. Interestingly enough, there are fairly reliable correlations for Newtonian and non-Newtonian fluids so long as the constraints of the correlations are observed. It is important to note that some impellers that have high power numbers under unaerated conditions can "unload" considerably under aerated conditions.

For example, it is fairly typical for a Rushton turbine to deliver, under aerated conditions, only 40% of the power it can deliver when unaerated. On the other hand, the SCABA 6SRGT, which is a curved-blade disc turbine, does not tend to unload at all. Some hydrofoils also exhibit very little unloading. This is an important consideration when one is trying to rank impellers in terms of how much power they can deliver for oxygen transfer purposes. (Often the ranking tends to be done qualitatively or "semiquantitatively.")

Large impellers

Most fermenters are equipped with more than one impeller. The reasons for this are to improve bulk mixing and power distribution and to avoid the need for very large impellers and/or very high agitator speeds. Impeller size is limited practically not only by vessel internals but also by the following:

1. *Torque transmitted to the drive shaft*. The larger the torque, the stronger and thicker the shaft must be. This also translates to higher torque and more expensive gear boxes.
2. *The size of the vessel manway*. This is particularly important if the impellers must be single piece for better cleaning/sterilization characteristics.

Agitator speed is limited by the natural frequency of the agitation system. This is because severe and potentially dangerous vibrations will occur if the rotational speed of the agitator approaches the natural frequency. We recommend that the maximum shaft speed not exceed 70% of the natural frequency. Having said all this, we are left with the problem of doing the power calculations. This usually involves an iterative calculation in which the parameters are (assuming we have selected impeller types) number of impellers, impeller diameter(s), and shaft speed. To do this we use the power correlations already noted. But these usually apply only to single impellers. In most cases, they apply to the lowest impeller, which is the one under which inlet air is introduced. Several approaches have been suggested. The simplest is to treat all the impellers as though each is the only one present. Among other things, this assumes that the impellers do not interact with each other.

There is empirical evidence that Rushton turbines will not interact substantively in Newtonian broths if they are spaced at least an impeller diameter apart. Obviously, the assumption of noninteraction is not meaningful for impellers that produce significant axial flow. Another approach used frequently, but not recommended by the authors, is to use the correlations for the lowest impeller but to use them for the upper impellers:

$$P_g/P_{ug} = (1 + HU)^{-1} \qquad ...(15)$$

where HU is the gas holdup. You guessed it: there are empirical correlations for HU. One example for Newtonian broths is

$$HU = 1.8(P_g/V_L)^{0.14}\ V_S^{\,0.75} \qquad ...(16)$$

The usual caveats apply. Typical values for gas holdup range from about 0.1 to 0.3; therefore, equation 15 predicts about 10-20% unloading for the upper impellers. Aeration number correlations predict around 50-60% unloading. Given the uncertainty in all this, and assuming that no other (reliable) information is available, we suggest the more conservative approach of applying the aeration number correlation to each impeller independently. The added cost for the larger drive will not be a major factor for vessels up to about 5,000 L; this is fairly cheap insurance. One should do pilot agitation studies at a meaningful scale for much larger vessels.

Characters of fluid forces

Finally, an additional constraint must be imposed if the organism is sensitive to fluid mechanical forces. This is seldom true of unicellular organisms. There are, however, some mycelial organisms that are sensitive. The extent of sensitivity and the nature of the forces that cause damage should be determined experimentally. One also should keep in mind that the character of fluid forces changes significantly with scale. For example, some turbulent forces that are negligible at small scale can be large and potentially destructive at large scale. Reliable guidance in this area is almost nonexistent. There is, however, a rough rule of thumb that states that damage will occur if impeller tip speed exceeds 1,500 ft/min. Bear in mind that this rule was developed almost 50 years ago and was based on information for organisms used at that time in vessels that had working volumes in the 30,000 to 50,000-gal range. It is clear that many factors (including broth rheology) will affect the potential for damage. Unfortunately, most of the information is seldom available when needed. In such cases, the "any port in a storm" approach to design usually becomes operable. Fortunately, we seldom encounter a circumstance in which the rule of thumb must be violated.

Condition for Homogeneity

Good bulk mixing is needed to insure homogeneity, which is required for reliable data acquisition and control. There are several important factors that affect mixing quality.

Good mixing

This is the primary factor affecting our ability to provide good mixing quality, but we seldom have any control over it. We simply must recognize that we will have to go to greater and greater lengths as the broth becomes more viscous and non-Newtonian. Also, one must bear in mind that the nonlinear behavior of most non-Newtonian fluids often results in nonobvious responses to various actions. For example, increasing agitator speed in a highly non-Newtonian, pseudoplastic broth often leads to a decrease in power input and a decrease in bulk mixing quality.

Categories of impeller

Impeller types fall into three basic categories: radial flow (e.g., Rushton turbine), axial flow (e.g., marine propellor), and mixed axial and radial flow (e.g., Lightnin' A-315). Radial flow impellers tend to deliver the high power required to enhance micromixing and mass transfer but do not promote top-to-bottom mixing; therefore, they are not the best choice for promoting homogeneity. Multiple radial flow impellers often are used in an attempt to compensate for poor bulk mixing. This can work fairly well for low-viscosity Newtonian broths when the ratio of liquid height to tank diameter does not exceed 1.7-2, but it is a very poor solution when the broth is very viscous and highly non-Newtonian. Anyone who has observed classical streptomycete or xanthan gum fermentations can attest to this fact. Pure axial flow impellers do not deliver much power but do tend to promote good top-to-bottom mixing and hence contribute significantly to bulk homogeneity. They do not, however, contribute much to mass transfer. Some mixed flow types appear to provide a good balance between bulk mixing and mass transfer requirements, particularly for non-Newtonian broths. We recommend that they be considered seriously for such cases. We also have had considerable success with combination systems, for example, those with turbines as the lower impellers and a hydrofoil on top. This appears to work well for Newtonian and non-Newtonian broths.

Diameter

In most cases, large-diameter impellers distribute power better and promote bulk mixing better than do small-diameter impellers. As a rough rule of thumb, based on considerable empirical evidence, we recommend D_i/D_t ratios of between 0.4 and 0.5 unless some other constraint (e.g., torque) is violated. Exceptions to this are some cases in which an axial flow impeller is the top impeller. In such cases, the ratio should not exceed 0.35 to avoid the risk of vortexing. This is particularly true for low- viscosity Newtonian broths. We suggest that the ratio of the aerated liquid height to the tank diameter be kept below 2.0 to minimize bulk mixing problems. This must be reconciled with impeller diameter and impeller spacing requirements and with architectural, shipping, and other constraints.

Spacing of turbine

As noted earlier, there is a body of empirical evidence that supports the spacing of Rushton turbines 1-1.5 impeller diameters apart in Newtonian broths. This spacing appears to provide a balance between preventing impellers from interfering with their neighbors and at the same time avoiding dead zones between them. The same sources of empirical data support placing the lowest impeller one impeller diameter from the bottom of the vessel (assuming a standard dished head) and placing the uppermost impeller one impeller diameter below the liquid surface. We have found these rules to work reasonably well for Newtonian broths and a few non-Newtonian broths, and for radial flow impellers other than Rushtons. Please note that these conclusions are based on limited observations. Finally, keep in mind that one cannot avoid interaction among impellers when at least one of them is an axial flow type or has a significant axial flow component, and that spacing effects in highly non-Newtonian broths are not well understood.

Transmittable power

Baffles are placed in vessels to minimize fluid swirling and vortex formation. The usual approach is to use four baffles on 90° centers, each baffle having a width equal to 0.1 of the D_t. Baffling tends to increase transmittable power and to improve mixing (except for the dead spots, which tend to form behind the baffles). Elimination of significant vortexing is important for safety as well as for improved bulk mixing. A vortex can reach down to the impeller in an unbaffled vessel. When this happens, a sizable portion of the impeller becomes air enshrouded, thereby decreasing resistance on the impeller. This tends to drive up the impeller speed, at least briefly (or the speed may be increased intentionally in an attempt to increase the power input). Unfortunately, vortices are unstable and can collapse shortly after formation, resulting in very large and potentially catastrophic stresses (impulse) on the impeller, particularly if the impeller speed is increased significantly while it is enshrouded.

Gas flow and holding

Gas flow has a complex effect on bulk mixing. The flow alone does tends to promote bulk mixing (as in bubble tanks), but it also tends to decrease the effect of the impellers, particularly at high values of gas linear velocity. Fortunately, this is not a major problem in most cases, but it can be for very large fermenters and for highly viscous non-Newtonian broths. There are several problems that can interfere seriously with a fermentation. Gas holdup decreases the effective volume of a fermenter. Foaming and aerosol formation can constrain operation, cause major cleaning and asepsis problems, and in the extreme cause termination of a fermentation. All three are dependent on broth characteristics, power input, and gas flow rate. There is some information in the literature, but these points have not been given the attention that even begins to reflect their importance. In general, all one can say is that all three become bigger problems for a given fermentation as gas flow and power input increase. Beyond that, each case is a new adventure. More will be said about the practical aspects later.

Heat Transfer

Heat transfer is required during fermentation to maintain constant temperature conditions, and at other times (i.e., sterilization, induction) to increase or decrease broth temperature. In most cases, cooling is required during most of an active, aerobic fermentation. A good approximation to the total heat load, $Q_{tot,}$ during such fermentations in which the primary carbon source is glucose or a similar carbohydrate is

$$Q_{tot} = Q_{metab} + Q_{mech}$$

$$Q_{tot} = 0.48(\text{OTR})\,V_L/Y_{x/o}, + 2545P_g \quad \text{...(17)}$$

where Q_{metab} is the heat generated by metabolism (Btu/h), Q_{mech} is the heat generated by power input (Btu/h), and $Y_{x/o}$ is the yield of cells on oxygen (g dry wt cells per g oxygen consumed). Heat transfer can be more of a limiting factor in aerobic fermentations than is oxygen transfer—particularly in large fermenters. The major heat transfer demand is usually during the maximum growth period (i.e., when μX is greatest).

It is always wise, however, to check other loads. For example, rapid cool-down after an induction phase could put enormous demands on the cooling system. Cooling a 10,000-L (wv) fermenter from 38°C (104°F) to 10°C (50°F) in 30 min would require an average heat transfer rate of approximately 2.4 x 10^6 Btu/h. The same heat transfer rate would support an oxygen transfer rate of approximately 495 mmol/L(h), which is very high by almost anyone's standards. Similarly, one should check other potentially high loads such as cool-down after sterilization. In any case, careful thought and solid, empirical facts (e.g., product thermal degradation rates) should form the basis of any specification that will require extraordinary transfer rates.

Rate of heat transfer

The rate at which heat can be transferred is dependent on (a) the driving force for heat transfer, (b) the area across which transfer must occur and, (c) the resistance to heat transfer:

$$Q = (1/\text{resistance}) \times (\text{area}) \times (\text{driving force}) \quad \text{...(18)}$$

The driving force for agitated vessels is usually taken to be the log mean temperature difference, ΔT_{LM}, defined as

$$\Delta T_{LM} = (T_{c,o} - T_{c,i})/LN[(T_f - T_{c,i})/(T_f - T_{c,o})] \quad \text{...(19)}$$

where $T_{c,o}$ is the coolant outlet temperature (°F), $T_{c,i}$ is the coolant inlet temperature (°F), and T_f is the fermentation temperature (°F). Note that use of equation 19 assumes that the broth is homogeneous and can be characterized by a single temperature.

The transfer area, A_{JA}, is the actual contact area (not necessarily the same as the available area) between the broth and the heat exchange surface, which is usually a jacket and/or an internal coil. The actual area can be found if the height of aerated liquid in the tank is known, along with the geometry of the jacket and any internal exchange area (e.g., a coil). The aerated liquid height measured along the centerline of the vessel, h_{LA} (ft), is

$$h_{LA} = [(1 + \text{HU})V_{LU} - V_D]/(0.785D_t^2) + \text{IDD} \quad \text{...(20)}$$

where V_{LU} is the unaerated liquid volume (ft^3), V_D is the lower dish volume (ft^3), IDD is the inside depth of lower dish (ft), and D_t is the tank inside diameter (ft). The resistance to heat transfer is a little more complicated. It is composed of five resistances in series and is expressed as

$$1/U = 1/h_i + 1/h_o + 1/h_{fi} + 1/h_{fo} + t/k \quad \text{...(21)}$$

where U is the overall heat transfer coefficient (Btu/[h °F ft^2]), h_i is the broth film heat transfer coefficient (Btu/[h °F ft^2]), h_o is the jacket fluid heat transfer coefficient (Btu/[h °F ft^2]), h_{fi} is the broth fouling factor (Btu/[h °F ft^2]), h_{fo} is the jacket fluid fouling factor (Btu/[h °F ft^2]), t is the fermenter wall thickness (ft), and k is the fermenter wall thermal conductivity (Btu/[h °F ft]). The resistance of the tank wall can be predicted accurately because both k and wall thickness are known accurately.

It should be noted that the k for stainless steel is very low (about 10 Btu/[°F h ft]). This means that the tank wall should be kept as thin as possible to minimize its contribution to heat transfer resistance. Correlations for h_o can be found in the literature but are seldom as reliable as those available from vendors for the specific jackets they fabricate. The problem in predicting U is in predicting reliable values of h_i, h_{fi}, and h_{fo}. Published correlations for h_i values for aerated, Newtonian fermentation broths have been published but are not very reliable. Correlations for non-Newtonian broths have also been published, but as was the case for mass transfer coefficients, h_i is strongly dependent on broth rheological characteristics.

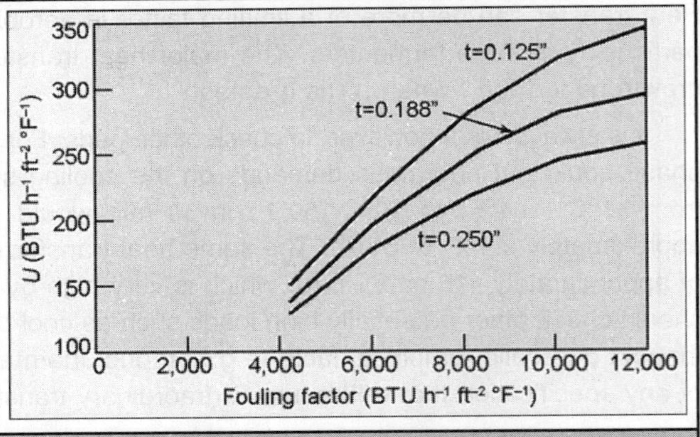

Figure 8.3 Effects of metal thickness and fouling resistances on the overall heat transfer coefficient.

Unfortunately, existing correlations do not do a very good job of accounting adequately for broth rheology, a problem exacerbated by the fact that good rheological information is seldom available. Finally, the fouling factors are essentially impossible to predict. The best one can do is keep fouling to a minimum by means of good cleaning practices and proper filtration of coolants. The following rough guides (based on many years of experience) are offered because the authors are unable to offer any collection of reliable prediction tools:

1. U values for simple, low viscosity Newtonian broths usually range from about 120 to 150 Btu/(h °F ft²) for new, clean stainless steel tanks.
2. U for viscous, non-Newtonian broths ranges from about 30 to 60 Btu/(h °F ft²)

It also should be noted that U's for heating usually are a bit higher than U's for cooling.

The heat transfer rate can now be expressed as

$$Q = UA\Delta T_{LM} \qquad \text{...(22)}$$

Similar equations can be developed for heat transfer via an internal coil. The only difference is that the heat transfer coefficients for coils are usually a bit higher than those for jackets. That having been said, we will try everything possible to discourage you from using internal coils. The reasons relate primarily to mechanical and cleaning considerations and are discussed later. In applying the preceding, we suggest that the reader consider the following suggestions:

1. Avoid the use of internal cooling surfaces (e.g., coils). They make cleaning difficult to impossible, are potential sources of contamination (leaking of coolant), put a lot of mechanical strain on the vessel, and can cause deterioration of bulk mixing (particularly for viscous non-Newtonian broths).

2. Avoid subfreezing coolant. Sooner or later, subfreezing coolant will cause severe valve freeze-up and worse.
3. Try to keep coolant flow rates low enough to avoid pipe diameters greater than 3 in.; cost increases considerably and availability of fittings and so forth decreases as size increases.

Finally, there are some cases in which high wall temperatures have been cited as causing problems with temperature-sensitive organisms. The wall temperature, T_w, depends on the temperatures of the broth and of the jacket fluid, as well as on the relative values of the heat transfer resistances. A simple energy balance across the heat transfer path yields

$$T_w = (T_f + \beta T_j)/(1 + \beta) \quad ...(23)$$

where

$$\beta = k/h_i/(t + k/h_o) \quad ...(24)$$

Note that this does not account for locally high wall temperatures that would be caused by steam applied to steam locks during fermentation.

Sterilization and Aseptic Operation

Sterilization and aseptic operation taken together (as they must be) have a much greater influence on fermenter design and operation than does any other requirement. In theory (i) sterilization destroys or removes all foreign organisms in all process equipment (including piping, seals, etc.) that might come into contact with the process fluid, and (ii) aseptic operation insures that *no* contaminating organisms enter the fermenter after sterilization. These ideals are not attainable in practice because (among other reasons) (i) there is a finite probability that a very low concentration of contaminating organisms will not be captured in a sample used to test for contamination, and (ii) there is a finite probability of false positives due to sample contamination and so forth. It also is important to note that there is a continuing debate concerning the definitions of pure culture and sterility. The driving forces for sterilization and aseptic operation range from minimizing product losses to insuring strict compliance with regulatory requirements. Among the many specific problems that contamination can cause are the following:

1. Production of a toxin that can't be removed by the purification system
2. Production of an enzyme that degrades the product
3. Decreased product yield due to use of substrate by contaminants
4. Production of toxins that inhibit the producer strain
5. Production of compounds (e.g., polysaccharides) that interfere with the operation of recovery and purification equipment

Which specific problems will exist and where in the spectrum a particular case will lie depend primarily on the product, its economic value, whether it is regulated, and how it will be used. Given all these variables and the fact that absolute sterility is an unachievable abstraction, the extents to which one should go should be considered on a case-by-case basis. As a practical matter, sterilization has to be interpreted as "effective sterilization," meaning that the design and procedures (including sampling and detection) are suitable for the specific case. For example, sterilization of ferm enters used to produce parenterals should be held to a much higher standard than sterilization of fermenters used to produce amylases for starch hydrolysis. Unfortunately, there remains a lot of room for disagreement among well-intentioned people.

Sterile piping system

Mechanical details and other practical considerations of sterilization and aseptic operation are discussed in the subsection "Sterile Piping Systems." Several methods have been developed to quantify sterilization. The simplest is based on the empirical observation that the death kinetics of many vegetative organisms and spores can be described by a simple first-order expression

$$dN/dt = -kN \qquad \text{...(25)}$$

where N is the number of viable organisms (or spores) at any time, t, and k is the thermal death constant. k is a function of most environmental variables but is most strongly affected by temperature. This dependence is very often expressible as an Arrhenius relationship

$$k = A \exp(-E_A/RT) \qquad \text{...(26)}$$

where A is the pre-exponential factor (min^{-1}), E_A is the thermal death activation energy (cal/mol), T is the absolute temperature (K), and R is the gas constant. The effect of temperature on k is pronounced for most organisms because their E_A's are so high (e.g., 65,000 cal/mol for *Bacillus stearothermophilus* spores). Among other methods of quantifying sterilization concepts are the following:

1. *Probability-based theories*. These are conceptually different from the first-order model, but give essentially the same results for the levels of kill that must be achieved practically.
2. *Methods based on the "decimal reduction factor," D*. These are used widely. It is important to note, however, that the nature of temperature dependence has no rational, physical basis and differs considerably from an Arrhenius relationship: it is based primarily on an empirical observation for relatively small temperature ranges.

We use the first-order model for the rest of the discussion.

Procedure

The primary design basis for sterilization systems and procedures is usually the level of sterility that must be achieved. This is most frequently expressed in terms of the probability of sterilization failure (i.e., the probability that a single organism will survive). In most instances probabilities of 10^{-3} to 10^{-4} are quite adequate. This means 1 failure in 1,000 or 10,000 sterilization operations, respectively. It also is important to note that the probability of failure is numerically very close to the value of N in the first-order model. The reader should also take note of the fact that in some quarters, sterilization criteria are stated in terms of "logs of kill." That is to say, a fixed number of logs is taken to be adequate. This is meaningless in that it does not take into account the initial contaminant loading. Now we consider prediction of requirements for batch sterilizing a fermentation medium as part of the fermenter design process to ensure that we'll have adequate heating and cooling capacity. It should also be done as part of the process design to ensure adequate timing. Batch sterilization involves heating the medium in the fermenter to sterilization temperature (T_{ster}), holding the medium temperature constant at T_{ster} for an adequate time, and then cooling the medium down to fermentation temperature. The medium is kept in the fermenter throughout. It is assumed in what follows that (a) heating and cooling are via the jacket only, and (b) the energy and condensate contributed by the small amount of steam that flows into the vessel as a result of sterilizing piping, air filters, and so on is negligible for energy balance purposes. The rate of kill at any instant during the sterilization process is

$$dN/dt = -kN = -A \exp(-E_A/RT)N \quad \ldots(27)$$

therefore,

$$N/N_o = \int \exp(-E_A/RT)dt \quad \ldots(28)$$

where N_o is the number of contaminating organisms initially present. The integration of equation 26 must be performed over heating, hold, and cooling phases of sterilization. This is done by numerical integration of equation 28, coupled with the equations that describe broth temperature as a function of time.

For heating

$$T_f = T_{ST} - (T_{ST} - T_{F,O}) \exp(-U * A_{JA} * t/M_F * C_{P,m}) \quad \ldots(29)$$

where T_f is the medium temperature (°F), T_{ST} is the steam temperature (°F), $T_{F,O}$ is the initial medium temperature (°F), A_{JA} is the active jacket area (ft²), t is the time since beginning of heating (h), M_F is the medium mass (lb), and $C_{P,m}$ is the medium heat capacity (Btu/[lb °F]).

For cooling

$$T_f = T_{C,i} + (T_{ster} - T_{C,i}) \exp(t(W_J C_{P,c}/(M_F C_{P,m})) \cdot (\exp(UA_{JA}/(W_J C_{P,c})) - 1) \quad \ldots(30)$$

where $T_{C,i}$ is the inlet coolant temperature (°F), T_{ster} is the sterilization temperature (°F), t is the time from beginning of cooling (h), W_J is the coolant flow (lb/h), and $C_{P,c}$ is the coolant heat capacity (Btu/[lb °F]). Equations 29 and 30 can be derived via energy balances.

One chooses times and temperatures such that the desired kill, N/N_o, is achieved. Unfortunately, one usually does not know the level of contamination or the nature of the contaminants. Indeed, it is very unlikely that contamination will remain constant from one fermentation to the next. The usual practice, therefore, is to design for what is thought to be the "worst case." This usually gives results that are quite satisfactory for design specifications. A "typical" worst case is a loading of 5×10^6 spores/mL of the organism *B. stearothermophilus*. Such spores are highly resistant and have the added advantage of being used widely to validate sterilization operations.

It is important to note that the purpose of the calculations is to provide a rational basis for obtaining reasonable design information and ensuring a high probability that the desired level of sterilization can be achieved over the whole range of anticipated operating conditions. Figure 8.4 illustrates the results of such calculations for the following case:

Medium volume = 15,000 L

Initial spore load = 5×10^6/mL

Initial medium temperature = 70°F

Steam temperature = 300°F

Coolant temperature = 35°F

Coolant flow rate = 100 gpm

Fermentation temperature = 86°F

Active jacket area = 225 ft^2

U = 150 BTU/(°F h ft^2)

$A = 2.1 \times 10^{36}$/min

E_A = 65,000 cal/mol

N (probability of failure) = 0.0001

As seen in this example, heating and cooling phases usually contribute only a small fraction of the total sterilization kill, but they do affect turnaround time significantly. Long heating and cooling times can also have other effects:

1. Cause damage to the fermentation medium to the extent that the fermentation can be compromised. The seriousness of this depends on some combination of economics and regulatory compliance.
2. Cause medium changes that have negative effects on recovery and purification without causing fermentation problems. This includes the possibility of introducing foreign materials that may pass undetected into the final product.

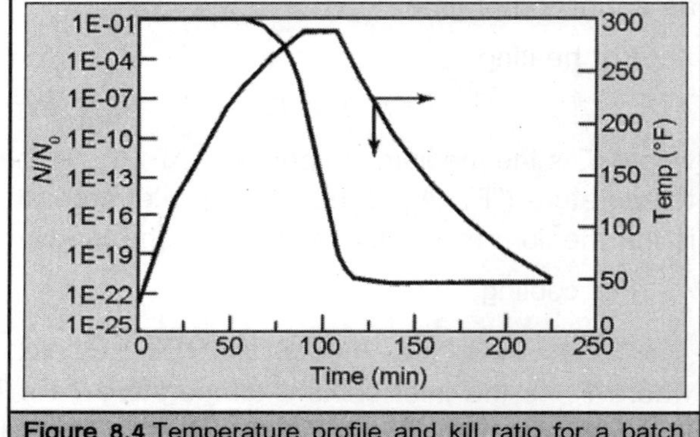

Figure 8.4 Temperature profile and kill ratio for a batch sterilization.

The example also shows that cooling time is considerably longer than heating time. This results primarily from the lower temperature-driving forces during cooling. The problem is exacerbated considerably when the heat transfer coefficient is very low as is the case for viscous non-Newtonian broths. Another method used for batch sterilization is direct steam injection: live steam is injected directly into the medium as the main source of thermal energy. This decreases heating time as well as the overall steam requirement. It also increases the medium volume by about 20% (as a result of steam condensation), which can cause some problems, including the following:

1. *Increased cooling time.*
2. *Medium dilution.* This will be a significant problem if the initial medium cannot be made concentrated enough to account for the dilution. Some reasons include low solubility of medium components, increased viscosity, and increased reaction rates among medium components at elevated temperatures.
3. *Introduction of impurities.* This depends primarily on the quality of the injected steam. In some cases plant steam is acceptable if boiler cleaning agents do not cause problems. At the other extreme is the requirement to use clean steam (WFI quality). It should be noted with regard to

this point that some steam will be injected directly even when jacket heating is used as the main energy source; therefore, one will always be in the position of having to evaluate the effects of contaminants carried by the steam.

For cases in which sterilization times are too long or medium alterations cause too many problems, one might consider continuous sterilization.

Problems of heat transfer

Heating and cooling times are not usually significant issues for sterilizing piping; however, there can be some serious heat transfer problems related to piping length, diameter, and orientation. Some of these are discussed in the subsection "Vessel Design."

Theoretical aspects of air sterilization are not discussed here. Suffice it to say that modern filters, if sized, installed, and maintained properly, will provide very reliable results. We'll discuss some of these practical aspects in the section "Mechanical Design."

Scrupulous Cleaning

Scrupulous cleaning is necessary to decrease nonbiological contamination and prevent cross-contamination of batches. Experience also has shown that reliable sterilization is difficult or impossible to achieve in the absence of rigorous cleaning. As noted earlier, fermenter cleaning is not mentioned specifically in 21 CFR, but the basis is provided elsewhere, which requires equipment be cleaned as per a defined plan and by trained people. Furthermore, the FDA has made clear the importance it attaches to cleaning not only in various agency publications, but also in more stringent enforcement. This is particularly true in multi-product facilities (a key issue for most modern biotech plants). This agency initiative and the desire to generally improve operation have caused the industry to develop more reliable cleaning systems and protocols and to move to fully automated cleaning-in-place (CIP) systems, which provide greater assurance of successful validation and long-term compliance. Developing and designing reliable fermenter cleaning systems is not as straightforward as it might appear, and considerable controversy continues.

Among the reasons for this is that while much is known about the basic science of cleaning in general, very little is known about the basic science of cleaning fermenters and little has been published. The approach taken is based primarily on experience derived from the dairy, food, and beverage industries. Such information is useful but is not directly applicable in general to pharmaceutical and biotech processes where soils are different and the cleaning requirements are far more stringent. What meager information there is concerning pharmaceutical and biotech soils has been obtained from experiments done on single soil components and/or studies done under conditions not truly representative of the process conditions. There have been no significant attempts to develop systematic analyses based on experiments done with complex mixtures typical of real fermentation soils under conditions found in real processes.

As a result there are no reliable general methods, tools, or correlations on which to base the development of cleaning agents, protocols and CIP system design; decisions tend to be made based on arbitrary criteria (e.g., coupon bake on studies). In keeping with our general design philosophy, we think it important that such arbitrariness be avoided and that cleaning protocols be developed along with the fermentation. It is clear that this will help to ensure not only proper design but will also minimize cleaning validation studies and any questions concerning the presence of contaminants and/or cleaning residues in the commercial product that could not have been present in clinical trial material. The selection of fermenter cleaning agents and protocols should be based not only on the specific soil but also on the materials of construction, surface finishes, and so on.

In addition, one should consider the issues of (i) compatibility of each material with the cleaning agents and protocols, and (ii) potential materials interaction during cleaning. These are decisions that usually are made during the design phase, but really should be defined much earlier. Fermenter cleaning validation is beyond the scope of this article; however, the design must be done so as tominimize validation problems and ensure maximum ease in ongoing compliance. The design team must therefore be aware of the procedures that will be used for validation and recognize that this is an issue that has not yet been resolved in the general community.

Interactive Processes

We now begin to see some of the interactions that can arise among the transport processes, and between them and the operating variables. This is a necessary step in developing sound designs, and a precursor to preliminary design calculations. As an example, consider a case in which we are trying to find ways to increase the OTR above some base value. One approach would be to increase the gas flow rate (v) keeping other variables constant:

- The overall driving force tends to increase (equation 10).
- V_s increases (equation 7).
- Increased V_s tends to increase K_ga (equation 11).
- Increased v also increases N_A (equation 12).
- Increased N_A tends to decrease power input. Decreased power input tends to decrease K_ga (eq. 11).

As another example, consider increasing the vessel pressure while keeping other variables constant:

1. The overall driving force tends to increase (equation 10).
2. The actual gas volumetric flow rate (v) tends to decrease (assuming constant molar flow rate) (equation 6).
3. V_s tends to decrease (equation 7).
4. The decrease in V_s tends to decrease K_ga (equation 11).
5. The decrease in m also tends to decrease NA (equation 12).
6. This tends to increase power input.
7. Increased power input tends to increase K_ga (equation 11).

Such complex interrelationships develop regardless of what steps are taken to increase OTR. The nature of these will change depending on the approach taken and the system being considered. There will also be additional pluses and minuses related to factors such as cost, safety, cleaning, and regulatory compliance. For example, using pure oxygen to increase the driving force carries with it very distinct cost and safety problems. Now let's carry the second approach a little further. We do this because increasing pressure is usually the easiest and seemingly least expensive way to increase OTR.

The decrease in V_s caused by the increased pressure has the positive effects of decreasing foaming, aerosol formation, and gas holdup. But it has the negative effect of requiring thicker tank walls, thereby increasing heat transfer resistance in the face of an increased heat load due to increased metabolic activity. This will become a significant problem when the pressure desired is greater than that required for vessel sterilization. The larger the tank diameter, the greater the possible problem. It should also be noted that thicker walls can make it more difficult to get a good surface finish and therefore increases the possibility of cleaning problems. In addition, the higher the pressure and thicker the walls, the shorter the list of qualified vendors. This type of reasoning process is applicable to any of the many permutations possible. We extend it after we consider some additional basic concepts, mechanical design practices, and a few more important practical constraints. We now consider two examples of preliminary design calculations, the objectives of which are to, first, define vessel geometry to ensure that it will not violate architectural constraints and be able to satisfy general rules of thumb for good bulk mixing, impeller spacing, and so forth. Some of this will require checking calculations done later and may require some iteration. The second objective is to establish ranges of air flow, pressure, oxygen enrichment, and power that will accomplish the following:

1. Satisfy the design basis maximum OTR requirement.
2. Keep V_S low enough to avoid foaming, holdup, and aerosol problems. We have found a limit of 200 cm/min to be a safe guide; however, there are cases (Newtonian broths) in which considerably higher values up to 300–350 cm/min tolerable and others (very viscous, highly non-Newtonian broths) in which it is not wise to go beyond 75–100 cm/min.
3. Not violate compressor and other component constraints.
4. Not require heroic efforts (e.g., extremes in pressure, power, etc.).

Please note that we are trying to establish reasonably broad ranges for each of the variables within which we can satisfy the requirements with a reasonable degree of comfort (based on experience). This is important because the correlations usually available have questionable accuracy, and it allows some breathing room for process improvements and so forth.

The third objective is to define impeller type(s), sizes, and speed that will do the following:

1. Provide the power required
2. Provide good bulk mixing
3. Satisfy impeller spacing rules of thumb
4. Not violate constraints related to the sensitivity of the organism
5. Make possible single-piece impeller construction and so forth (see the section "Mechanical Design" for consideration of mechanical constraints)
6. Not violate any other constraints specified

The fourth objective is to define heat exchange area and coolant flow and temperature that achieve the following:

1. Satisfy peak heat transfer requirements
2. Not violate coolant temperature and flow constraints and such
3. Avoid internal heat exchange area
4. Not require subfreezing coolant or flow that will require very large pipe diameters

(*Note*: One should check to make sure that other heat loads [e.g., cooling after fermentation] are not greater than the load during maximum growth activity. We assume that to be the case here.) The fifth objective is to use the results of the calculations not only to define mechanical details, but also as a basis for recommendations for rational changes in design basis. Please note that experimentally derived correlations for the subject cases are used in the examples (this is a design gift as rare as hen's teeth). All were obtained at 500-L scale, and we are just within our comfort range concerning their scalability to 15,000 L (design scale). The reader is cautioned, however, that there is no basis to believe that the correlations would be applicable to other cases, nor would we recommend extrapolation to much larger scales. It also is worth noting that there are some good examples in the literature that might be applicable to the problems at hand or to others; however, one must exercise considerable caution if the conditions specified in the reference differ significantly from those for the case being analyzed. Unfortunately, it is often not possible to determine if significant differences exist.

Case 1.

Product	Intracellular protein
Organism	*E. coli*
Final broth volume	15,000 L
Maximum OTR	234 mmol/(L h)
Dissolved oxygen requirement	10% relative to atmospheric conditions
Broth rheological type	Newtonian
Maximum broth viscosity	5 cp
Regulation	CBER
Containment	BL2-LS
Maximum coolant flow	250 gpm
Minimum fluid temperature	35°F
Internal coils	No
Compressor max press/max flow	50 psig/25,000 SLPM
Minimum heat transfer coefficient	150 Btu/(h °F ft^2)
Ceiling height	336 in.
Floor space	1000 ft^2, unobstructed

The following experimental correlations are applicable:

$$K_g a = 108(P_g/V_L)^{0.6}\ V_S^{0.2}$$

$$P_g/P_{ug} = 1.0 - 18.56N_A + 215.64N_A^{2} - 1082.66N_A^{B} + 1859.68N_A^{4}$$

$$HU = 0.0133(P_g/V_L)^{0.4}\ (60V_S/^{0.5}) - 0.0333$$

These were developed under the following conditions for single Rushton turbines:

$0.8 < P_g/V_L < 2.9$ hp/100 gal

$100 < V_S < 200$ cm/min

$0 < N_A < 0.1$

$0.33 < D_i/D_t < 0.45$

Vessel geometry calculations are the same for both cases. We start by assuming a vessel diameter and wall thickness. (Note that the thickness will probably be changed a bit later to satisfy code requirements. This will not have a significant effect on height and volume calculations to follow.) In this case we assume a 96-in. outside diameter and a 0.25-in. wall thickness. By straightforward geometry and using tabulated values of volumes and inside depth of dish (IDD) for standard ASME heads, we get the following:

Liquid height	134.2 in.
(Liquid height)/tank diameter	1.41
Total tank height	192.5 in.
Totalvolume	21.1L
Percent fill (unaerated)	71.1%

This satisfies rules of thumb concerning liquid height-to-diameter ratio and fill percent. If we add about 72 in. for legs and 30 in. for opening the manway, we get a total height requirement of 294.5 in. We should be able to satisfy the maximum height limit.

Table 8.3 Oxygen transfer calculation results (single cell)

Pressure (PSIG)	35	20	20	2
O_2 trans eff. (%)	21	30	28	28
O_2 (MOL%in)	21	21	40	40
Gas flow (in, SLPM)	29,762	20,833	11,719	11,719
VVM	1.98	1.39	0.78	0.78
Average V_s (cm/min)	186	183	103	200
hp/100 gal	1.3	2.7	1.0	2.6
Theor. P_g	52.8	105.6	41.1	103.8

From these results we conclude that the following will satisfy the OTR required and give us a reasonable amount of "breathing room."

Gassed power	100 hp (includes transmission efficiency)
Maximum operating pressure	40 psig
Maximum air flow	10,000 SLPM

We do not recommend oxygen enrichment at this point, but we do suggest that the "holes be punched" to facilitate future installation of an oxygen system. The agitation system calculations are based on the assumption of using two 40-in. diameter Rushton turbines. A maximum speed of 150 rpm should be adequate. Note also that the impellers can be spaced to satisfy the rules of thumb for bulk mixing rules discussed previously. Although this should give satisfactory results for such a low viscosity broth, we recommend consideration of a slightly larger turbine as the lower impeller and a hydrofoil (e.g., A-315) as the top impeller.

Table 8.4 Citation system calculation results (single cell)

Shaft speed (rpm)	114	141	88	141
Turb imp dia (in)	40.00	40.00	40.00	40.00
Turb D_i/D_t ratio	0.42	0.42	0.42	0.42
Ungassed hp (turb)	123.1	233.0	56.6	233.0
N_a, aeration numb	0.071	0.057	0.051	0.062
P_g/P_o ratio	0.43	0.46	0.48	0.45
Gassed hp	53	106	42	104

A minimum coolant temperature of 35°F was used in all cases. The results show that the heat loads can be handled, but the conditions are uncomfortably tight. For example, in no case will increasing the coolant flow to the maximum make up for a 10% decrease of the heat transfer coefficient. The potential consequences of inadequate heat transfer in this case should be considered seriously before final design commitments are made. Alternatives (e.g., lower coolant temperature) should be explored.

Table 8.5 Heat transfer calculation results (single cell)

Heat load (MMBtu/h)	1.808	1.943	1.779	1.938
Hold-up	0.199	0.271	0.123	0.282
Jacket area (ft²)	297	317	276	320
UJKT(BTU/ft² h °F)	150	150	150	150
Coolant temp in (°F)	35	35	35	35
Coolant flow (GPM)	190	210	235	200

Case 2.

Product	Antibiotic
Organism	*Streptomyces* sp.

Final broth volume	15,000 L
Maximum OTR	50 mmol/(L h)
Dissolved oxygen requirement	10% relative to atmospheric conditions
Broth rheological type	Non-Newtonian (pseudoplastic)
Maximum broth viscosity	1,000–1,500 cp
Regulation	CDER
Containment	BL1-LS
Max coolant flow	250 gpm
Min fluid temperature	35°F
Internal coils	No
Compressor max press/max flow	50 psig/25,000 SLPM
Min heat transfer coefficient	40 btu/(h °F ft²)
Ceiling height	336 in.
Floor space	1000 ft², unobstructed

The following empirical equations are applicable:

$$K_g a = 8.0(P_g/V_L)^{0.5}\ V_S^{0.4}$$

$$P_g/P_{ug} = 1.0 - 26.99N_A + 417.37N_A^2 - 2789.43_N^3 + 6643.30N_A^4$$

$$HU = 0.012(P_g/V_L)^{0.3}\ (60V_S/100)^{0.52} - 0.029$$

These were developed under the following conditions for single Rushton turbines:

$1.0 < P_g/V_L < 2.3$ hp/100 gal

$60 < V_S < 130$ cm/min

$0 < N_A < 0.1$

$0.36 < D_i/D_t < 0.42$

From these results we conclude that the following will satisfy the OTR required.

Gassed power	100 hp (includes transmission efficiency)
Maximum operating pressure	40 psig
Maximum air flow	25,000 SLPM

Oxygen enrichment should be considered in this case despite the fact that the calculations show that it is not necessary. The reason is that the gas flow rate, and hence V_S, is quite high without enrichment. This results primarily from the very low oxygen transfer efficiency typical of very viscous non-Newtonian broths. Although it is true that we haven't broken the rule of thumb for holdup, foaming, and aerosol we should take note of the fact that high air flows in pseudoplastic broths can exacerbate impeller unloading and cause flooding more readily than for non-Newtonian broths. It also is generally true that our correlations become less reliable as conditions become more extreme and further removed from the conditions under which they were developed. Note in particular that our correlations were not developed for the high values of V_s found in the calculations. Additional calculations show that 40–45 psig would bring the V_s into a comfortable range. We suggest, therefore, the following:

Gassed power	100 hp (includes transmission efficiency)
Maximum operating pressure	45 psig
Maximum air flow	25,000 SLPM
Oxygen enrichment	40% oxygen in inlet stream

Table 8.6 Oxygen transfer calculation results (mycelial)

Pressure (PSIG)	35	25	20	10
Oxygen transfer eff(%)	6	6	6	6
O_2 mol% in	21	21	40	40
Gas flow IN (SLPM)	22,222	22,222	11,667	11,667
VVM	1.48	1.48	0.78	0.78
Average V_s (cm/min)	154	172	102	140
hp/100 gal required	1.7	2.0	1.0	1.6
Theoretical P_g	60.9	79.9	41.7	61.8

The agitation system calculations are based on the assumption of using two 40-in.-diameter Rushton turbines. A maximum speed of 140 rpm appears to be adequate. Note also that the impellers can be spaced to satisfy the bulk mixing rules of thumb discussed previously.

Table 8.7 Agitation system calculation results (mycelial)

Shaft speed (rpm)	124	137	107	124
Ungassed hp	159	214	102	159
Aeration number	0.049	0.055	0.042	0.049
Gassed hp	61	80	42	62

That's very nice, but we have significant reservations about the results. An alternative should be given serious consideration because of the rheological nature of the broth, the profound effect this can have on mixing, and the issues of scalability for such broths. At the very least, the lower turbine should be replaced by one having a D_i/D_t approaching 0.5, and the upper turbine should be replaced by a hydrofoil. The reader should understand that this is based on experience only: we have no solid correlations for this case. The best route to take would be to do a few pilot mixing studies in as large a vessel as possible with the real broth. The minimum coolant temperature of 35°F was used in all cases.

The results show that the heat loads can be handled. It also appears that we are well below the maximum available coolant flow. This may be comforting, but the reader should realize that there is not a lot of breathing room for process improvement. In particular, an increase in OTR up to 60-65 mmol/(L h) would put us over the top even if there were no deterioration in heat transfer coefficient. Again, one should give this serious consideration before committing to a final design. We trust that the discussion and examples in this section have helped the reader to understand not only the many interactions that must be considered, but also the nature of preliminary design calculations and their limitations. The results obtained here must now be translated into a mechanical design that must also take into consideration the items to be discussed in the next section.

Table 8.8 Heat transfer calculation results (mycelial)

Heat load (MMBtu/h)	0.533	0.560	0.463	0.515
Gas hold-up	0.202	0.230	0.124	0.182
Jacket area (ft^2)	298	306	276	292
UJKT [(Btu/ft^2 h °F)]	40	40	40	40
Coolant temp in (°F)	35	35	35	35
Coolant flow (GPM)	190	210	235	200

MECHANICAL DESIGN

Much of the mechanical design follows from the results of calculations discussed in the previous section; however, these must be tempered by other considerations that should be included in the process and/or facility design basis. Some of these (e.g., regulatory compliance, containment) have already been discussed and are revisited in greater detail in this section. Other items that should be included and will affect fermenter design include the following:

1. *Extent of automation, nature of plant wide control system, etc.* Such items affect not only the instrumentation (beyond our scope) and similar factors, but also valving, piping, vessel ports, and so forth. Among some of the major issues here are identification of critical (as defined by cGMP requirements) valves and other components. It must be kept in mind that failure to identify critical instrumentation and control components frequently leads to very complex valving systems in which the failure of a single switch can bring everything to a grinding halt. Unfortunately, there are too many people who do not think this through (as a system) before final design begins. Much of this results from the human tendency to rationalize and/or push "problems" downstream, the kind of thinking that results in the thought "Why worry? We'll get it whether or not we need it." Such thinking is in no one's best interest (buyer or seller).
2. *Staff requirements and the anticipated nature of the operating staff.* This can influence the complexity and physical layout of (among other things) the piping system(s).
3. *Plant location.* This will have some affect on overall design and component selection if for no other than service issues.
4. *Available utilities.* This can influence such things as the designs of the cooling and aeration systems.
5. *Architectural constraints.* These can have profound effects on the ways in which process requirements will be satisfied. Floor space and ceiling height constraints often require that a

fermenter be designed in such a way that it violates some or most or even all of the rules of thumb mentioned in the previous section. This is not good but is better than building a vessel that does not fit into the plant.

6. *Transportation constraints*. The fermenter has to be moved from the fabricator's shop to the plant—at a cost not greater than the total project budget. This may require some thumb bending or breaking. In extreme cases it may be better to fabricate the fermenter on site; however, this can carry large penalties, particularly in licensed facilities.

7. *Maintenance*. Many design decisions can have major effects on the ease of maintenance. It is also often true that some of the design features that ease maintenance are more costly than those that do not. In most cases, the savings in capital expenditure will not come near paying for losses that will result later because of maintenance problems. This is particularly true in licensed facilities where regulators view good, facile maintenance as an integral, indispensible part of cGMP operation.

8. *Definition of standards*. Standards for welds, finishes, and so on should be standardized for all parts of the equipment that will contact process fluid. It does not, for example, make sense to call for a high-quality vessel finish (e.g., 320 grit, EP) and at the same time accept unpolished tubing and valves in inoculation and medium addition lines.

Finally, there are a few recurring themes that are encountered in fermenter design:

1. *Building-in "versatility."* The types of versatility desired range from wanting a fermenter that can operate well over a very wide range of conditions but with a narrow range of organisms, to wanting a convertible bioreactor that can handle microbes, mammalian cells, and maybe (someday) transgenic animals. Most people understand the value of versatility, but many do not understand the attendant problems and costs. As a general rule, converting laboratory glassware directly into large, stainless steel equivalents usually costs more than any rational person should be willing to pay. Also as a general rule, versatility should decrease as a process goes from the lab to the production floor. Versatility in the laboratory is almost a requirement because change is in the nature of laboratory work. Versatility on the plant floor, however, leads to more procedures, more paper work, more testing, and more confusion, particularly when equipment modifications are required to achieve the versatility; change is not in the basic nature of most plant work. Equipment capital savings can justify the added costs of all the preceding for many cases in which the equipment is designed to handle multiple, similar fermentations on a campaign basis and without significant equipment modification. This statement becomes less true as the fermentations become less similar, as more modifications are necessary, and as regulatory scrutiny increases. Bottom line: analyze very carefully any inclination to want versatility built into plant equipment (or even pilot equipment, in some cases).

2. *Retrofitting existing equipment*. The usual thinking is that capital and time savings can be had by refurbishing "old faithful." Just how true this is depends not only on the condition of the existing equipment but also on the intended application(s) of the reborn version. Comments similar to the ones made for versatility apply here. The chances of success decrease as one goes from a lab to licensed production facility. There are ample, expensive corpses to prove this point.

Vessel Design

The major choices that must be made are (i) the type of metal to be used for the vessel and nozzles, and (ii) the type of elastomer to be used for static seals. The selections should be based on compatibility with the organism, compatibility with the product, corrosion resistance, cleanability (also related to finish), welding characteristics, and cost and durability. All of these should be determined during process development but seldom are. In almost all cases, the metal selected will be some grade of stainless steel (usually SS304, SS304L, SS316, or SS316L). The choice is usually associated with the nature and value of the product, although some of the other factors noted earlier may be considered. SS304 is usually good enough for lower-value, unlicensed products, whereas SS316L is the material of choice for high-value, licensed products. L-grade is selected when better corrosion resistance and good multipass welding characteristics are required; it adds

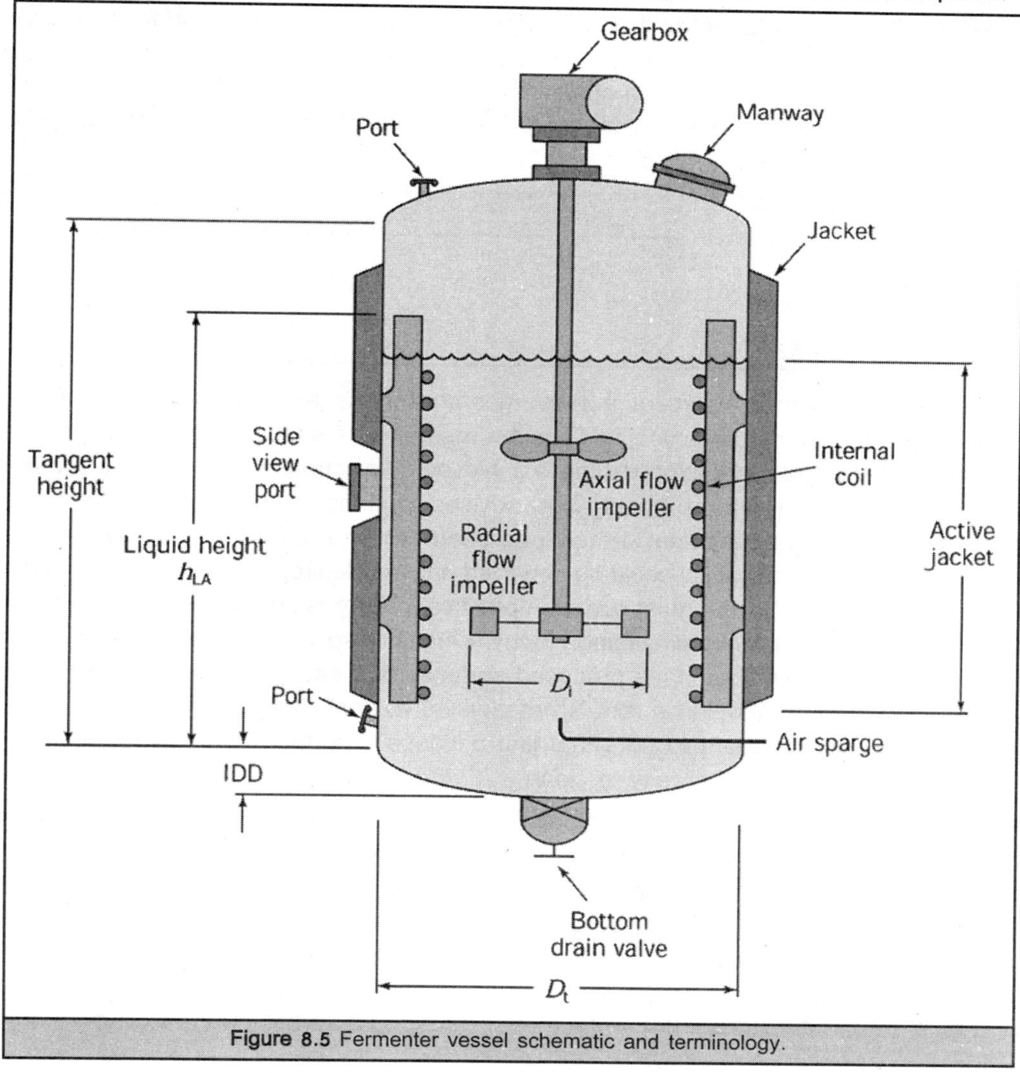

Figure 8.5 Fermenter vessel schematic and terminology.

about 15% to the cost of the vessel. It is important in making metal selection to keep in mind that clean steam, purified water, and WFI are all highly corrosive.

Static seal (O-rings and gaskets) materials are usually limited to silicone, EPDM, Viton, or Teflon. Silicone and EPDM are the materials of choice for headplates and elsewhere, respectively. Teflon has much better temperature resistance than either silicone or EPDM, but it doesn't stretch, and it cold flows. Viton hardens on use. Finally, one should be aware of that some O-ring/gasket-forming processes can leave very small quantities of metals in the seals. These may not be enough to cause fermentation or product problems per se, but they can be the cause of considerable corrosion.

High quality cleaning

In most instances, the need for a high-quality finish is closely related to the need for high-quality cleaning: there is little question that CIP systems function better when surfaces are smoother. Smoothness is expressed on several scales, two of the more common being grit number and surface R_A. Grit number is related directly to the abrasive used to achieve the finish (mechanical); surface R_A is the actual surface roughness as measured by means of a profilometer and expressed in microns:

Grit	R_A (μm)
120	3.2
220	0.8
320	0.4

The two most common ways of achieving smoothness are mechanical polishing and electropolishing. Mechanical polishing is done by means of a series of continually decreasing abrasive particle sizes (grits). Electropolishing is achieved by electrochemical removal of metal from high points on the surface. One is usually best advised to electropolish only after getting a good mechanical polish (240-grit minimum). It has long been held that the very good finishes required for vessels where cleaning is critical could be achieved only by electropolishing a 320-grit mechanical finish (although 240-grit EP is the most common practice). There is some evidence in the literature to support this, but it is not overwhelmingly convincing. It also is important to note that recent advances in mechanical finishes have produced surfaces that easily rival the appearance of any electropolished surface. It should be (but is not always) evident that cost increases with surface smoothness. It is also important to note that just because a surface is almost perfectly smooth does not guarantee that it will be easy to clean.

Cleanability will also depend on the nature of the interaction between foulants and the surface material. The only way to know for sure is to try your broth on various surfaces and then make a specification based on rational evaluation of the experimental information. Surface material and finish, cleaning agents, and cleaning protocols must be considered together—preferably before design. Passivation must also be considered. This is a treatment that restores oxides that protect stainless steel from corrosion but can be removed during vessel fabrication (primarily welding). The process is rather simple, but the surface chemistry is not and is not completely understood;

hence, there is controversy as to which method is best and whether it is really as necessary for fermenters as it is for high-purity water systems where rouge formation is a serious problem. Two methods are in common use: (i) treatment with sodium hydroxide, citric acid, and nitric acid; and (ii) treatment with chelating agents. The second is much milder and does not generate the noxious waste of the first. Both appear to be equally effective; hence, the chelating method is becoming more popular.

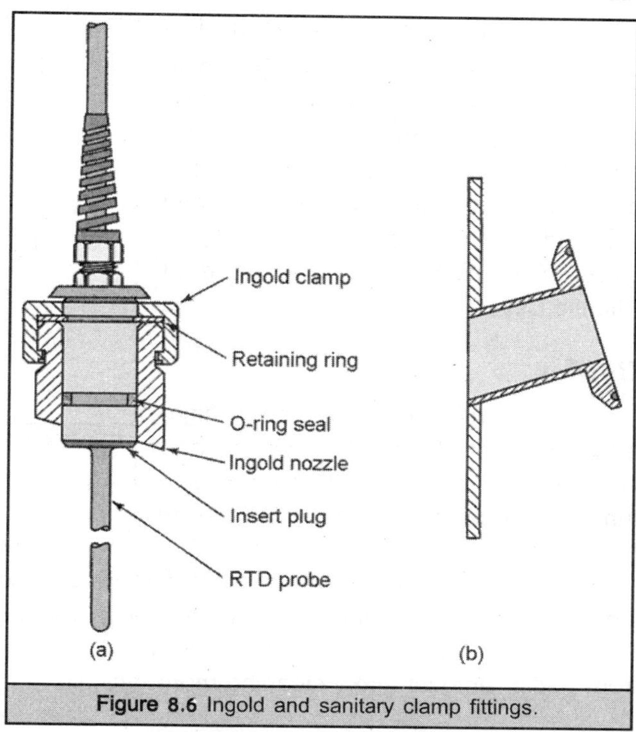

Figure 8.6 Ingold and sanitary clamp fittings.

Ports and nozzles

Ports for additions and probes must be designed to be sterilizable and cleanable. Among other things, this means they must have reliable seals and be free draining. The major debate here usually focuses on the choice between Ingold ports and those designed for sanitary clamp connection. There have been sterility problems with Ingolds, most of which have been traced to the ports being out of round as a result of distortions caused by welding. Such distortion causes obvious problems with the O-ring seals. This problem can be corrected quite easily by making the nozzle thick enough to avoid distortion. Our experience has been that such a correction yields nozzles that are just as reliable and more easily cleaned than are sanitary fittings when both are mounted at the usual 15°. The first word that comes to mind is "don't." Side view ports let the bugs see out more than they let you see in. The little that's gained by being able to see less than 1% of the total action does not seem to justify the expense, the jacket coving, and the cleaning and sterility problems caused by these little gems.

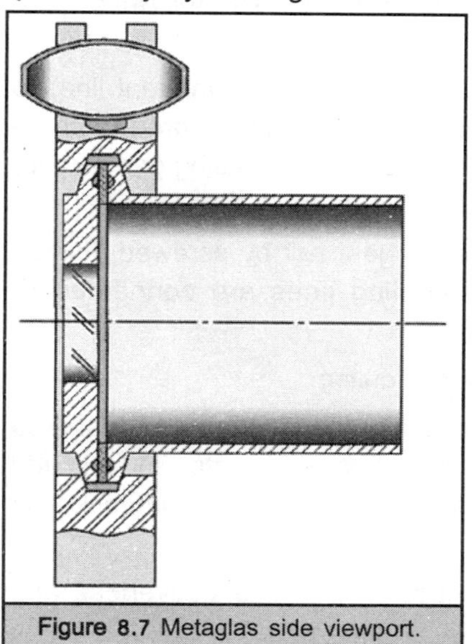
Figure 8.7 Metaglas side viewport.

If you must have one, we suggest the circular Metaglas type attached to the vessel with sanitary clamps; the glass is bonded directly to the metal, which avoids the sealing problems encountered with other types. It does, however, have the disadvantage of having a smaller viewing area than does a standard, circular view port, and it is more expensive. We recommend avoiding long rectangular view ports, which are very difficult to keep clean and free draining, and types in which the glass has to be sealed with gaskets and/or O-rings on both sides. Keep in mind that tolerances for glass are measured in fractions of an inch, not thousandths. That, coupled with the fact that the gaskets will compress means that the probability of repeatedly achieving a good seal and good cleaning is lower than your odds of winning more than once at Atlantic City.

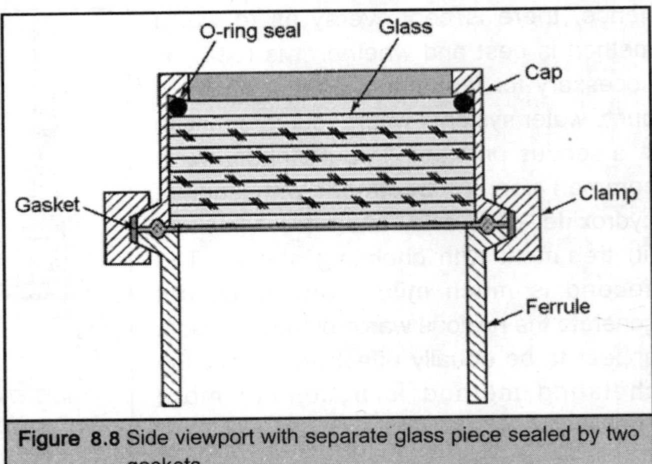

Figure 8.8 Side viewport with separate glass piece sealed by two gaskets.

Baffles

Baffles are usually required in high-power systems to prevent swirling and vortexing, thereby increasing the power that can be delivered to the fluid. The usual practice is to use four baffles on 90° centers welded directly to the wall. Each baffle should have a width equal to 10% of the tank diameter and should have long slots cut out of the edge facing the wall so as to prevent solids build up and to make cleaning easier. Removable baffles are used in some cases; however, we discourage this practice, particularly in cases where cleaning is a critical issue, simply because unsealed joints resulting from baffle removal make cleaning more difficult.

Role of jackets in fermentation

Several types of jacket are used on fermenters; the choice of type is usually not critical for heat transfer purposes and is best left to the vessel fabricater. In some cases, however, a jacket type (e.g., half pipe) may be chosen to increase the vessel pressure rating. The following have proven to be useful practices:

1. The jacket should extend from the probe ring (about 2 in. above the bottom tangent line) to the top tangent line and should be zoned. This allows additional active surface area to come in contact with the broth as the volume increase due to additions and to increasing gas holdup, minimizing cooling-loop pressure drop, and minimizing medium bake-on.
2. Connections to the jacket should be via sanitary clamps or flanges, not by screwed fittings. This minimizes possible damage to jacket welds when cooling lines are connected or disconnected.
3. Coolant should be filtered to avoid solids buildup and/or jacket fouling.
4. Bottom jackets should be considered only for vessels larger than 5,000 L. The additional area will be about 10% of the straight side jacket area (allowing for bottom drain, etc. and will add about 5% to the cost of the vessel). In cases where there is an internal coil, bottom jacketing isn't necessary or desirable.
5. Jacket coving to accommodate view ports, decreases jacket area and effectiveness and increases cost considerably. It should be avoided.

Coils

Coils are the most common internal cooling surfaces, although there are other types. They can easily double the available heat transfer area and tend to be more effective than jackets; however, they can have big disadvantages:

1. They make cleaning very difficult.
2. They can cause additional bulk mixing problems, particularly for very viscous non-Newtonian broths.
3. They will eventually leak nonsterile coolant into the broth.
4. They can add up to 25% to the cost of the vessel.

We recommend that every other alternative (including slight decreases in growth rate and/or cell mass) be considered before yielding to the temptation to use coils. Keep in mind that, once installed, they will be there for the life of the vessel. If you absolutely, positively must use a coil, there are several points to remember:

1. Mount it in a way that will insure minimum stress during heat-up and cool-down.
2. Weld cladding over the butt welds used to join the coil pipe sections.
3. Leak test after construction and build in design features that will simplify leak testing on a regular basis thereafter.
4. Space coil turns at least 3 in. apart. Anything closer will insure major cleaning problems.

Single orifice spargers

There has been a lot of discussion concerning the pros and cons of ring and single-orifice spargers, and the details of design of each. We have seen both work well and have not found any evidence for the validity of claims concerning the importance of hole size (for example) for oxygen transfer per se. The primary focus should be on gas distribution (which will depend on other aspects of the agitation system) and on aseptic operation. We usually favor single-orifice spargers.

There also have been advocates of porous (frits) spargers. The rationale presented has focused primarily on the small bubble size such spargers produce. There is some basis for these claims in cases where little mechanical energy is available for bubble breakup (e.g., in mammalian cell systems). There might also be some value in cases for which bubble coalescence is not a problem (far and few between in practical systems). They can be extremely difficult to clean, particularly for mycelial organisms. We usually suggest they not be used except in very special cases, as just noted.

Finally, it is important that the sparger be made and mounted in such a way that it can be removed easily for cleaning.

Piping System

Design and construction details of the piping system depend on process, sterility, cleaning, and containment requirements, taken together.

Contact process

All lines providing fluids that cannot contact process fluid surfaces (e.g., coolant lines, plant steam lines for heating only) fall into this category. Satisfactory service is provided by either copper or stainless steel piping along with a rational combination of welded and compression fittings. Durability, servicability, cost, and corrosion resistance are major considerations. Ball valves are satisfactory on lines not turned on and off frequently (diaphragm valves tend to withstand a greater number of on-off cycles prior to failure).

System of sterile piping

Before we discuss design of sterile piping systems, we reiterate the importance of a systems approach to integrating vessel and piping design. Independent design almost always results in a lot of aggravation, as well as higher cost and lost time. A sterile piping system can usually be divided into five major subsystems: process, steam/condensate, air, harvest, and seal lubrication. Each has specific requirements dependent on its unique functions as well as the specific demands of the process. None of these systems is completely independent of the others, and all share common components. Details of each of these subsystems follow a simplified description of sterilization procedures. The reader is advised that the procedure is one of several commonly practiced; there are, for example, different opinions as to when live steam to the air inlet line should be turned on.

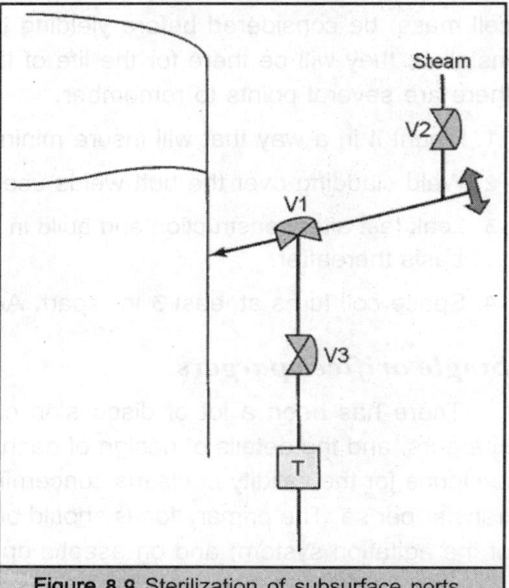

Figure 8.9 Sterilization of subsurface ports.

Phase 1

After the vessel is filled with set medium, steam flow is started through the jacket. The agitator is run at low speed, and particulate free steam lubricates the shaft seal. The exhaust line is open for venting to the atmosphere. All other valves are closed. During this time not only are the vessel and the set medium being heated, but air is being purged from the medium. If the air is not purged, the vessel pressure during sterilization can become dangerously high, and there will not be a reliable correlation between pressure and temperature. Phase 1 continues until the medium temperature reaches 100°C.

Phase 2

Steam continues to flow through the jacket. The air exhaust line is closed; the only path out is via steam traps. Steam is now admitted through the air inlet line to sterilize the inlet filter; steam leaving the vessel through the air exhaust line sterilizes the exhaust line piping and the exhaust filter. Steam flow is started through subsurface ports via steam lock assemblies. Ports above the liquid surface are sterilized by steam escaping through them from the vessel and then to condensate lines. Steam flow also is started through the bottom drain valve. All steam flows continue until the cool-down begins. Phase 2 continues until the medium temperature reaches 121°C.

Phase 3

Steam flow to the jacket is throttled to hold sterilization temperature constant until cool-down begins.

Phase 4

Cool-down. This is not discussed here.

Objectives

There are several general objectives that should be applied to all sterile piping:

1. Make all piping system components free draining. This requires special attention to the details of pipe pitches, valve orientations, and so forth
2. Design all components and the piping layout so as to eliminate nooks and crannies where contamination (biological and nonbiological) can hide so as to escape cleaning and/or sterilization. Frequently overlooked problems include such things as dead legs and ridges formed by welding operations.
3. Make steam and condensate piping at least 3/8 in. diameter to insure proper steam flow and condensate draining.
4. Make all piping lengths as short as possible without interfering with good fabrication practices (GFP), operability, and maintenance.
5. Use check valves only when no other solution is possible (e.g., in over-pressurized lines).
6. Do not use sight glasses in condensate drain lines except in the seal lubricant drain line.
7. Pay careful attention to piping orientation to avoid the possibility of air traps and inadequate heating.
8. Make sure that bottom drain valves are flush-mounted diaphragm valves capable of being steamed in the closed position. They must be free draining.
9. Avoid dip tubes unless there is no other way (there almost always is).

Finally, the design should take into consideration the methods that will be used for validation of sterilization operations. These will vary considerably with the product and the organism. At one extreme is virtually no validation; at the other extreme is temperature mapping of the vessel and all the trap lines, as well as the use of spore strips and/or spore ampules in the vessel and in the trap lines. In some cases, tee sections have been included in drain lines for insertion of spore strips.

Orbital welds

Orbital welding of the piping system produces the most reliable results with regard to aseptic operation and cleanability; it is essentially a requirement for many classes of regulated products, particularly those regulated by Center for Biologics Evaluation and Research (CBER). It also is the most expensive to build and to maintain. There are, however, many processes for which an economic combination of welded and compression fittings is acceptable. These can be operated quite satisfactorily if proper attention is paid to cleaning, sterilization, and maintenance protocols. The fact is that such systems have been used for many years to manufacture regulated products safely and efficaciously.

Valves

Diaphragm valves have become essentially required for most classes of regulated fermentation products. There is little argument that their design provides the greatest reliability for clean, aseptic operation. They have the added advantage of integral sterile access ports, which permit very short piping runs, particularly in valve clusters. Historically, diaphragm valves have been more expensive than have ball valves, but this has changed recently, particularly for some materials. It therefore makes sense to use diaphragms for most applications. The primary exceptions are steam lines or other applications in which diaphragm life is a problem; in those cases one should consider ball valves or plug valves, as appropriate. Another important consideration is the extent to which one uses valve actuators and position indicators. Obviously, this will depend in large measure on the nature of the control systems employed. In any event, one should consider very seriously the problems associated with valve "extras" while control system decisions are being made.

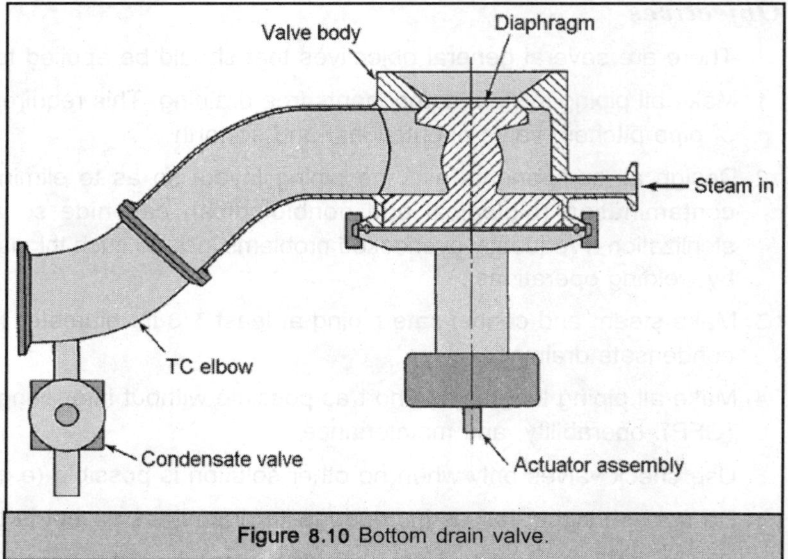

Figure 8.10 Bottom drain valve.

Significance of steamlock

The primary purpose of steam- locks (also known as block and bleed) is to allow sterilization of ports and process piping at any time during a fermentation in such a way that connections can be made and broken without risk of breaching the sterile barrier and/or the containment barrier. In this case, vessel T1 contains presterilized nutrient that must be added at some time during the fermentation. T1 is connected to the fermenter by means of sanitary clamps (other means are possible). Steam flow is then started (open V1, V3, and V5) so as to sterilize the connecting hose and all the valving not previously sterilized. Condensate goes to drain (sanitary, if necessary) via V3 and V5. The steam and condensate valves are closed after sterilization is complete; process valves V2 and V4 can be opened anytime thereafter to make the sterile transfer. Note that the diaphragm valve sterile access ports make it much easier to sterilize the steam side of the valves than is possible with other valve types. Variations on this theme are better suited to cases involving long transfer lines or having other special requirements.

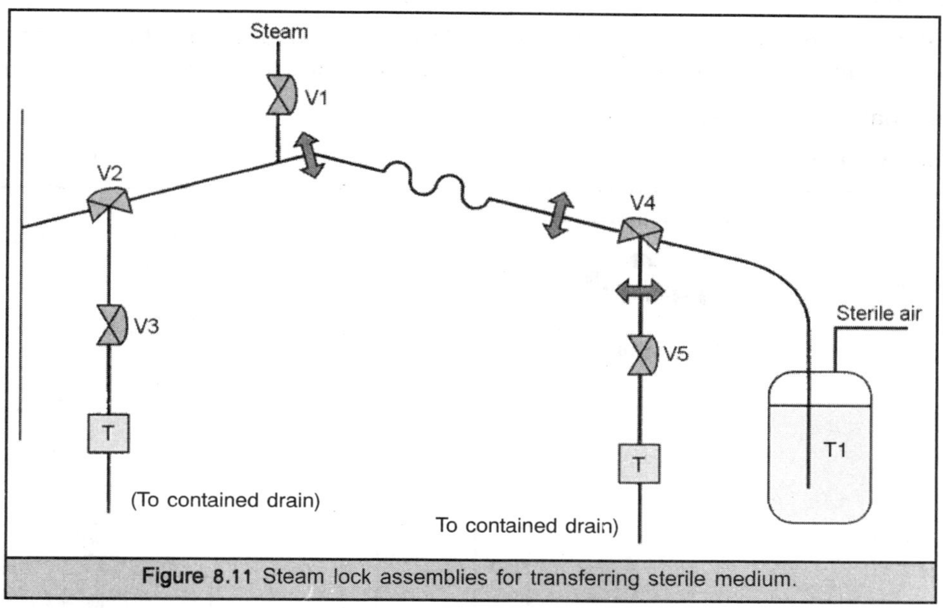

Figure 8.11 Steam lock assemblies for transferring sterile medium.

Wetting and clogging

The major problems for this system are related mainly to the filters. Chief among these (other than outright failure) is wetting of both filters caused by condensate during sterilization, and wetting and clogging of the exhaust filter by condensate and/ or aerosol during fermentation. These problems are not all easy to resolve and should be dealt with most seriously during design. Wetting and/or clogging of the inlet filter by condensate and/or compressor oil during fermentation also can be a problem but is usually the result of not using proper quality air, having no or improper prefilters in the air inlet line, and/or poor maintenance (e.g., of the compressor). All of these are easily remedied. The extent of condensate problems during sterilization can be remedied best by ensuring that whatever condensate does form can drain freely. This can be complicated by the fact that it is undesirable to have the sterile side of the filter connected directly to the drain line.

Another approach is to steam heat the filter housing such that condensation cannot occur. This is effective but has the disadvantages of adding cost and decreasing filter life. Other factors that can affect condensate problems are the positioning and orientation of the housing. There are differing opinions concerning these; one is best advised to consider the advice the filter and the fermenter vendors for specific cases. Plugging of the exhaust filter by condensate and aerosols during fermentation require special consideration. Air leaving the fermenter will be essentially saturated with water vapor at fermentation temperature. The exhaust line, filter, and so on usually are colder than the fermenter; therefore, condensation is inevitable. The amount of condensation will depend on temperature differences, air flow rate, and the nature of the surfaces of the components in the exhaust line. Approaches that have been used to deal with these problems include:

1. Condensers
2. Heat exchangers; used before the exhaust filter to avoid condensation from the fermenter off gases and after the pressure control valve to prevent reflux from the exterior exhaust line
3. Steam-heated filter housings
4. Heated exhaust lines

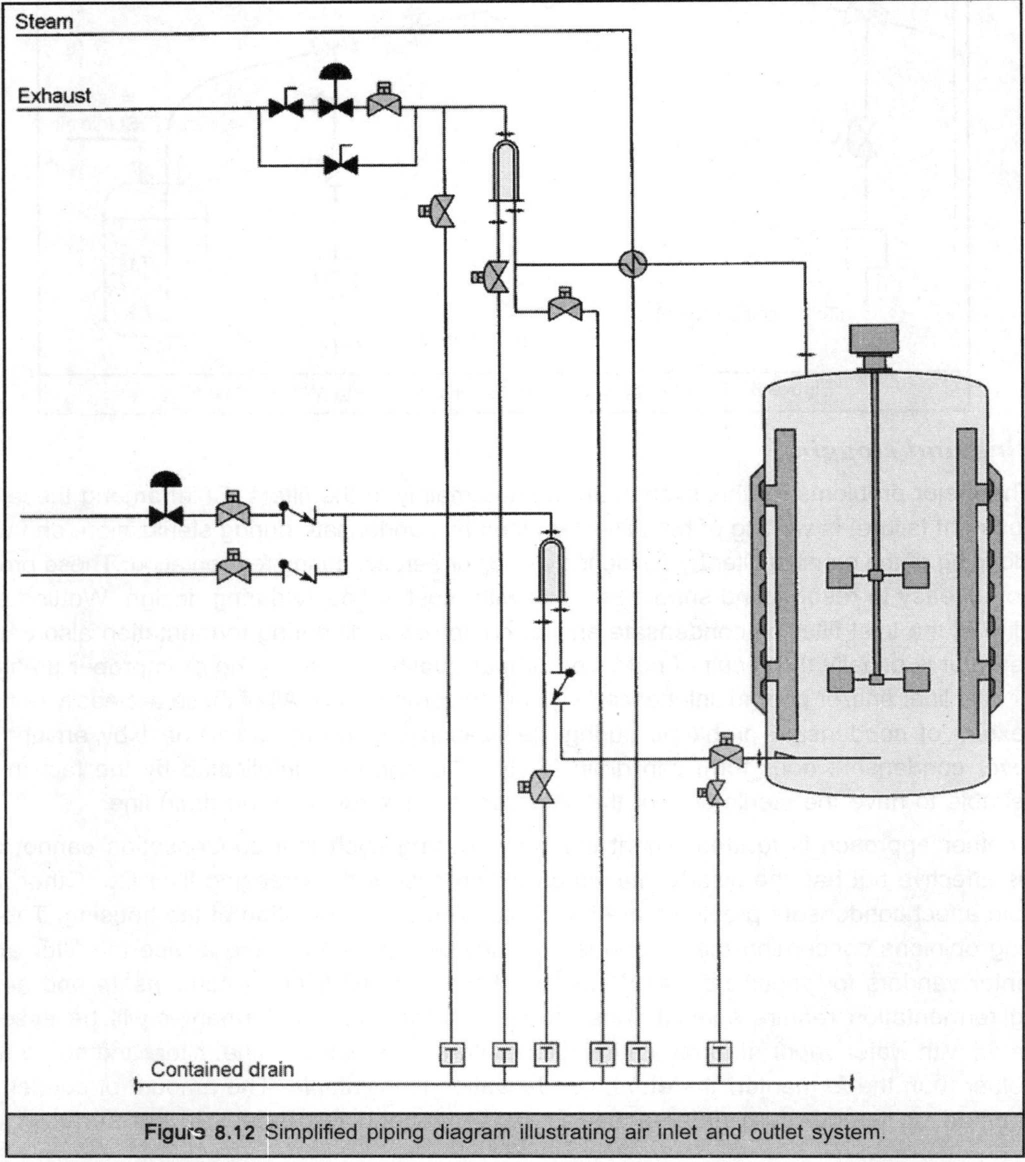

Figure 8.12 Simplified piping diagram illustrating air inlet and outlet system.

5. Coalescers
6. All of the above

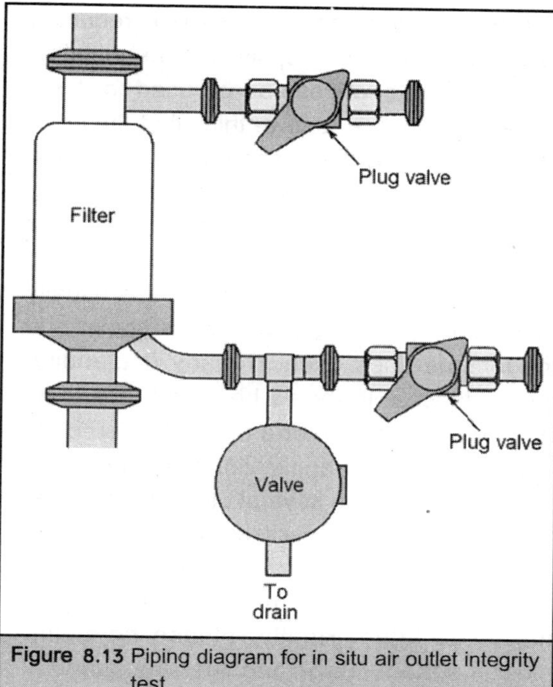

Figure 8.13 Piping diagram for in situ air outlet integrity test.

None of these is completely successful, and each has its proponents and detractors. Suffice it to say that each case should be considered on its own merits (e.g., fermentation temperature, duration, and air flow) and that combinations of the approaches should not be ignored. The condensate problem is exacerbated by the aerosol problem. Aerosols will form in any aerobic fermentation and will carry liquid along with dissolved solids and particulates (including organisms) to the exhaust line. The only questions are how much and what problems they will cause. Among the problems are deposition of dissolved and particulate solids, not only on the filter, but also on heated surfaces or condenser surfaces.

Among the problems deposition causes is reduction in heat exchange effectiveness, which reduces the capabilities of the previously mentioned devices to eliminate condensation. In severe cases, exhaust line blockage can lead to other problems related to pressure buildup and/or higher linear velocities in the exhaust line. The problem should be dealt with on a case-by-case basis during process development and fermenter design.

Provisions should be made for in situ integrity testing of sterile filters. This can be done by means of any of several commercial electronic testing devices that are based on liquid intrusion, diffusion, or some other well-established method for testing membrane filter integrity. Finally, two pressure control valves are required in the case of constant pressure sterilization (continuous flow of steam into the vessel and out to atmosphere). The reason is that the steam flow will be very much lower than the air flow during fermentation; hence, the valve C_v required to maintain steam pressure accurately will be far too small to control air pressure accurately.

Methods of Sampling

The sample vial, vent filter, and valve V1 are sterilized as a unit in an autoclave. The unit then is connected to the fermenter piping via a sanitary clamp. After the connection is made, steam is introduced via V3 and passes across the nonsterile sides of V2 and V1 and then to trap via V4. Once the valves and line are sterile, the steam and condensate valves are closed. Samples may be taken any time after the sampling system cools. Samples are taken via V1 and V2. Lines and valves are resterilized before the sanitary connection is broken. The valves selected for this system will depend on the class of service required. We suggest that vessel sampling valve V3 be a flush-mounted diaphragm valve designed in such a way as to minimize dead volume and piping lengths;

it is an excellent choice for service requiring very high levels of asepsis and cleanability. Less expensive valves can be used for less-demanding service, but we think that the initial extra cost will more than pay for itself. Diaphragm valves are suggested for V1, V3, and V4 in cases of demanding cGMP or containment requirements. This is not required for all classes of service; indeed, it is usually satisfactory to use plug valves for steam and condensate lines even in many demanding circumstances. It should be noted, however, that you may encounter a perception problem if you use them for applications deemed to be sensitive. There are several other designs that have been used successfully. A few have been designed specifically for systems requiring BL3-LS containment or higher. These are discussed later. Finally, please note that addition system design and operation are very similar to those for sample systems and are not discussed here.

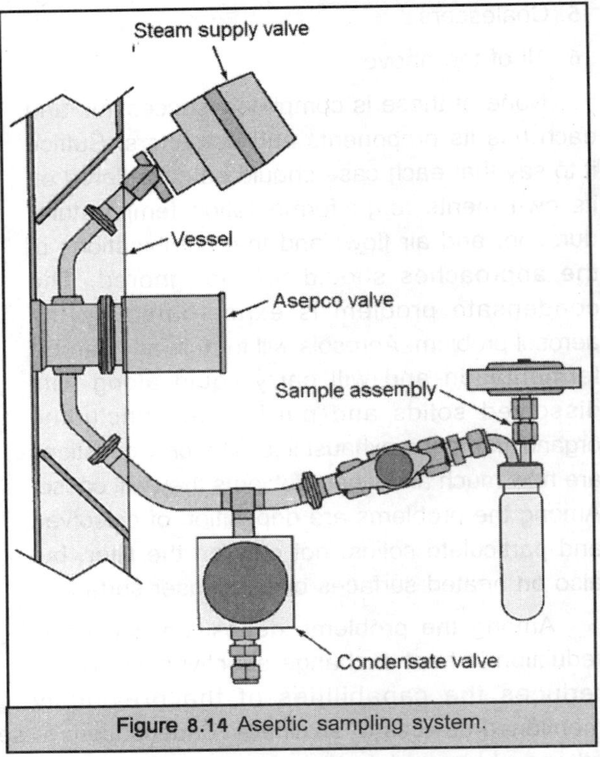

Figure 8.14 Aseptic sampling system.

Mechanical Coupling

Agitation can be done by direct mechanical coupling of the shaft to the drive or by magnetic coupling. The latter has been recommended for cases where high levels of containment are required. There also is the perception that magnetic drives are more suitable for maintaining more stringent levels of asepsis and cleanliness than is possible with direct drive. Neither claim has any substantive basis; indeed, there is good reason to believe that existing magnetic drives may present greater cleaning difficulties because of the manner in which the driven magnet must be mounted at the bottom of the vessel. In addition, power transfer by magnetic drive is quite low. Direct drive is the predominant current choice at almost any scale of operation.

The major mechanical decisions that must be made in this case (other than power) have been discussed at length elsewhere. The major arguments for bottom drive are ease of maintenance, shorter shafts, less support structure, and lower overall height. The arguments against bottom (and for top drive) focus primarily on the potential of catastrophic spills resulting from bottom seal failure, seal grinding as a result of broth particulates working into a bottom seal, and greater cleaning difficulties. If the seals are designed and maintained properly, none of these is a problem. We have seldom seen any real difference in aseptic operability between top and bottom drives, and we are reasonably certain that the organisms don't care. The choice probably will continue to be driven primarily by personal preference. There are very few cases in which double mechanical seals are not (or should not) be used in fermenters. The major debate focuses on seal orientation and the details of individual seal designs. There are basically two orientations used: inline and back-to-back. We recommend in-line design because we have found it to operate more cleanly and require a simpler sterilizing/ lubricant system.

Seal lubrication is usually provided by means of sterile steam condensate. It is extremely important that this condensate be free of particulates: their presence guarantees rapid seal failure, contaminated fermentations, and a hyperactive maintenance program. It is also important to note that during sterilization live steam flows through the seal housing. There are some who insist on keeping the steam flowing throughout the fermentation. (Obviously, they have no faith in the seals.) The one thing this will guarantee is much more rapid wearing of the seals (perhaps supporting the lack of faith in the seals). One must also decide on the means for controlling lubricant flow rate. Most use a valve for this purpose. We suggest an orifice sized to deliver the proper flow. This avoids the cost and maintenance of a valve and insures fiddle-proof operation. It does, however, require the use of particulate-free condensate, but then so does proper operation of the seals.

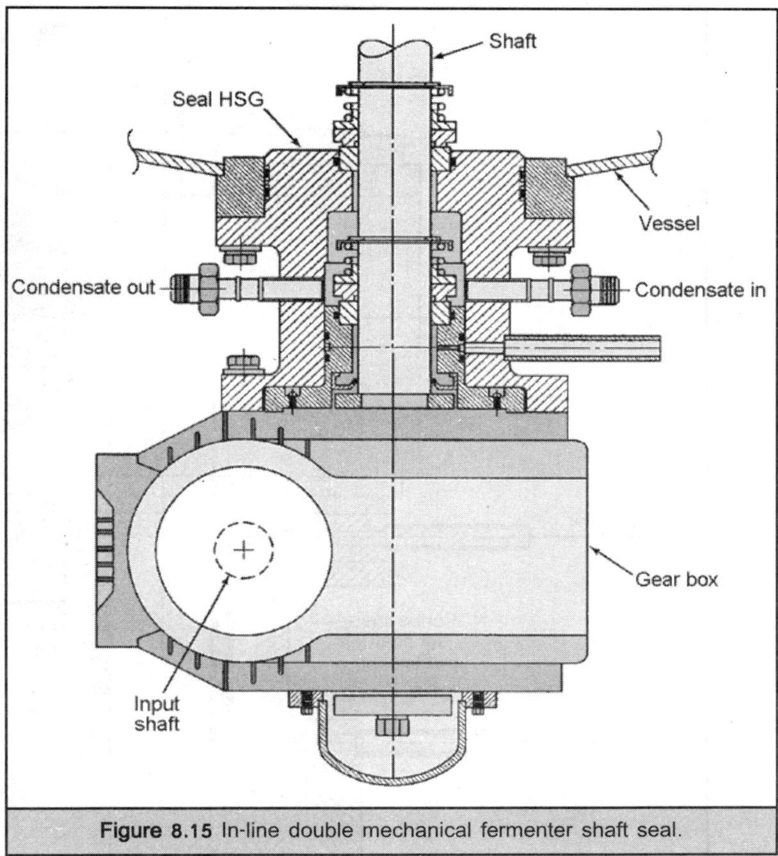

Figure 8.15 In-line double mechanical fermenter shaft seal.

We also recommend including a sight glass in the lubricant drain line as well as a seal leak detector. Other more complex detection systems might be considered for specific circumstances (e.g., severe containment requirements). Finally, preventing shaft vibration is another important factor in agitation system design. Shaft vibration is a safety hazard. It will also cause premature seal failure and other costly mechanical damage.

Requirements for Cleaning

As noted earlier, specific requirements for cleaning depend on several factors including the nature of the fermentation broth. There is a major difference between most microbial broths and

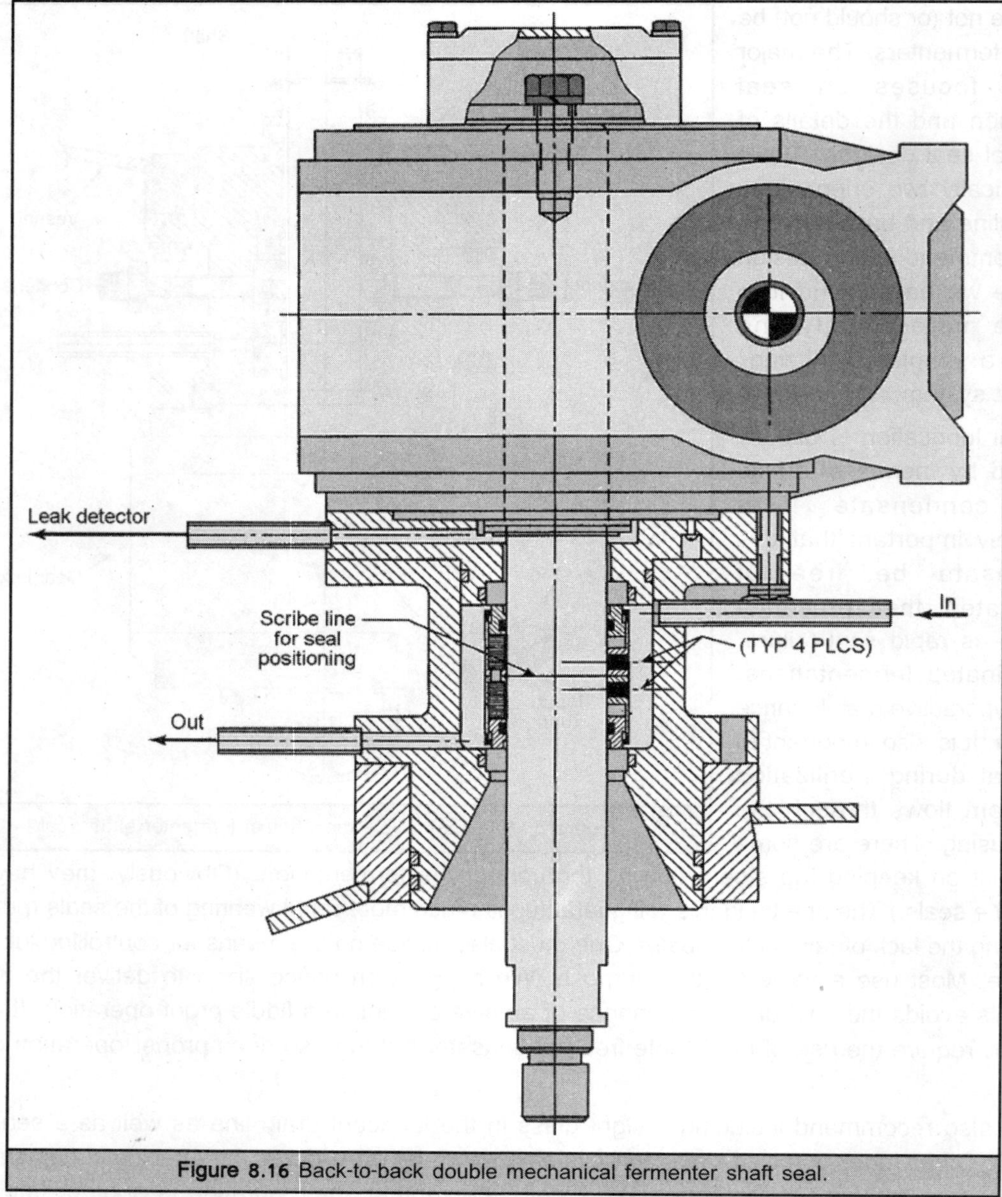

Figure 8.16 Back-to-back double mechanical fermenter shaft seal.

most mycelial broths. Single-cell microbial broths, with the exception of those containing a lot of undissolved particulates and high viscosity components (e.g., xanthan broths), tend to be free draining and readily amenable to cleaning based primarily on the physicochemical action of the cleaning agents. Many fungi and mycelial bacteria, on the other hand, tend to cling to fermenter internals and may require mechanical action (e.g., high-velocity jets) in addition to cleaning agents. The following guidelines are applicable for most systems. These principles must be applied in light of the actual cleaning agents and protocols to be used:

1. Eliminate internals and nooks and crannies to the greatest extent possible in the vessel and throughout the piping system. This practice is consistent with design for aseptic operation.
2. Drill and position spray balls to insure complete coverage of all surfaces inside the vessel. This usually requires an empirical approach. Coverage can be tested by means of the riboflavin test. The reader is cautioned, however, that complete coverage is a necessary but not sufficient condition for cleaning. Also, please note the following:
   (a) Spray balls designed for sanitary operation are self-draining and self-cleaning. They can be sterilized in situ. There is, however, considerable debate concerning whether they should be removed prior to fermentation.
   (b) High-velocity, rotating devices that may be necessary when large clumps of sticky residue must be removed (e.g., as with a fungus) are not designed to be inherently self-cleaning, self- draining, or sterilizable. They should not be mounted permanently.
3. Eliminate dead legs in piping and dead spaces in valves, fittings, and other system components. This is also consistent with design for aseptic operation.
4. Specify the same materials and finishes for process and CIP piping as are specified for the fermenter. Obviously, these must be compatible with the cleansers and conditions used.
5. Avoid threaded joints. A completely welded system is best, but compression fittings can be satisfactory when cost is a major consideration, particularly for nonregulated products.
6. Make CIP piping as simple as possible. For example, make dual use of process and other piping as much as possible.
7. Avoid complex, expensive transfer panels wherever possible. Use swing elbows wherever practical.
8. Design and construct the system to facilitate validation and on-going testing for removal of contaminants, including the cleaning agents. This design should provide for swabbing, obtaining rinse samples, or whatever else the cleaning protocol requires.

There are many cases for which a portable CIP system is preferable to an integral system. For such cases, the fermenter and the portable unit should be designed to allow simple mechanical attachment of the necessary hoses between the two units and with utilities, and a straightforward means for interfacing the instrumentation, logging data, and controlling the systems of the two units.

Finally, we note that classical CIP systems rely on continuous cleanser flow and the maintenance of a shallow puddle in the bottom of the fermenter. To achieve this they rely on special pumping devices such as eductors. Aside from the design, control, and other operating problems this causes, it has been observed in some facilities that the stable pool leads to the formation of "cleaning rings" at the bottom of the fermenter.

These rings may not be real problems (additional evidence is still required), but they do cause perception problems. We suggest the use of a pulsed-flow system to overcome not only the need for special pumping devices but also the "cleaning ring" problem. Such pulsed systems have been found to accomplish both objectives in practice.

Containment

Our scope is limited to containment of fermenters; however, we reiterate here the importance of making certain that the equipment, the facility, and the operating protocols be considered simultaneously to ensure compatibility with regard to containment (just as in the cases of sterility and cleaning). As noted earlier, most new facilities for regulated products are being built to satisfy BL2-LS facility requirements and (in most cases) BL3-LS fermenter containment requirements. With few exceptions (about which there is no agreement) there is no difference with regard to the equipment containment requirements at the two levels—at present. The guidelines require that BL2-LS/BL3-LS fermenters be designed and constructed to prevent organism release (primary containment) from the vessel and from any of the subsystems noted earlier; hence, they must be built as closed systems. (Note that it also is possible to enclose a fermenter in a biosafety "cabinet," but this is not usually practical or necessary.) Some areas that have been subject to greater scrutiny than has been required for cGMP compliance are as follows. In evaluating the discussion, the reader should bear in mind that there is not much information readily available concerning containment reliability data. Some progress is being made via the Industrial Biosafety Project in the United Kingdom.

1. *Static seals*. It has been suggested that double static seals (e.g., in headplates) be used for BL2-LS and that double static seals with a steam trace between the seals be used for BL3-LS. There has been little to demonstrate that these are either needed or desirable. Indeed, there have been substantive arguments made against using such devices. There is as yet no consensus.

2. *Rotating seals*. Arguments have been made in favor of using magnetic drives at anything higher than BL2-LS. This was discussed earlier. There are many who argue that only top drive be considered for BL2-LS and above. Their reasons have mainly to do with avoiding the possibility of large spills in the event of catastrophic seal failure. Such a circumstance is possible but highly unlikely. Leak detectors (which would normally be installed for cGMP requirements) would almost certainly indicate a leak long before there was the remotest possibility of catastrophic failure. One also should consider that a leak in top drive seal might be more difficult to detect and could lead to release of a more insidious nature. Again there is no consensus. Finally, it has been suggested that a low-pressure sensor or flow sensor be used to detect lubricant flow failure. This could help to avoid seal failure.

3. *Sampling Systems*. Note that it is similar to the system recommended for cGMP compliance. It is a bit more complicated and may provide some incremental benefit; however, this might be outweighed by the greater likelihood of operator error. There have also been more complex systems suggested that involve more intricate valving and/or biosafety cabinets. It is apparent that this is an area that needs considerably more development work.

4. *Air exhaust system*. It has been suggested that two in-series sterile filters or a sterile filter followed by an incinerator be used at BL2-LS and higher. All other aspects are covered by the earlier discussion of exhaust lines.

5. *Pressure relief*. This is an area of some controversy because of the need to satisfy physical and biological safety requirements simultaneously. It has been suggested by many that relief

venting be done via a large kill tank protected by a HEPA filter; however, this could be in conflict with physical safety codes that require there be no devices in the relief path. We do not have anything approaching universal agreement here; however, it does seem fairly apparent that limiting the supply pressure would help to minimize the risk. There does appear to be agreement on the use of rupture disks rather than pressure relief valves; this is consistent with accepted practice for cGMP compliance. Rupture disks are cleaner and are not prone to sticking.

Finally, HAZOP evaluations for fermenter containment have been recommended. Such analyses would evaluate potential hazards that might be caused by a wide range of incidents (e.g., fire) not limited to normal operation.

ROLE OF rDNA TECHNOLOGY

Today high-performance production strains to be used in fermentation processes are often constructed using rDNA technology. However, the concept of metabolic pathway manipulation is by no means new. Thus, for decades better strains of *Saccharomyces* to be used for beer fermentation have been obtained through classical breeding and crossing of different strains, and in the production of penicillin productivity has increased by more than 500 times through repeated rounds of mutation and selection of new strains of *P. chrysogenum*. With the introduction of rDNA technology it has become possible to apply a more rational approach to strain improvement–namely by the introduction of targeted genetic changes resulting in strains with a phenotype that gives a better process. This rational approach has been named cellular and metabolic engineering, and different definitions of the approach have been given.

The term *metabolic engineering* was first introduced by Bailey, who defined it as "improvement of cellular activities by manipulation of enzymatic transport and regulatory functions of the cell with the use of recombinant DNA technology." His definition of metabolic engineering includes the following:

1. Inserting new pathways in microorganisms with the aim of producing novel metabolites (e.g., production of polyketides by *Streptomyces*), with the aim of degrading toxic compounds (e.g., in bioremediation), or with the aim of constructing a novel biotransformation system
2. Production of heterologous peptides, such as production of human insulin, erythropoitin, and tPA, or industrial enzymes, such as lipases and cellulases
3. Improvement of pathway fluxes leading to higher yields of metabolites (e.g., increasing the flux toward antibiotics or primary metabolites) or yields of biomass (e.g., increasing the cell mass yield in baker's yeast production using industrial media containing mixed sugars)

What characterizes metabolic engineering is the rational approach to performing genetic changes, and as with all other fields of engineering it consists of two steps: analysis and synthesis. As a consequence of the difficulties in performing detailed analysis of cellular metabolism there has mainly been focus on synthesis in the past, such as expression of new genes in various host cells, amplification of endogenous enzymes, and deletion of genes or modulation of enzyme activities. With modern experimental techniques it has, become possible, however, to perform detailed analysis of cellular function through both in vivo and in vitro measurements, and in the following some of these techniques are described.

PRODUCTION OF HIGH PERFORMANCE STRAINS

Whereas the development in rDNA technology has enabled the design of novel high-performance strains that are tailor-made to specific needs, the revolution in computer technology and in analytical chemistry has allowed a much more fundamental characterization of cellular function than previously possible. Thus, flow cytometry allows quantification of population dynamics, while advances in microscopy and image analysis have allowed studies of cell growth and function at the single cell level.

In studies of fermentation processes it is important to study the function of whole cultures, and here the development of high-performance bioreactors has been an important contribution. When microbial or cell cultures are grown in these bioreactors, it is ensured that all the cells experience the same conditions, and it is therefore possible to perform quantitative physiological studies, such as study of growth and production kinetics, study of gene expression under well-controlled conditions, quantification of metabolic fluxes, and study of the control of flux through the different cellular pathways.

Bioreactors for High-Performance

High-performance bioreactors are characterized as being practically ideal bioreactors with a very low mixing time and a very high gas–liquid mass transfer. In these bioreactors the cells are subjected to an unchanging environment when they are circulated throughout the liquid medium, and the response to imposed variations in the environment is therefore a consequence of the cellular behavior only. High-performance bioreactors are equipped with a large number of in situ sensors, and to control operating variables a flexible direct digital control (DDC) is used, rather than classical single-purpose controllers. This allows precise control of many operating variables, including temperature, pH, dilution rate, and dissolved oxygen concentration, which therefore can be taken to be culture parameters.

Many commercially available laboratory bioreactors (volumes less than 10 L) normally fulfill the requirements for being high-performance bioreactors. Several bioreactor companies offer systems with flexible DDCs that enable precise control of the culture variables, and when these bioreactors are designed with the proper stirrer and sparger it is possible to ensure a very low mixing time (less than 1 s) and a high gas–liquid mass transfer (k_1a values above 0.1 s^{-1}) . In addition to sensors for the culture parameters, modern high-performance bioreactors are normally equipped with in situ sensors and on-line analyzers that allow frequent monitoring of key culture variables. The bioreactor is equipped with on-line flow injection analyzers for monitoring of important culture variables, exhaust gas analysis, and an automatic sampling system. Hereby the most important culture variables can be measured at a high frequency (30-1,000 h^{-1}), whereas other culture parameters can be measured in the automatically withdrawn samples at a lower frequency (1-3 h^{-1}).

The bioreactor system can be operated as a constant mass, continuous culture with measurement of the feed flow. It is therefore possible to vary the dilution rate (with a precision of

0.005 h^{-1}), and through proper design of the feed medium it is possible to operate the bioreactor such that there is a single limiting substrate component, such as glucose or ammonia. Finally, the bioreactor is equipped with gas blending, which allows variations in the composition of the inlet gas to the bioreactor, so the influence of, for example, the oxygen concentration on the growth and production kinetics can be studied at the same specific growth rate.

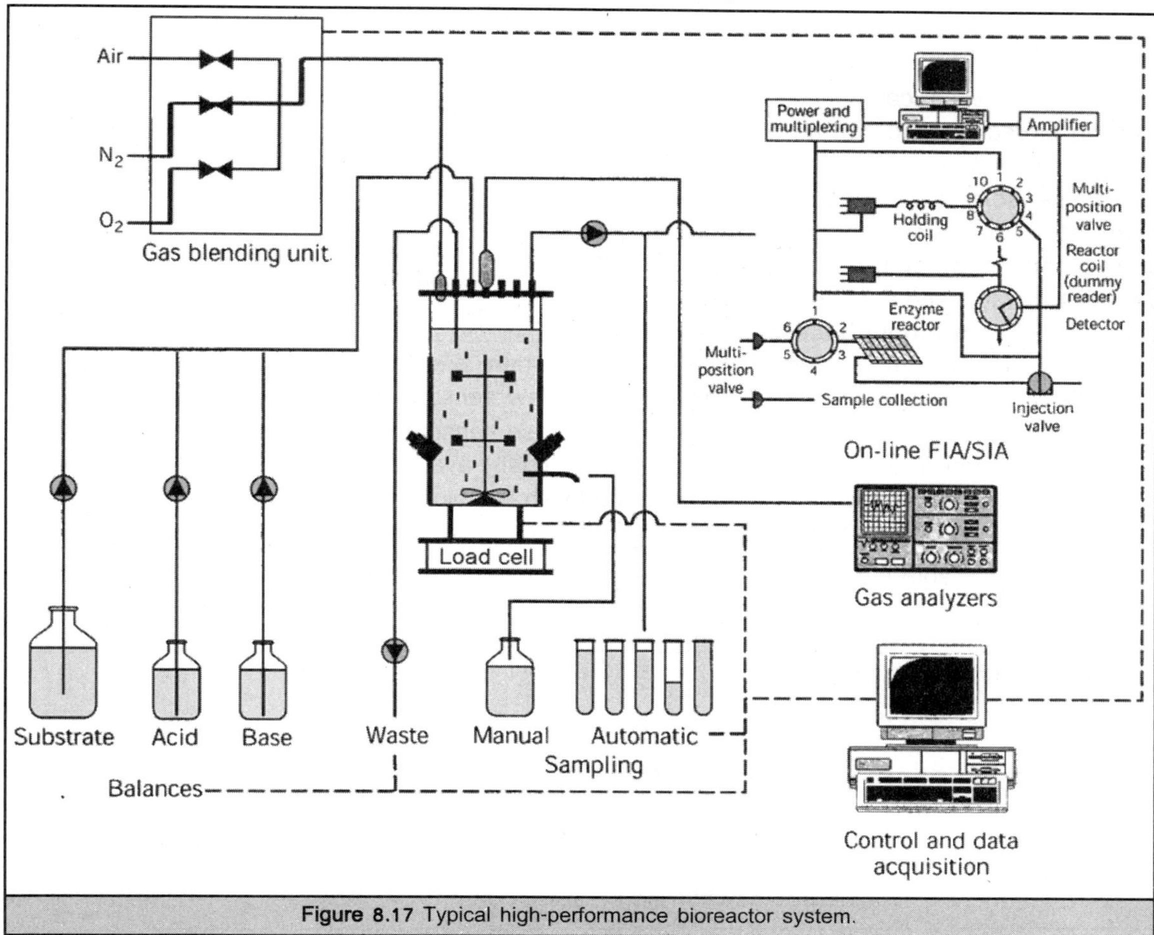

Figure 8.17 Typical high-performance bioreactor system.

With advanced computer control of culture parameters it is possible to perform very reproducible experiment, which is of paramount importance for a fundamental understanding of the underlying cellular reactions. This is illustrated in Figure 8.18, where the carbon dioxide evolution rate is shown for two different batch cultivations of the filamentous fungi *P. chrysogenum*. With remarkable reproducibility the CER signal can obviously be used to map the different phases of the batch fermentation, which were carried out with sucrose as the carbon source.

In the first phase sucrose is hydrolyzed to glucose and fructose and glucose is metabolized. When glucose is exhausted after 45 h there is a distinct shift in the metabolism, and when fructose is exhausted after 60 h there is another shift in the metabolism (i.e., the CER decreases rapidly).

Even when both sugars are exhausted there is still some cellular activity due to metabolism of gluconic acid, which was formed in the initial phase during growth on glucose.

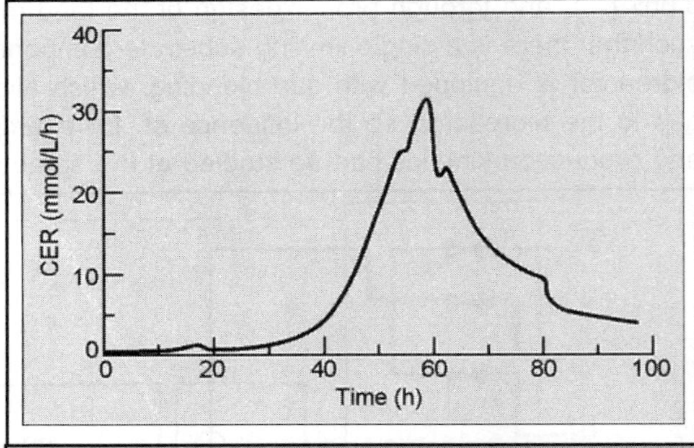

Figure 8.18 On-line measurement of the carbon dioxide evolution rate (CER) during two consecutive batch cultivations of *P. chrysogenum.*

In situ sensors

An in situ sensor can be inserted directly into the bioreactor and can be sterilized in place. Table 8.9 gives the characteristics of in situ sensors used to monitor culture parameters in a typical high-performance bioreactor. Most of these sensors are extremely reliable and are used routinely not only in research laboratories but also in industrial plants. For monitoring of culture variables there are, however, only a few commercially available in situ sensors available, and besides a few sensors based on enzymatic analysis of medium components (e.g., for glucose analysis), these sensors measure biomass using different measurement principles. There are several commercially available sensors for in situ monitoring of the optical density, which is linearly correlated with the biomass concentration. These sensors have a wide linear range, but there is interference from gas bubbles and solids in the medium.

Table 8.9 In situ sensors used for measuring culture parameters

Culture parameter	*Sensor*	*Range*	*Accuracy*
Temperature	Pt-100	0-150°C	0.1°C
Pressure	Piezoresistor	0-2 bar	20 mbar
Gas flow	Thermal mass flow meter	0-20 L min^{-1}	20 mL min^{-1}
pH	pH electrode	2-12	0.02
pO_2	Polarografic Clark electrode	0-400 mbar	2 mbar
pCO_2	Membrane-covered pH electrode	0-100 mbar	2 mbar

In some sensors special designs are used to reduce this interference such as by passing the sample into an internal measurement chamber. Sensors measuring the culture fluorescence (normally the NAD(P)H-related fluorescence) were first introduced as biomass sensors, but despite high sensitivity and correlation of the culture fluorescence with the total cellular activity, the signal from these sensors is difficult to interpret, and there is generally a poor correlation with the biomass concentration.

Measurement of dielectric properties (or capacitance) is one of the later contributions to measurement of biomass, and it has the advantage of a wide measurement range. However, the signal depends on the cellular activity, which although it is an interesting variable, may not necessarily correlate with the biomass concentration. Even though none of the commercially available in situ biomass sensors stands out as the best biomass sensor, several of the sensors are very useful additions to the sensors normally applied in high-performance bioreactors.

Table 8.10 Advantages and disadvantages of different measurement principles for in situ biomass sensors

Principle	*Advantages*	*Disadvantages*
Optical density	Wide linear range	Some interference
Culture fluorescence	High sensitivity	Much interference
	Measurement of cellular activity	Signal interpretation difficult
Capacitance	Wide measurement range	Interferance from aeration and agitation
	Measurement of cellular activity	Signal interpretation difficult
Ultrasonic	Wide linear range	Interference from aeration and agitation
	Self-cleaning	Temperature sensitive

On-line analyzers

With the lack of reliable in situ sensors for measurement of important culture variables, including glucose and other medium components, on-line analysis is frequently used. Here a sample is automatically withdrawn and analyzed. For this purpose flow injection analysis (FIA) is well suited due to its high speed, good precision, and good reliability. The major drawback of FIA is that it is a single component analysis, and even though a novel technique termed *sequential injection analysis* (SIA) offers the possibility to measure several components using the same hardware, other analytical systems such as mass spectrometry (MS), high-performance liquid chromatography (HPLC), and gas chromatography (GC) may sometimes be preferred. These analytical systems offer the ability to measure many components in a single run, but, in general, it is difficult to match the high measurement frequency and reliability of FIA. Aprerequisite for on-line monitoring is automatic sampling, and the sampling system normally serves as the sterile barrier between the bioreactor and the analytical system. Automatic sampling can be performed either by withdrawing a sample directly or by using a membrane module. In direct sample withdrawal, a nonfiltered sample is pumped directly from the bioreactor through a syringe or catheter. Direct sample withdrawal is a very simple solution that allows for on-line measurement of the biomass concentration using FIA and measurement of intracellular components. However, a construction with a syringe is not sufficiently robust to be used in an industrial environment. For sampling via a membrane module there are several commercially available designs, which can be divided into two groups:

1. *Membrane modules placed in a recycle loop connected to the bioreactor*. The advantage of these modules is that it is possible to change the membrane while the process is running. However, the continuous pumping of the medium through the membrane loop may have a significant effect on the cellular behavior (especially for shear-sensitive organisms).
2. In situ *membrane modules, which are inserted into the bioreactor and sterilized in place*. The advantage of these modules is that the modules do not influence the bioreactor operation and

the withdrawn sample is a good representation of the medium in the bioreactor. However, to ensure mechanical breakage of the membrane it is necessary to apply rather thick membranes, which influences the response time.

With a reliable membrane module for automatic sampling it is possible to perform routine on-line monitoring of important medium components, and if FIA is used it is especially possible to obtain high frequency measurements, which allows precise quantification of the fluxes in and out of the cells (e.g., the substrate uptake rates and the product formation rates). These fluxes form the basis for quantitative physiological research, as discussed later, and high-performance bioreactors equipped with on-line analyzers for the most important culture variables are therefore indispensable tools in such research.

Quantitative analysis

For quantitative physiological studies, the measurement of intracellular variables is extremely important. These variables should ideally be performed in vivo, while cellular function is maintained, but presently there are no techniques that allow for precise in vivo measurements of intracellular variables; therefore they are taken in vitro. To ensure that the in vitro measurements are good representatives of the in vivo conditions, it is critical to have techniques that allow for rapid sampling and quenching of the cellular activity. For the measurement of intracellular metabolites and cofactors, such as glycolytic intermediates, ATP, and NADH, which have a turnover time of a few seconds, it is extremely important to have an almost instantaneous quenching of the cellular metabolism.

Using a specially designed sampling module and rapid cooling of the sample in precooled test tubes containing small glass beads, Theobald et al. were able to measure the intracellular level of many metabolites of the EMP pathway and the cofactors ATP, ADP, and AMP in *S. cerevisiae*. Furthermore, using a freeze–quench method, where the sample is mixed with a reagent and then quenched by pumping it into a reservoir of liquid methanol at –40°C, de Koning and van Dam studied the uptake kinetics of glucose by *S. cerevisiae* at the scale of 5 ms to 1 s. Using these sampling techniques, together with HPLC and enzymatic assays, it is possible to obtain a fundamental understanding of the metabolite levels inside the cells at different growth conditions, which, together with knowledge of the metabolic fluxes, will give the ultimate physiological characterization.

Morphology and product formation

For filamentous microorganisms there is in many cases a relation between morphology and product formation. In the past, detailed studies of the growth mechanisms of these organisms have been hampered by the lack of good experimental techniques. However with the rapid development in image analysis it has become possible to study the growth of these organisms in much detail. Yet, because of the many different processes that influence morphology, it is important to combine experimental work and mathematical modeling to extract information about the underlying mechanisms. When the morphology is measured for a given number of hyphal elements in a sample taken from a fermentation with filamentous microorganisms, there will be a certain distribution of properties, governed by the underlying mechanisms for growth and hyphal breakage. It is difficult to extract information about the different mechanisms from these measurements directly. However, using small growth chambers, where the developing morphology can be monitored on-line, it is possible to obtain fundamental information about the growth mechanisms. When this information is combined with computer simulations of hyphal element populations, it is possible to obtain some insight into the mechanisms behind hyphal breakage during submerged fermentations.

Physiochemical Studies

Despite many years of extensive analysis of glycolysis, all the details in regulation of glycolytic flux have not been elucidated. Even in *S. cerevisiae*, where studies of a large number of mutants in the glycolysis have resulted in a meticulous mapping of regulation patterns, all the control structures have probably not been revealed. Thus, it was only a few years ago that fructose-2,6-bisphosphate was discovered not only as an important regulator of phosphofructokinase but of other enzymes as well.

Further physiological studies are therefore extremely important to identify yet-unknown regulatory patterns and regulatory compounds. With the recent advances in genome characterization it is now possible to rapidly construct many different mutants by introduction of gene disruptions or over- expression of key enzymes, and this is a good basis for further studies. However, it is important to combine genetic work with quantitative physiological studies in which gene regulation is studied under well-controlled growth conditions, and metabolic fluxes and their control are quantified.

Expression of gene

To illustrate how gene regulation can be studied in submerged fermentations carried out in high- performance bioreactors, two examples are considered. The first example is regulation of the strong TAKA (or α-amlyase) promoter of *Aspergillus oryzae*, which is often used as expression promoter for production of industrial enzymes by filamentous fungi. In the past there has been much speculation on the regulation of this promoter, but most studies were based on shake-flask experiments, in which it is difficult to distinguish between repression and induction. However, through the use of continuous cultures it can clearly be demonstrated that there is both glucose repression and maltose induction. Thus, by adding a glucose pulse to a steady-state chemostat operated at a low dilution rate, when the glucose concentration is very low (about 5 mg L^{-1}; i.e., derepressed conditions), it was found that there is an immediate decrease in the α-amylase synthesis. This clearly indicates that there is glucose repression.

Furthermore, by shifting from a glucose-based medium to a maltose-based medium when the cells were grown under derepressed conditions (i.e., at a low glucose concentration) it was possible to show that there is maltose induction. Besides allowing qualitative conclusions to be drawn, measurements of the glucose concentration at the milligram per liter level even allowed quantification of the degree of glucose repression. These findings are consistent with mapping of the TAKA promoter, which indicates the presence of several *creA* binding sites (creAp is a regulatory protein associated with glucose repression in filamentous fungi) and a binding site for a protein that is induced by starch (or maltose). In a classical approach to analyze the promoter further, one would construct promoter fragments and identify the function of the individual binding sites. In this analysis shake-flask cultures may be useful in drawing qualitative conclusions (i.e., whether glucose repression has been eliminated). However, to draw quantitative conclusions these experiments should be followed by another set of continuous fermentation experiments. The other example is on glucose control in the yeast *S. cerevisiae*. In the production and application of baker's yeast, glucose repression is a serious problem.

An often-used carbon source is molasses, which consists of glucose, fructose, sucrose, and raffinose (the trisaccharide galactose–glucose–fructose), and in the uptake and metabolism of the different sugars glucose repression results in lag phases. Furthermore, in the application of baker's yeast in break making, maltose is the typical carbon source and glucose repression of maltose uptake and metabolism results in a reduction in the gassing power. One approach to alleviate glucose repression is to delete the genes coding for key regulatory proteins, for example, *MIG1*, a general repressor that influences the transcription of a number of genes, including the *SUC*, the *GAL*, and the *MAL* genes.

Analysis of the effect of deletion of *MIG1* can only be done properly using well-controlled fermentation experiments, and the effect of *MIG1* deletion on sucrose and maltose utilization has been studied using batch cultures. Two different approaches were applied to silence *MIG1*: (i) gene disruption and (ii) introduction of high copy number plasmids with antisense fragments of *MIG1*. Use of antisense is especially interesting in silencing genes in polyploid industrial yeast strains, but even though the antisense fragments were expressed, there was no change in the degree of glucose repression. However, with gene disruption of *MIG1* a significantly higher activity of invertase is obtained during glucose consumption, resulting in an almost parallel uptake of glucose and sucrose.

Disruption of *MIG1* was also found to partly alleviate glucose repression on maltose uptake and metabolism in laboratory strains; there was an almost parallel uptake of glucose and maltose, whereas there was no effect of *MIG1* disruption on maltose uptake in a polyploid industrial strain. This was explained to be a consequence of strong catabolite inactivation of the maltose permease in the industrial strain. Based on the physiological characterization of the recombinant strains it was concluded that silencing of MIG1 may alleviate glucose repression, but other effects including catabolite inactivation, also have an important influence. Only through careful physiological characterizations was it possible to identify the different effects.

Table 8.11 Invertase activity in different haploid laboratory strains of *Saccharomyces cerevisiae* during growth on glucose

Strain	*Genetic construct*	*Invertase activity [U (g DW)$^{-1}$]*
X2180-1A	Wild-type	61
TA8	Antisense, plasmid	55
TA25	Antisense, chromosomal	55
T301	Disruption	263

Optimization of metabolic products

In connection with optimization of metabolite production, where the aim is to direct as much carbon as possible from the substrate into the metabolic product to ensure a high yield, it is of prime importance to quantify how the carbon fluxes are distributed through the various cellular pathways. A very powerful technique for calculation of the fluxes through various pathways is the so-called metabolic flux analysis, where the intracellular fluxes are calculated using a stoichiometric model for all the major intracellular reactions and applying mass balances around the intracellular metabolites. A set of measured fluxes as input in the calculations is used, and these fluxes typically are the uptake rates of substrates and secretion rates of metabolites. If one performs experiments with ^{13}C-enriched carbon sources and measures the fractional enrichment of ^{13}C in intracellular metabolites, it is possible to apply an additional set of constraints in the model, and a better estimation of the intracellular fluxes may be obtained. Finally, if one applies information about the isotopomer distributions for the intracellular metabolites much more information is supplied, and it may even be possible to quantify the net fluxes through reactions that are practically reversible.

Besides quantification of the various pathway fluxes, and therefore determination of the carbon flows inside the cell, metabolic flux analysis is useful for the following:

1. *Identification of possible rigid branch points (or nodes) in the cellular pathways*. Through comparison of the distribution of fluxes at different operating conditions and in different mutants it is possible to identify whether a pathway node is rigid or flexible. Thus through metabolic flux analysis of various mutants of *C. glutamicum,* Vallino and Stephanopoulos concluded that in lysine production the glucose-6-phosphate node is flexible and the pyruvate node is weakly rigid.
2. *Identification of the existence of different pathways*. Formulation of the reaction stoichiometry, which is the basis for metabolic flux analysis, requires detailed information of the biochemistry. However, for many microorganisms certain details in the pathway stoichiometries are unknown, and it may not be known whether a given pathway is active. Furthermore, in many cells there are several isoenzymes whose functions are not known in complete details. By calculating the metabolic fluxes with a different set of cellular pathways it may be possible to identify which pathway is most likely to exist or to obtain indications of the functions of different isoenzymes and/or pathways. Thus, in analysis of anaerobic growth of *S. cerevisiae,* we found that alcohol dehydrogenase III, a mitochondrial enzyme whose physiological function has not been identified, plays an important role in maintaining the redox level inside the mitochondria. However, this type of analysis should always be followed by enzyme assays in which the presence of enzyme activity in cell extracts is tested.
3. *Calculation of nonmeasured extracellular fluxes*. Normally the number of fluxes that can be measured is larger than what is needed for calculation of the intracellular fluxes. In this case it is possible to calculate some of the extracellular fluxes, such as the rate of production of various by-products, by use of the stoichiometric model and the measured fluxes.
4. *Examination of the influence of alternative pathways on the distribution of fluxes*. In connection with optimization of metabolite production it may be possible to identify one or several constraints

for increasing the yield of a metabolite on the substrate or the flux leading to the desired metabolite. In these cases it is possible to envision various scenarios, such as examining whether insertion of a new pathway or an isoenzyme (or perhaps deletion of an isoenzyme) can have a positive effect on removing this constraint, leading to an increased flux toward the desired metabolite. In a study of penicillin production we calculated that the yield of penicillin on glucose is likely to be higher if cysteine, one of the precursors for penicillin production, is synthesized by direct sulfhydrylation rather than by the trans-sulfuration pathway claimed to exist in *P. chrysogenum*.

5. *Calculation of maximum theoretical yields*. Based on a stoichiometric model it is normally possible to calculate the maximum theoretical yield of a given metabolite if a given set of constraints is specified. This value is of interest because it specifies an upper limit for the yield in the process. This has been illustrated both for lysine production by *C. glutamicum* and for penicillin production by *P. chrysogenum*. In these calculations it is necessary to introduce a constraint, for example, that reactions do not run in thermodynamically unfavorable directions.

Regulatory mechanism

One of the most important aspects of quantitative physiology is control of flux. As discussed, the concept of metabolic flux analysis is useful for studies of interactions between different pathways and for quantification of flux distributions around branch points, but it does not allow for evaluation of how the fluxes are controlled (i.e., how the rates of synthesis and conversion of metabolites are kept in close balance over a very wide range of external conditions without catastrophic rises or falls in the metabolite concentrations).

The discoveries in the 1950s of feedback inhibition, co-operativity, covalent modification of enzymes, and control of enzyme synthesis introduced a number of molecular effects that may play a role in control of fluxes. With these many different mechanisms it is not surprising that disputes often arises over how the flux through a given pathway is controlled. Since the many reports on enzyme regulation typically are verbal and qualitative (e.g., phosphofructokinase is the major flux-controlling enzyme of glycolysis in muscle), it is difficult to discriminate between different findings, and in many cases different findings may seem conflicting. Furthermore, one often finds terms such as *rate-limiting steps* and *bottleneck enzyme* when control of flux is discussed. Thus, the early findings that it is the first enzyme in a pathway or the first enzyme after a branch that is under some type of control (e.g., by feedback inhibition) often results in statements such as "the first step in a pathway is the rate-limiting step."

The concept of metabolic control analysis (MCA) tells us that these kinds of qualitative statements have no meaning, since flux control is distributed over all the steps in a pathway, but that some steps may exert a higher degree of flux control than others. The concept of MCA was developed from the landmark papers of Kacser and Burns and Heinrich and Rapoport. Its basis is a set of parameters, called elasticity coefficients and control coefficients, that quantify the control within a reaction network. The elasticity coefficients are given by

$$\in_{x_j}^{i} = \frac{X_j}{\nu_i} \frac{\partial \nu_i}{\partial X_j} \qquad \ldots(31)$$

and they express the sensitivity of the rates of the enzyme- catalyzed reactions (v_i) toward the metabolite concentrations (X_j). The most often used control coefficients are the flux control coefficients (FCCs), which are given by

$$C_i^{J_j} = \frac{\nu_i}{J_j} \frac{\partial J_i}{\partial \nu_i} \qquad \text{...(32)}$$

The FCCs express the fractional change in the steady-state flux through the pathway (J_j) that results from an infinitesimal change in the activity of the enzymes (or reaction rates). The FCCs and the elasticity coefficients are related to each other via the summation theorem, which states that the sum of all the FCCs is 1, and the connectivity theorem, which states that the sum of the product of the elasticity coefficients and the FCCs is zero.

Three different groups of experimental methods are available for determination of the FCCs:

1. Direct methods, where the control coefficients are determined directly
2. Indirect methods, where the elasticity coefficients are determined and the control coefficients are calculated from the theorems of MCA
3. Transient metabolite measurements, where the metabolite concentrations are measured during a transient and this information is used to determine the control coefficients

Whereas the elasticity coefficients are properties of the individual enzymes, the FCCs are properties of the system. The FCCs are therefore not fixed but change with the environmental conditions. This has been illustrated in analysis of the penicillin biosynthetic pathway. Based on a kinetic model for the enzymes in this pathway, the FCCs were calculated at different stages of fed-batch fermentation, and a drastic shift in the flux control was found. During the first part of the cultivation the flux control was mainly at the first step in the pathway, the formation of the tripeptide LLD-ACV by ACV synthetase (ACVS). Later in the cultivation flux control shifted to the second step in the pathway, the conversion of LLD-ACV to isopenicillin N by isopenicillin N synthetase (IPNS). This shift in flux control is due to intracellular accumulation of LLD-ACV, which is an inhibitor of ACVS. Obviously, it makes no sense to talk about a rate-limiting step or a bottleneck enzyme in this process. Besides the shift in flux control, it is interesting to note that most of the flux control is by the first two steps in the pathway. Furthermore, through analysis of the kinetic model it was found that the value of the FCCs depends on the dissolved oxygen concentration, which is a substrate in the IPNS catalyzed reaction.

CONCLUDING REMARKS

Our intent has been to present a balanced view of stirred-tank fermenter design and construction for microbial and mycelial organisms. There are, however, several important aspects that space and scope considerations prevent us from considering in the depth they warrant. In particular, we would like to have included more information related to overall process considerations, instrumentation, and control; however, we believe that these are addressed elsewhere in this volume. We trust that we have conveyed the sense that fermenter design is very much an art and that we have convinced you to consider each case separately.

9

COMMON MOLD (ASPERGILLUS)

There are about 1.5 million species of fungi on earth, of which only approximately 100,000 have been described. Of these described species, the approximately 200 taxa that comprise the genus *Aspergillus* are among the most commonly encountered molds. Adapted to many habitats, exhibiting a wide range of metabolic activities, and capable of the production of astronomical numbers of spores, *Aspergillus* species are found throughout the biosphere. A few species are economically important as biotechnological resources, whereas others have gained notoriety for their negative impact on human health and commerce. For example, some aspergilli are used in industrial fermentations to produce enzymes, pharmaceuticals, and bulk chemicals; others are important components of Asian food fermentations; still others are major agents of biodeterioration; and a few common species cause allergenic, pathogenic, and toxigenic diseases in humans and other animals.

SYSTEMATIC POSITION

The filamentous fungi that comprise the genus *Aspergillus* reproduce mitotically through the production of spores borne on a specialized structure called the conidiophore. The morphology of this spore-bearing structure, with its distinctive foot cell, long stalk, swollen apex, and protuberant phialides, is the defining feature of the genus. It was first described in 1729 by Micheli, an Italian priest, who was reminded of a Roman Catholic device used to sprinkle holy water, the aspergillum. All members of the genus *Aspergillus* produce an aspergillum-like conidiophore. The conidiophore wall is usually thicker and more hydrophobic than that of the other hyphal cells. The stipe of the conidiophore attaches to the vegetative mycelium in a T- or L-shaped base that has been traditionally called a foot cell, even though it is not a separate cell. The presence of the foot cell is an important diagnostic criterion of the genus. The apex of the stipe swells into the vesicle. Spore-producing phialides are borne directly (uniseriate) or indirectly on metulae (biseriate) on the surface of the vesicle, another important characteristic for distinguishing species within the genus. The phialides are sporogenous. Repeated mitotic divisions in the phialide nucleus lead to the formation of a chain of uninucleate or multinucleate conidiospores, also called conidia. The phialides also extrude primary wall layers and pigments that form part of the mature spore. Conidiospores are spherical or elliptical, smooth or echinulate, extremely hydrophobic, and easily airborne when mature.

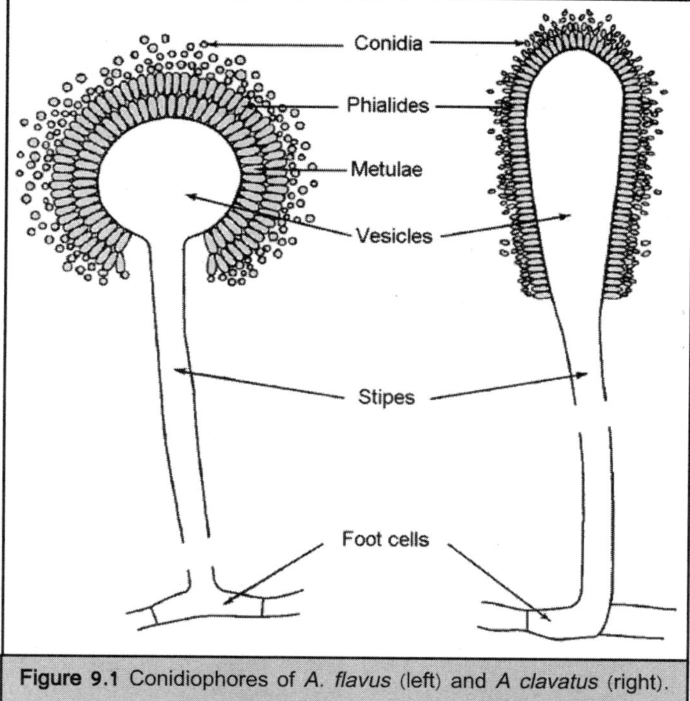

Figure 9.1 Conidiophores of *A. flavus* (left) and *A clavatus* (right).

Charles Thom (1872–1956), most famous for his work on penicillin and the other great mold genus, *Penicillium*, was also a major figure in the development of *Aspergillus* taxonomy. With Margaret Church in 1926 and Kenneth Raper in 1945 he produced the first two complete monographic treatments. Kenneth Raper went on to expand this work and with Dorothy Fennell published *The Genus Aspergillus* in 1965, still considered by many to be the most authoritative source of *Aspergillus* taxonomy. Highly sensitive to substrate, temperature, moisture level, and other environmental variables, potential morphological plasticity is controlled in modern taxonomic schemes by the use of standard media and culture conditions. Pure cultures are grown on standardized agar media for specific lengths of time with defined temperature, light, and other environmental parameters.

Species within *Aspergillus* are distinguished by cultural characters such as colony color, growth rate, and texture and by microscopic traits such as phialides and metulae, shape of the vesicle, spore size, and spore ornamentation. The ability to produce certain pigments, sclerotia, or cleistothecia is highly dependent on culture conditions, but can also be useful in distinguishing species. Modern taxonomists place increasing emphasis on secondary metabolite profiles, isoenzyme electrophoresis, and molecular genetic characters such as ribosomal DNA sequences to supplement the traditional characters. After adequate characterization, identification of species depends on careful comparison with published descriptions. The last monograph of the entire genus was Raper and Fennell's *The Genus Aspergillus*, but several more recent books provide identification systems for the more common species. Medically important species are described in Kwon-Chung and Bennett. Some aspergilli possess a sexual stage.

All the sexual stages result in the formation of ascospores inside of cleistothecia and are classified formally in the Ascomycete family Trichocomaceae. According to the rules of botanical nomenclature, which govern the naming of all fungi, the sexual (teleomorphic or perfect) state requires its own name. Because *Aspergillus* by definition refers only to the asexual (anamorphic or imperfect) state, many species within the genus have two names. For example, the asexual species *Aspergillus tetrazonus* produces a teleomorphic state called *Emericella quadrilineata*, and the teleomorph of the well-known genetic model *Aspergillus nidulans* is called *Emericella nidulans*. This dual nomenclature mystifies workers not trained in the rules of botanical nomenclature.

Table 9.1 Glossary of *Aspergillus* systematic position

Term	*Definition*
Anamorph	The nonsexual form of a fungus, loosely synonymous with *imperfect*
Ascus	A baglike sexual structure containing ascospores
Cleistothecium	A closed ascus-containing structure; the type of ascocarp (ascoma) found in the sexual stages of *Aspergillus*
Conidium (*pl.* conidia)	A thin-walled asexual spore borne on a conidiophore (also called *conidiospore*)
Foot cell	Lower portion of conidiophore, not a separate cell
Heterokaryon	A mycelium containing genetically dissimilar nuclei
Metullae	Sterile cells subtending the phialides in biseriate species; also called *secondary sterigmata*
Phialides	Conidiogenous cells; also called *primary* sterigmata
Sclerotium	Multicellular structure composed of hardened, thick-walled hyphae containing no spores inside or outside
Stipe	Stalk of the conidiophore
Teleomorph	The sexual stage of the fungal life cycle, loosely synonymous with *perfect stage*
Vesicle	The swollen apex of the conidiophore

Moreover, Thom, Church, Raper, and Fennell all resisted these rules and applied the generic name *Aspergillus* to both the conidial and ascosporic stage of these organisms. Much common practice is also blind to the rules of botanical nomenclature. Revisions of the Botanical Code in 1981 have added another layer of complexity by endangering many well- known specific epithets. Strict constructionists renamed the asexual state of *A. nidulans* as *Aspergillus nidulellus*. Rules of priority also endanger the name *Aspergillus niger*, one of the most widely used industrial species. Although the change to *A. nidulellus* is not followed, and although most workers continue to use the name *A. niger*, it demonstrates the potential for nomenclatural confusion in both the traditional literature and with on-line data retrieval. For a more detailed description of the nomenclatural conundrum; the contemporary rules of taxonomy are outlined in Samson and Pitt. A list of the Ascomycete genera (e.g., *Emericella, Eurotium*) with *Aspergillus* anamorphs are presented in Samson. Additionally, a list of *Aspergillus* species names in current use has been published in an attempt to help stabilize nomenclature in the genus.

BIOPHYSIOLOGICAL DIVERSITY

Members of the genus *Aspergillus* are found almost everywhere. They are truly ubiquitous and have been isolated from penguin dung in the Antarctic, binocular lenses in the tropics, and spoiled foodstuffs all over the planet. Aspergilli can utilize an incredible variety of substrates, including plant debris, foods and feeds, fabrics, feathers, leather, and dung. Aspergilli have been associated with the biodeterioration of kerosene, paper, pesticides, plasticisers, and rubber and have been isolated from unlikely sources that range from biblical scrolls to sauna bath boards. They often grow on the surface of humid walls in cellars and stables, and their spores are found in high concentrations in attics, crawl spaces, air conditioning ducts, house dust, and insulation. Although the majority of species are saprophytic, the genus also includes opportunistic pathogens of mammals, insects, and plants. The major natural habitats for aspergilli are soil and decaying vegetation.

Water and temperature are the major factors influencing mycelial growth. Within the soil environment, aspergilli occur most frequently in subtropical/warm temperate zones; however, certain species such as *Aspergillus niveus, A. nidulans,* and *Aspergillus tamarii* are more common in tropical zones. Most aspergilli prefer moist habitats, although several species are xerophilic. Whatever the optimal conditions for growth, all species of *Aspergillus* form abundant conidiospores that are resistant to harsh environments. These propagules bridge survival until hospitable conditions for germination prevail. *Aspergillus* colonies are visible macroscopically as black, green, yellow, or brown mold. The colony color reflects the spore color of the particular species. Pigment production is also influenced by trace elements. In fact, the color of *A. niger* spores was once used as a bioassay to detect copper. As with all filamentous fungi, the hyphae grow by apical extension, allowing for rapid exploitation of new habitats and facilitating the colonization of solid and semisolid substrates.

The ramifying hyphal filaments are profusely branched and have a high surface-to-volume ratio in intimate contact with their substrate. As mycelia colonize, they extrude enzymes that break down complex carbon sources such as cellulose, hemicellulose, pectin, and starch, not only changing their substrate, but literally becoming part of it. Simultaneously, mycelia influence the growth of other organisms, especially bacteria, and create complex degradative consortia. Because of their robust physiology, aspergilli are easy to grow in the laboratory and can be cultured with simple carbon and nitrogen sources, mineral salts, water, and oxygen. Mutations blocked in various catabolic and anabolic pathways are also easily isolated. Biochemical and molecular genetic analyses in *A. nidulans* have revealed considerable detail about the physiology and genetics of these processes. In addition, the weedy nature of these molds means they often cause trouble as common contaminants in tissue culture and bacteriology laboratories.

HEREDITY

Aspergillus nidulans is an excellent model organism for heredity. If the work by Raper and Fennell is the bible of *Aspergillus* taxonomy, then the paper of Pontecorvo et al. is often referred to as the old testament. This paper "contained not only the foundations of *Aspergillus* heredity, but most of the superstructure as well". The parasexual cycle was first elucidated in *A. nidulans*. The first step of the parasexual cycle is the production of heterokaryons, a mycelium with genetically dissimilar nuclei, usually forced between different auxotrophic mutants. Somatic diploids are formed that are identified by their ability to grow on minimal media.

Parasexual recombination occurs through mitotic crossing over and vegetative haploidization. After the parasexual cycle was elucidated in *A. nidulans*, it was found in a large number of other fungi, allowing formal genetic analysis in anamorphic species. Because most species of *Aspergillus* do not form a sexual stage, these artificially produced diploids provide "an alternative to sex." For example, early genetic studies on the aflatoxin-producing species *Aspergillus flavus* and *Aspergillus parasiticus*, both of which lack a sexual stage, were conducted by means of the parasexual cycle. Even more importantly, parasexual cycle genetics was adopted to mammalian tissue culture, where it was usually referred to as somatic cell genetics. For several decades before the recombinant DNA revolution, somatic cell genetics was the major method for mapping human genes. *Aspergillus nidulans* remains an important genetic model in its own right. The steps of conidiation display morphological phenotypes amenable to genetic dissection and have been exploited as a model for developmental biology through molecular analysis of successive stages of the defined developmental sequence.

The ability of the species to degrade all kinds of natural materials and to utilize a wide range of inorganic nitrogen sources has also been exploited for the genetic dissection of carbon metabolism, alcohol metabolism, purine degradation, proline utilization, sulfur metabolism, and many other metabolic pathways. The genetic regulation of nitrogen metabolite repression and carbon catabolite repression, both tightly controlled by induction of appropriate enzymes, has also received considerable attention. In addition, *A. nidulans* has been a valuable model for the study of mitochondrial and population genetics, for penicillin biosynthesis, and for the molecular biology of gene expression. The formal genetics of *Aspergillus* has been reviewed by Smith and Pateman and Martinelli and Kinghorn. The small size of fungal chromosomes, long a disadvantage for cytogenetic analysis, has become an advantage for doing electrophoretic karyotyping by means of pulsed gel electrophoresis.

Electrophoretic karyotypes allow workers to bypass conventional genetic and cytological maps. Any DNA sequence for which a probe is available can be mapped to a chromosome by Southern analysis. The fungal genomics community recently selected *A. nidulans* as a premier target for genomic analysis. The genetic versatility and well-elucidated genetic systems have made *Aspergillus* a suitable genus for industrial application of the latest recombinant DNA techniques to large-scale fermentation processes. These industrial processes are discussed below.

COMMERCIAL VALUES

Aspergilli have been used since prehistoric times as components of several Asian food fermentations. Members of the genus have exceptional protein- and metabolite-secreting capacity, and these physiological activities have been adapted to human enterprise. From a contemporary perspective, the most important economic impact is the manufacture of fungal products such as enzymes, organic acids, and bioactive secondary metabolites. Modern fermenters are sterilizable and operate under aseptic conditions. They are frequently controlled by computers that monitor temperature, agitation, pH, dissolved oxygen, and other parameters.

In general, wild-type strains do not work well in industrial fermentations. Yield of the product is low, and the strain usually is not adapted to the conditions of large-scale fermentation. Genetic strain improvement has traditionally been conducted by means of brute strength mutagenesis and screening. Although repeated rounds of mutagenesis and screening are generally successful, they are also laborious. Recombinant DNA techniques have revolutionized strain improvement programs. *Aspergillus* gene expression systems out perform and are cheaper and simpler than analogous animal or insect systems.

Citric Acid

Citric acid is the most important acid produced by fungi and the only organic acid produced almost exclusively through fermentation. Widely used in the food and beverage industry, its pleasant acid taste enhances products such as soft drinks, fruit juices, jams, jellies, candies, prepared desserts, and frozen fruits. Citrates are also efficient buffering and chelating agents and are used by the cosmetics industry and in blood transfusion products, effervescent tablets, detergent manufacturing, electroplating, printing, inks, leather tanning, and a host of other applications. With increasing emphasis on nonpolluting chemical products, the market for nature's acidulant is increasing. Many molds and yeasts accumulate large amounts of citric acid from glucose. The biochemical basis of this metabolic overproduction has been studied for a century.

Table 9.2 Economically useful low molecular weight products of *Aspergillus*

Product	*Producing species*	*Use*
Organic Acids		
Citric acid	*A. niger*	Flavoring agency, antioxidant, chelating and cleaning agent, component of effervescent powders
Gluconic acid	*A. niger*	Manufacturing of toothphaste
Itaconic acid	*A. terreus*	Copolymer in resin manufacture
Pharmacologically active secondary metabolites		
Asperillomarasmine	*A. oryzae*	Angiotensin-converting enzyme inhibitor
Asperlicins	*A. alliaceus*	Cholecystokinin antagonists
Lovastatin	*A. terreus*	Cholesterol inhibitor
Echinocandin	*A. nidulans/A. rugulovalvus*	Antifungal agent

The first commercial fermentation was perfected by Currie at Pfizer and Company in Brooklyn, New York, during the 1920s and involved surface fermentations of *A. niger*. This species is still used in many industrial processes worldwide, although submerged fermentations have generally replaced surface and solid-state systems. Production strains are selected not only for their high yield of citric acid, but also for their ability to use cheap raw materials as substrates. Optimal yields of citric acid are associated with low pH (below 2.5), high carbohydrate, and manganese deficiency. Although a great deal is known about the environmental cues involved in maximizing citric acid production, the exact biochemical basis and the crucial regulatory control points are the subject of continuing research. Mycelium is removed by filtering or centrifugation; citric acid is then precipitated. Subsequent recovery steps can include treatment with activated carbon, cation and anion exchanges, and final crystallization. The whole process is highly optimized, with *A. niger* capable of producing about 90% of the theoretical yield from a carbon source.

Extremely Diverse Metabolites

The families of low molecular weight compounds that have no obvious role in the life cycle of producing species, often produced after active growth has ceased, are called *secondary metabolites*. Both biologically and chemically, secondary metabolites are extremely diverse; however, they arise from just a few biosynthetic pathways based in primary metabolism. Acetate-derived polyketides and amino acid–derived compounds are the most important biosynthetic routes in filamentous fungi, including aspergilli. Members of the genus *Aspergillus* are prolific producers of these natural products, and the structures of a large number of secondary metabolites from *Aspergillus* have been elucidated by chemists. In general, the most famous natural products are bioactive and are classified on the basis of this activity as antibiotics, chemotherapeutic agents, toxins, and so forth. The single best-known fungal secondary metabolite is the antibiotic penicillin.

Although several species of *Aspergillus* make penicillin, and although *A. nidulans* has been developed as a model system for studying its biosynthesis, high-yielding industrial fermentations all use strains of *Penicillium*. Economically speaking, toxic secondary metabolites of *Aspergillus* are more important than the medically useful ones. Aflatoxins and sterigmatocystins, for example, cause millions of dollars of damage in agriculture each year. Currently, the most important pharmacological agent is lovastatin (mevinolin) a potent inhibitor of cholesterol synthesis, derived from *Aspergillus terreus*. Lovastatin is a polyketide. Several amino acid–derived metabolites also show pharmacological activity. For example, echinocandin B (cilofungin), produced by *A. nidulans* and *Aspergillus rugulovalvus* (also called *E. rugulosis*), is under development as an antifungal agent.

Microbial Fermentation

Foods fermented with microorganisms have interesting flavors and improved digestibility. In Western culture, the majority of food fermentations involve yeasts (e.g., bread and wine) or bacteria (e.g., cheese and pickles). Aspergilli play only minor roles in Western cuisine (e.g., ripening salamis and country-cured hams). This is in sharp contrast to Asian food traditions, where filamentous fungi have been used for millennia, especially in complex two-stage fermentations. The best known of these processes include shoyu (soy sauce), sake (rice wine), and miso (fermented soybean paste). The initial step in each case is the production of a *koji*, a Japanese word that roughly translates as "bloom of mold." To produce a koji, *Aspergillus oryzae* or the closely related species *A. sojae* is grown on steamed rice or other cereal and incubated in warm, humid conditions. The hyphae infiltrate the grain, secreting hydrolytic enzymes that partially degrade the substrate, producing the koji, a fragrant, crumbly mass of mold and substrate. Perhaps because of the paucity of Western fungal fermentations, the English language does not have a word that similarly denotes a substance thus composed of living mycelia, extracellular enzymes, and partially degraded substrate. This mold–substrate mixture has different properties than either alone and constitutes a form of matter not captured by the term solid-state fermentation. Kojis are classified according to the degree of mycelial development and sporulation, their intended use, and the cereal substrate.

Kojis are usually used as inocula for further fermentations. For soy sauce, the koji is mixed with soybeans and wheat. After fungal amylases and proteases have partially degraded the mixture, a salt solution is added. An anaerobic yeast–lactobacilli fermentation ensues; the final pressed liquid is clarified as soy sauce. For sake, a koji inoculum is mixed with steamed rice and water and incubated for 3 to 4 days in the moto stage. As the fungal enzymes partially saccharify the rice, a predictable change in the microbial flora takes place, with yeasts becoming the predominant microorganisms, to yield the moroni stage. After appropriate incubation, yeast metabolic activities yield alcohol. Miso is a fermented paste with the consistency of peanut butter, often made in Japanese homes. Widely used as a soup base, miso varies according to substrate (rice, barley, soybeans, mixtures), length of fermentation, amount of salt, etc. Similar products, based solely on fermented soybeans, are known as *chiang* in China, jang in Korea, *tao-jo* in Indonesian and Thailand, and *tao-tsi* in the Philippines. *Aspergillus oryzae* and *A. sojae* are close relatives of the toxigenic species *A. flavus* and *A. parasiticus*. These domesticated species do not produce toxins; *A. oryzae* is on the GRAS list by the U.S. Food and Drug Administration. The first industrial enzyme patented in the United States was an amylase mixture purified from a koji by the early Japanese-American biotechnologist Jokichi Takamine in 1894. Contemporary scientists have subsequently identified more than 50 different enzymes from koji.

Biological Catalysts

Enzymes are *biological catalysts*. Long before they were purified by scientists, their properties had been used in baking, brewing, tanning, and other processes. Because of their high specificity, their biodegradability, and their ability to work in watery solutions at moderate temperatures, enzymes are becoming increasingly important in today's environmentally conscious industries. In many processes, enzymes have replaced the strong chemicals and high temperatures formerly used in commercial catalysis. The major industrial enzymes from fungi are hydrolytic, and the major producing species are *A. oryzae* and *A. niger*. Amylases and glucoamylases are used for turning starches into sugars and oligosaccharides.

Pectinases are applied in fruit juice clarification. *Aspergillus* proteases are used in bread making and in chill-proofing beer, and a thermostable phytase is used widely in the animal feed industry as an additive. With the advent of modern molecular techniques, the genes for many native *Aspergillus* enzymes have been cloned, sequenced, amplified, and engineered to improve yields, stability, and other properties. Genetic manipulation also has allowed *Aspergillus* to become the host for the production of various nonnative (foreign) proteins, the best known of which is chymosin. Rennet is a substance extracted from the stomachs of slaughtered calves that is used in cheese making. The principle active ingredient of rennet is chymosin, a protease that coagulates milk. Structural genes for the calf stomach enzyme can be transferred to fungi. The mammalian enzyme is then expressed and secreted by the fungus in fermentation broths. The yeast *Kluyveromyces lactis* and *A. niger* have both been developed as production hosts for chymosin production. Given the economic advantages of fungal fermentations, it can be predicted that *Aspergillus* will be the production host of choice for many other nonfungal enzymes that traditionally have been extracted from higher plants and animals.

Table 9.3 Commercially important *Aspergillus* enzymes

Enzyme	*Applications*
Amylases	Hydrolysis of starch; bread and beer production; removal of sizing from fabrics; high-fructose syrups
Invertase	Confections; soft center in chocolates
Lactase	Hydrolysis of lactose; production of syrups for sweetening agents
Pectinases	Pretreatment of fruit juices to remove turbidity, reduction cloudiness in wines
Phytase	Animal feed additive for liberation of phosphate from plant material
Proteases	Meat tenderizing; removal of bitter flavors and chill proofing of beer; reduction elasticity of gluten proteins in bread

Descriptions of the use of enzymes in fermentation, diagnostics, baking, brewing, cheese manufacturing, chemical biotransformation, detergents, effluent and waste treatment, leather, textile, starch conversion, and the like are presented in Godfrey and West's *Industrial Enzymology*. This work also gives detailed information on the legislation and regulation of commercial enzymes, toxicological and regulatory considerations, and the scientific and regulatory aspects surrounding the use of genetically engineered production strains. The long history of safe use and the GRAS status of *A. oryzae* and *A. niger* will ensure that they will remain the species of choice for industrial enzyme production, especially in the food industry.

DISORDERS

Although *Aspergillus* species are very common in the human environment, only a limited number of taxa (approximately 17) are capable of initiating human disease. The usual reservoir for an *Aspergillus* infection is not another infected person but, rather, a nonliving site where the fungus is growing saprophytically. Air conditioning vents, attic insulation, cellar walls, stables, and even house-plants in hospitals have been implicated in various disease states. Because *Aspergillus* conidia are among the most common of airborne molds, the diseases they cause are usually pulmonary. The single species most commonly associated with human disease is *Aspergillus fumigatus*. In addition, *A. flavus, A. nidulans, A. niger,* and *A. terreus* are regularly implicated.

The term *aspergillosis* has been broadly used to describe any disease caused by an *Aspergillus* species, excluding mycotoxicoses. Because this term encompasses so many different clinical entities, with different researchers applying different emphases, and because some conditions defy simple categorization, the medical mycology literature can be confusing. The classification of Kwon-Chung and Bennett and Dixon and Walsh recognizes three broad categories of aspergillosis: (i) allergy, (ii) saprophytic colonization, and (iii) invasive mycosis.

Allergens

Like other fungal spores, *Aspergillus* conidia are potent allergens, especially associated with respiratory allergies. Spore inhalation may cause hypersensitivity reactions in susceptible individuals. For example, allergic bronchopulmonary aspergillosis occurs in patients with preexisting asthma who develop bronchial sensitivity reactions to inhaled conidia.

Occupational exposure is a risk factor. Industrial-scale fermentations can cause sensitization of workers to high concentrations of airborne spores or to culture fluid materials, resulting in asthma or other adverse health effects. Malt worker's lung is a hypersensitivity pneumonitis caused by *Aspergillus clavatus*. Some medical mycologists classify respiratory allergies as extrinsic (inhalation of conidia) or intrinsic (*Aspergillus* growing in air-ways). Intrinsic conditions are an example of saprophytic growth.

Fungus Ball

In contrast to extrinsic respiratory allergies caused by spore inhalation, intrinsic aspergillosis is a localized hyphal growth associated with colonization of air spaces. Aspergilloma, or fungus ball, is formed when *Aspergillus* grows in a preexisting lung cavity. Fungus balls have also been reported to occur in the paranasal sinus.

Pathogenesis

Sometimes *Aspergillus* species are opportunistic pathogens causing frank mycosis. These mycoses are collectively called invasive aspergillosis and are usually classified by the portal of entry. Invasive pulmonary aspergillosis is typically manifested as acute pneumonia. Nonpulmonary portals of entry include the eye, paranasal sinus, or gastrointestinal tract or may be associated with vascular surgery and drug addiction. Invasive aspergillosis is largely a disease of a compromised host. Those at high risk include patients with diabetes, drug addicts, chronically ill patients, transplant patients receiving immunosuppressive drugs, and cancer patients receiving chemotherapy. Occupational exposure is also a factor.

Historically, the earliest reports of aspergillosis were among wig cleaners and pigeon fanciers. *Aspergillus fumigatus* and *A.flavus* are the most common species associated with invasive and noninvasive aspergillosis. It is hypothesized that these species possess special virulence factors that facilitate their role as opportunistic pathogens.

Invasive disease is treated with antifungal chemotherapy, but existing antimycotic drugs have toxic side effects, and the therapeutic outcome is often poor. As the use of immunosuppressive drugs increases in modern medicine, invasive aspergillosis is becoming more common. There is an important need for effective systemic antifungal drugs. Several pharmaceutical companies have active drug discovery programs in this area.

Toxins of Filamentous Fungi

Filamentous fungal metabolites that cause human or veterinary diseases are called mycotoxins. The mycotoxins that have received the most attention are the aflatoxins, a family of highly toxic, mutagenic, and carcinogenic compounds produced by certain strains of *A. flavus, Aspergillus nomius,* and *A. parasiticus*. These species regularly contaminate peanuts, corn, cottonseed, rice, tree nuts, and other agricultural commodities. A large body of epidemiological and molecular biological data implicates aflatoxin with human liver cancer. Acute human aflatoxicosis, however, is rare and usually only occurs when famine or poverty forces people to subsist on moldy food-stuffs. On the other hand, veterinary aflatoxicosis is a major problem, with poultry and trout being particularly susceptible. Other mycotoxins produced by members of the genus *Aspergillus* tend to be produced at toxicologically less important levels. These include cyclopiazonic acid (*A. flavus*), patulin (*A. clavatus*), and sterigmatocystin (*A. nidulans* and *Aspergillus versicolor*). The toxigenic potential of *Aspergillus ochraceus* is particularly varied, with different strains reported to make ochratoxin, citrinin, xanthomegnin, viomellein, and penicillic acid.

Like aflatoxin, patulin and sterigmatocystin are carcinogenic; citrinin and ochratoxin A are nephroxic; cyclopiazonic acid is a neurotoxin; and xanthomegnin and viomellein have been associated with photosensitization and liver damage. Many of the compounds classified as cytotoxic or neurotoxic in vertebrates also show toxicity against insects. Sclerotia have been a particularly rich source of metabolites with insecticidal properties. Mycotoxin contamination varies from year to year based on weather, agricultural practices, storage conditions, and other factors. Aflatoxin is usually worst in drought years. International agencies have set levels for permissible standards of contamination. Among the most commonplace and versatile of all fungi, the aspergilli play an important role in natural degradative cycles. Although a few species have gained notoriety as animal pathogens and mycotoxin producers, the vast majority are benign. In processes ranging from traditional Asian food preparation to massive contemporary industrial fermentations, the capacity of aspergilli to secrete enzymes and metabolites is exploited for human benefit. In basic science, *A. nidulans* is a model system for genomics and molecular biology as well as classical genetics. The increasing societal emphasis on environmentally safe processes and commodities guarantees that the metabolic capabilities of this genus will continue to be used in modern industry.

10

ENZYMATIC REACTIONS

Enzymatic reaction has increasingly been applied to organic synthesis, and there are several reports of the attempted in vitro synthesis of the D-amino acid–containing peptides and amides. However, because these enzymatic reactions are not stereospecific for substrates with D-configurations, they are at an innate disadvantage. Some peptidases and amidases act on peptides and amides containing D-amino acids. Soluble *Streptomyces* carboxypeptidase DD catalyzes not only the transpeptidation reaction on the peptide intermediate in peptidoglycan biosynthesis but also the hydrolysis of N_{α},N_{ε}-diacetyl-L-lysyl-(D-Ala)$_2$ in water. A D-peptidase has been purified and characterized from an actinomycete, although it is not strictly specific toward peptides containing D-amino acids. In *Enterococcus*, the *vanX* gene product (D-Ala)$_2$ hydrolase plays a role in vancomycin resistance. The chemically synthesized D-enzyme of an HIV-1, in which all the amino acids were replaced with the corresponding D-amino acids, displays D-stereospecificity.

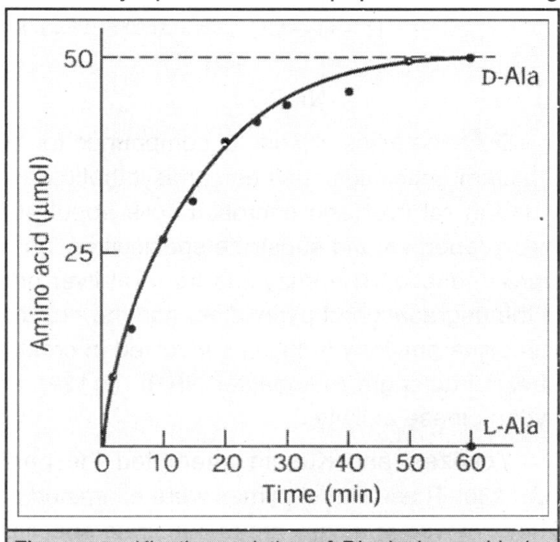

Figure 10.1 Kinetic resolution of DL-alanine amide by *O. anthropi* D-aminopeptidase.

D-Stereospecific amino acid amidases (amidohydrolases) were isolated from some microorganisms. Of the enzymes, a D-alaninamide–specific amidohydrolase from *Arthrobacter* has been used in the industrial manufacture of D- and L-alanine. Production of D-amino acids by use of D- or L-specific enzymes such as D-amino acylase, D-hydantoin hydrolase, *N*-carbamoyl-D-amino acid hydrolase, D-amidase, D-transaminase, and amino acid racemase was recently reviewed. A new enzyme, D-aminopeptidase, was isolated from *Ochrobactrum anthropi* and characterized. We describe its function, structure, and application to the D-stereospecific hydrolysis of racemic amino acid amides and the formation of peptide bonds. A new enzyme, alkaline D-peptidase, acting on a synthetic peptide, (D-Phe)4, was also isolated from *Bacillus cereus*. We describe its structure and function, as well as its application to the synthesis of D-phenylalanine oligomers. We propose that these two enzymes are new members of the group of penicillin-recognizing enzymes.

AMINOHYDROLASES

N-Carbamoyl-D-amino acid amidohydrolase (hereinafter abbreviated as DCase) is the another enzyme that catalyzes the following reaction:

$$\mathrm{R{-}C(H)(NHCONH_2){-}COOH} + H_2O \xrightarrow{\text{Dcase}} \mathrm{R{-}C(H)(NH_2){-}COOH} + CO_2 + NH_3$$

The DCase converts *N*-carbamoyl-D-amino acids to D- amino acids with strict D-form specificity and is expected to be applicable to the industrial D-amino acid production process instead of chemical diazonation, in which *N*-carbamoyl-D-amino acids are converted by DCase after hydrolysis of the corresponding 5-substituted hydantoins by the hydrolyzing enzyme, D-hydantoinase. These reactions can be performed separately as two serial reactions or as one batch reaction with microorganisms possessing both enzymes.

$$\mathrm{R{-}C(H){-}CO{-}NH{-}CO{-}NH}\ (\text{ring}) \xrightarrow{\text{D-Hydantoinase}} \mathrm{R{-}C(H)(NHCONH_2){-}COOH} \xrightarrow{\text{DCase}} \mathrm{R{-}C(H)(NH_2){-}COOH}$$

D-Amino acids are useful compounds for the preparation of physiologically active peptides and β-lactam antibiotics such as semisynthetic penicillins and cephalosporins. DCase activity was first found in rat liver and microbial cells about 40 years ago. These enzymes were examined as to their properties and substrate specificities, and their reaction mechanisms and biological functions were deduced. The enzymes from rat liver and *Clostridium uracilicum* were found to be involved in the degradation of pyrimidine, and the enzyme from *Pseudomonas putida*, which is N-carbamoyl-sarcosine amidohydrolase, is involved in creatinine metabolism. Olivieri et al. found DCase activity of Agrobacterium radiobacter NRRL B11291 in intact cells or a cell-free extract, together with D-hydantoinase activity.

Yokozeki and Kubota attempted the partial purification of DCase from *Pseudomonas* sp. AJ11220. These two enzymes were examined as to their substrate specificities and reaction profiles. They showed relatively broad substrate specificities such as for aliphatic and aromatic compounds, with strict D-form specificity, and almost the same optimal conditions, that is, pH 7 and 55–60°C.

Ogawa et al. screened enzyme producers by means of enrichment cultures with the *N*-carbamoyl-D-amino acids citrulline and ornithine as sole nitrogen or carbon sources. They isolated alkaliphilic and thermotolerant enzymes, which were produced by strains classified as *Comamonas* sp. and *Blastobacter* sp., respectively, purified them to homogeneity, and determined their properties.

These enzymes also showed strict specificity toward N-carbamoyl-D-amino acids and hydrolyzed *N*-carbamoyl-amino acids having a hydrophobic side chain very efficiently; on the other hand, those having a polar group or short-chain alkyl group were only weakly hydrolyzed. However, these enzymes showed that the *N*-carbamoyl compounds involved in the metabolism of pyrimidine and purine are not hydrolyzed at all. On characterization of these enzymes, the reactive molecular weights of the native enzymes were found to be about 120,000 and those of the subunits to be about 40,000. Enzymochemical analysis was also performed. Recently, Louwrier and Knowles purified a DCase from *Agrobacterium* sp. which was composed of genetically engineered self-cloning cells, and characterized it. This enzyme was able to cleave a variety of N-carbamoyl substrates but was strictly D-form specific. The active enzyme was suggested to be present as a dimer with a subunit molecular weight of 38,000 Da, differing from the trimer enzymes of Ogawa et al.

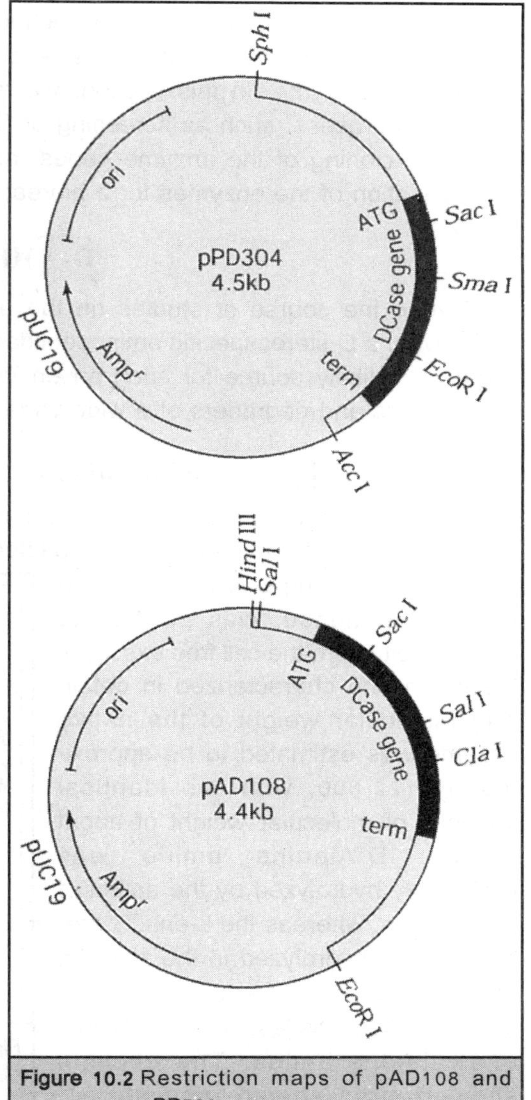

Figure 10.2 Restriction maps of pAD108 and pPD304.

Concerning the enzyme genes, some DCase genes have been cloned and analyzed. The DCase gene from *Agrobacterium radiobacter* NRRL B11291 was cloned, analyzed, and expressed in *Escherichia coli*, and the recombinant enzyme was then characterized. By means of a site-directed mutagenesis experiment, the relationship between activity and amino acid substitutions was examined, and some mutations concerning enzyme stability were found. Neal et al. also isolated the DCase gene of *Agrobacterium* sp. and expressed the gene in *E. coli* and *Agrobacterium* sp. We also screened some strains producing the enzyme from mesophile and thermotolerant strains and cloned two enzyme genes.

We tried to improve the native DCase to obtain a practical DCase that exhibits both high reactivity and sufficient stability for repeated use in a bioreactor system by means of amino acid substitutions using recombinant DNA technology. We succeeded in creating a practical DCase by the substitution of three amino acids, and applied it to an industrial production process as an immobilized enzyme. In this section, we report our attempts to establish a new D-amino acid production process, such as screening of DCase-producing bacteria from soil, purification of the enzymes, cloning of the enzyme genes, mutagenesis to obtain thermostabilized enzymes, and immobilization of the enzymes for a bioreactor system to produce D-amino acids.

D-AMINOPEPTIDASE

During the course of studies on the enzyme-catalyzed organic synthesis of D-amino acid derivatives, a D-stereospecific aminopeptidase was needed. It was speculated that microorganisms might be a likely source for such an enzyme, given their important role in the environment as synthesizers and degraders of a wide variety of substances.

Extraction of the D-Aminopeptidase

An enrichment culture in a medium containing a synthetic substrate (D-Ala-NH2) as the sole nitrogen source led to isolation of a bacterial strain, *Ochrobactrum anthropi* SCRC C1-38. The enzyme hydrolyzing D-Ala-NH_2, named D-aminopeptidase, was purified to homogeneity from the cell-free extract of the strain and characterized in detail. The molecular weight of the native enzyme was estimated to be approximately 122,000, with two identical subunits of molecular weight of about 59,000. D-Alanine amide was completely hydrolyzed by the action of the enzyme, whereas the L-enantiomer remained unhydrolyzed in the reaction mixture. The rate of the hydrolysis of L-alanine amide was less than 0.01% that of D-alanine amide. The enzyme showed strict chemo- and stereo-specificities toward D-amino acid amides, peptides, and esters, as shown in Table 10.1. Each substrate (100 mM) was incubated under the standard enzyme assay condition.

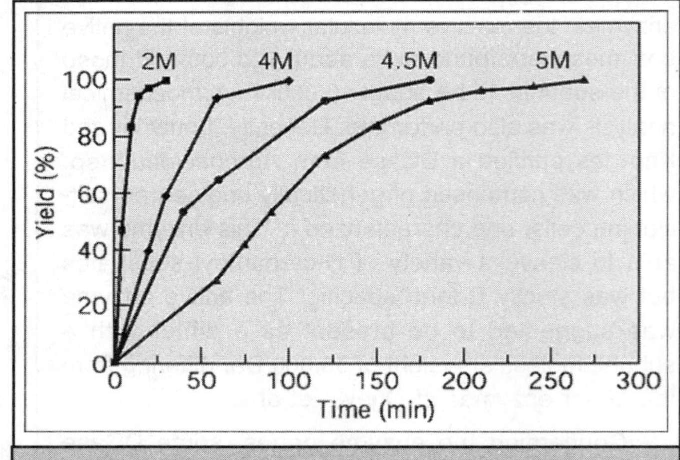

Figure 10.3 Synthesis of D-alanine from amide HCl by the cells of *E. coli* JM109/pC138DP.

The substrates include D-alanine amide (relative velocity: 100%, K_m value: 0.65 mM); glycine amide (44%, 22.3 mM); D-α-aminobutyramide (30%, 18.3 mM); D-serine amide (29%, 27.0 mM); D-alanine 3-aminopentane amide (32%, 2.27 mM); D-alanine-*p*-nitroanilide (96%, 0.51 mM); D-alanine methyl ester (75%), dimer (21%, 10.2 mM), trimer (92%, 0.57 mM), and tetramer (89%, 0.32 mM) of D-alanine; D-alanylglycine (95%, 0.98 mM), D-alanylglycylglycine (45%, 0.37 mM); and D-alanyl-L-alanyl-L-alanine (100%, 0.65 mM). These results show that the enzyme has higher affinity toward peptide substrates than amino acid amides.

Table 10.1 Substrate specificity of D-aminopeptidase from *O. anthropi* SCRC C1-38

Substrate	*Relative activity (%)*	K_m *(mM)*	V_{max} *(U/mg)*
D-Alanine amide	100	0.65	600
Glycine amide	44	22.3	365
D-*a*-Aminobutyric acid amide	30	18.3	576
D-Serine amide	29	27.0	22.0
D-Threonine amide	9	100	60.3
D-Methionine amide	2		
D-Norvaline amide	1.8		
D-Norleucine amide	0.8		
D-Phenylglycine amide	0.7		
D-Alanylglycine	95	0.98	1,000
D-Alanylglycylglycine	45	0.37	799
D-Alanyl-D-alanine	21	10.2	326
D-Alanyl-D-alanyl-D-alanine	92	0.57	866
D-Alanyl-D-alanyl-D-alanyl-D-alanine	89	0.32	702
D-Alanyl-L-alanine	46	1.03	312
D-Alanyl-L-alanyl-L-alanine	100	0.65	730
DL-Alanyl-DL-serine	27		
DL-Alanyl-DL-methionine	20		
DL-Alanyl-DL-phenylalanine	9		
DL-Alanyl-DL-asparagine	7		
DL-Alanyl-DL-leucine	1		
DL-Alanyl-DL-valine	0.5		
Glycine methyl ester	229		
D-Alanine methyl ester	75		
D-Alanine-*b*-naphthylamide	32		
D-Alanine benzylamide	72	0.51	768
D-Alanine anilide	73		
D-Alanine-*p*-nitroanilide	96	0.51	696
D-Alanine *n*-butylamide	66	0.73	670
D-Alanine-3-aminopentane amide	32	2.27	288
D-Alanine *n*-laurylamide	19		
D-Threonine benzyl ester	3.2		

The enzyme showed neither endopeptidase nor carboxypeptidase activity. The mode of action of the enzyme toward a peptide substrate was studied with D-alanylglycylglycine. The reaction was followed over time, and it was observed that alanine and glycylglycine were released until the substrate tripeptide was nearly completely consumed, whereupon glycine release began, and finally alanine and glycine were produced. This result shows that the enzyme catalyzes the hydrolysis of a single amino acid from the N-terminus of a peptide. Because the mode of action is typical of an aminopeptidase, we named the enzyme D-aminopeptidase.

Microbial Synthesis

The D-aminopeptidase from *O. anthropi* SCRC C1-38 was utilized for a stereoselective synthesis of D-alanine *N*-alkylamide from an amine and D-amino acid amide or D-amino acid methyl ester. In water, D-alanine *N*-alkylamide once formed by the enzyme was successively hydrolyzed to yield D-alanine. The enzyme was immobilized by urethane prepolymer PU-6. When the aminolysis reaction was performed in water-saturated organic solvents such as butylacetate, benzene, and 1,1,1-trichloroethane with the immobilized enzyme as a catalyst, it progressed well quantitatively in a highly D-stereoselective manner, to give optically pure D-alanine *N*-alkylamides. The acyl donor specificity of this reaction was rather limited, with D-alanine, glycine, and D-α-amino butyric acid as substrates, while the acyl acceptor specificity was relatively wide, with bulky, aromatic-containing, and straight-chain amines.

Table 10.2 D-Aminopeptidase-catalyzed synthesis of D-Amino acid derivatives in organic solvents

Acyl donor	*Acyl acceptor*	*Yield (%)*
In 1,1,1-trichloroethane		
D-Alanine methyl ester HCl	3-Aminopentane	99
D-Alanine methyl ester HCl	Neopentylamine	98
D-Alanine methyl ester HCl	*n*-Butylamine	66
D-Alanine methyl ester HCl	Benzylamine	15
In butyl acetate		
D-Alanine methyl ester HCl	3-Aminopentane	99
Glycine methyl ester HCl	3-Aminopentane	80
D-α-Aminobutyric acid methyl ester HCl	3-Aminopentane	44
In toluene		
D-Alanine methyl ester HCl	D-Alanine methyl ester HCl	58

The catalytic center activity (k_{cat}) of this reaction, 7,700 (min^{-1}), was much higher than in the case of the nonstereoselective synthesis of D-amino acid–containing peptides. Alkaline conditions are usually essential for peptide bond formation in kinetically controlled systems. Several attempts to oligomerize D-alanine methyl ester with D-aminopeptidase in an alkaline aqueous medium were unsuccessful, probably because the substrate is unstable under alkaline conditions. The immobilized D-aminopeptidase catalyzed the synthesis of D-alanine oligomers from D-alanine methylester in organic solvents. Dimer and trimer of D-alanine were obtained in 58% and 6% yield, respectively, when urethane prepolymer PU-6 immobilized D-aminopeptidase (1.5 U/mL) was incubated in water-saturated toluene with 250 mM of D-alanine methyl ester HCl and 750 mM of triethylamine. The k_{cat} of the reaction was calculated to be 19,500 (min^{-1}), which is several tens of thousands times greater than that of the known enzymatic syntheses of amino acid oligomers. The gene for the D-aminopeptidase was cloned in *Escherichia coli* JM109 to overproduce the enzyme.

An expression plasmid pC138DP was constructed. The amount of the enzyme in a cell-free extract of *E. coli* JM109/ pC138DP was elevated to 288,000 units per liter of culture, which is about 3,600-fold over that of the wild strain. The intact cells of *E. coli* transformant were used as a catalyst for the D-stereospecific hydrolysis of several racemic amino acid amides HCl. Complete hydrolysis of D-alanine amide was achieved in a short time (4.5 h) from 5.0 M of racemic alanine amide HCl using the cells of *E. coli* transformant. The concentration of D-alanine reached up to 220 g/L. The cells or the cell-free extract catalyzed the synthesis of D-α-aminobutyric acid, D-methionine, D-norvaline, and D-norleucine from their amides in a similar manner.

Resemblance between D-Aminopeptidase to Carboxypeptidase DD and β-Lactamases

The nucleotide sequence of the D-aminopeptidase gene was determined and analyzed to study the structural relationship to other proteins. The gene consisted of an open reading frame of 1,560 nucleotides, which specifies a protein of M_r of 57,257. The deduced primary structure is considerably similar to that of *Streptomyces* R61 carboxypeptidase DD (30) (27% identical over 287 amino acids) and class C β-lactamase from *E. coli* (22% identical over 234 amino acids). The enzyme is also structurally related to class A β-lactamases, penicillin-binding proteins, and class D β-lactamases. The sequence Ser-X_{aa}-X_{aa}-Lys is perfectly conserved among this class of enzymes. With carboxypeptidase DD and class C β-lactamase, the sequences at the highly similar area, Ser-Val-X_{aa}-Lys-X_{aa}-Phe-X_{aa}-Ala-X_{aa}-Val-Leu-Leu and Ser-Val-Ser-Lys-X_{aa}-Phe-Tyr, respectively, are well conserved. Furthermore, some of the residues scattered around the conserved region are similar among the enzymes.

The Met-57 located four residues upstream from the Ser-61 is found similarly in the carboxypeptidase DD and β-lactamases as a hydrophobic residue Phe. A common observation regarding D-aminopeptidase and the β-lactamase is that the consensus sequence is located around 60 from the N-terminus of the enzymes. The inhibition by *p*-chloromercuribenzoate (PCMB) and other sulfhydryl reagents suggested that the enzyme would be a thiol peptidase. However, the findings of the site-specific mutagenesis study showed that the amino acid sequences Ser-X_{aa}-X_{aa}-Lys, which had been conserved in the penicillin-recognizing enzymes, are essential in exerting the D-aminopeptidase activity. The sites Ser-61 and Lys-64 are essential in the catalysis, because the V_{max}/K_m value measured with the mutants modified to Cys-61 and Asn-64 have been reduced to 0.26% and 0.008%, respectively, of those of the natural enzyme, although these K_m values were hardly lowered. The mutant with the inert Gly residue in place of the likely active center Ser-61 had a lower activity (10^{-5}-fold that of the native enzyme). On the other hand, the mutations at other sites (Met-57 and Cys-68) did not greatly affect the enzyme activity. The mutations at Cys-60, which is adjacent to the likely active center Ser-61, gave notable effects on the kinetic profiles, indicating that the residue is important in the binding of the substrate. The substitutions at Cys-60 to Ser and Gly produced mutants with slightly altered V_{max}/K_m values. They were tolerant to inhibition by PCMB, which suggests that the inhibition of the native enzyme by PCMB would have been due to the steric hindrance by a mercaptide bond formation between Cys-60 of the enzyme and PCMB. A mutant of D-aminopeptidase with increased thermostability was obtained.

The enhancement of the thermal stability of mutant enzyme was attributed to the substitution of Gly-155 to Ser. Figure 10.4 shows similarities of the reactions catalyzed by D-aminopeptidase,

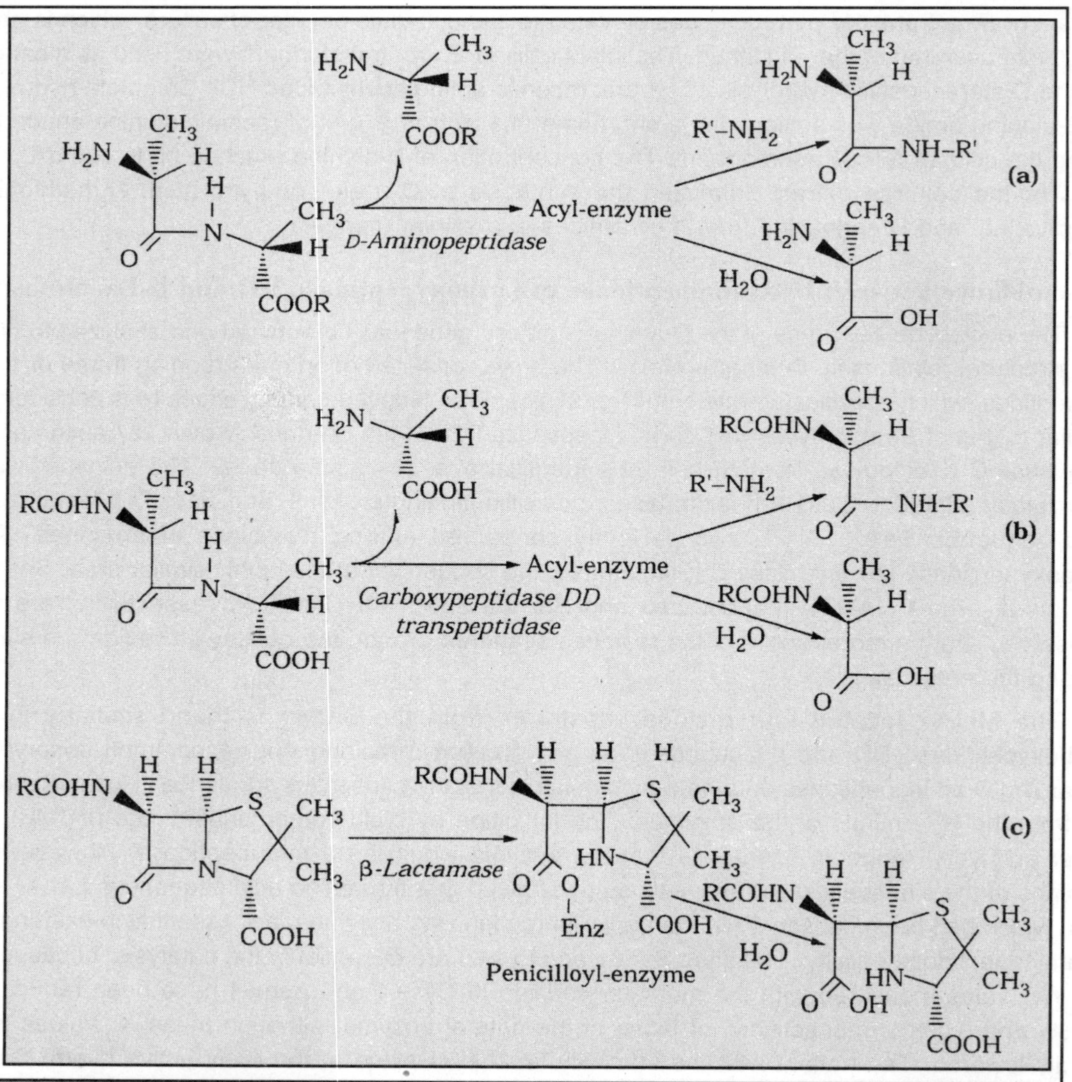

Figure 10.4 Reactions catalyzed by (a) D-aminopeptidase, (b) carboxypeptidase DD, and (c) β-lactamase.

carboxypeptidase DD, and β-lactamase. All three enzymes catalyze the hydrolytic cleavage of the amide bond between D-amino acids of the same configuration. The former two catalyze the transpeptidation reaction; carboxypeptidase DD catalyzes the cross-linkage of the peptideglycan, and D-aminopeptidase catalyzes the transpeptidation and aminolysis of D-amino acid ester and amides, respectively, in water and organic solvents. The aminolysis reaction catalyzed by D-aminopeptidase in butylacetate is efficient (k_{cat}: 7,700/min). The transpeptidation reaction can also be catalyzed in water, although the product D-alanine-3-aminopentane amide was easily hydrolyzed in water. Inheritance of the carboxypeptidase-like tertiary structure in the D-aminopeptidase would have made the transpeptidation reaction in organic solvents possible.

The enzyme is actually inhibited by β-lactam compounds, such as 6-APA, 7-ACA, benzylpenicillin, and ampicillin, although none of these is the substrate for the enzyme. The D-aminopeptidase is a new member of the penicillin-recognizing enzymes by virtue of the following characteristics: similarities in the primary structure by gene sequencing, similarities in the reactions catalyzed in water and organic solvents, and the findings obtained by kinetic studies of the mutants generated by site-directed mutagenesis and the inhibition by β-lactam compounds. The mutants generated by inhibition of β-lactam compounds may constitute evidence to rebut the contention that the β-lactamases evolve from the penicillin-binding proteins. The existence of a third enzyme with a similar structure, which does not appear to be a selective target of the β-lactam compounds, was shown. To our knowledge, this enzyme is the first example of an aminopeptidase with Ser at the active site.

N-CARBAMOYL-D-AMINO ACID AMIDOHYDROLASE

Screening

The screening and isolation methods used for soil microorganisms were as follows. For the isolation of microorganisms that can hydrolyze *N*-carbamoyl-D-amino acids to D-amino acids, soil samples were suspended in saline and the supernatants were inoculated into growth medium containing *N*-carbamoyl-D-amino acids as sole nitrogen sources for enrichment culture. After aerobic cultivation, the reduction of *N*-carbamoyl-D-amino acids and the production of D-amino acids were detected by silica gel thin- layer chromatography (TLC) with a solvent system of *n*-butanol:acetic acid:water (4:1:1). *N*-Carbamoyl-D-amino acids and D-amino acids were detected with p-dimethylaminobenzaldehyde in a 6 M HCl solution and ninhydrin, respectively. From the culture broth from which *N*-carbamoyl-D-amino acids disappeared or in which D-amino acids accumulated, microorganisms were isolated on agar plates. The substrate specificities of these DCases were examined by means of the resting cell reaction with detection by the TLC method.

We screened two good mesophile strains producing a lot of enzymes that were classified as *Agrobacterium* sp. or *Rhizobium* sp. A similar enzyme-producing strain, classified as *Agrobacterium radiobacter*, was also found by Olivieri et al. and characterized. We also screened thermotolerant strains by means of enrichment culture at 45°C and obtained some *Pseudomonas* sp. strains. The strain isolated by Yokozeki and Kubota was classified as *Pseudomonas* sp. but was not a thermotolerant strain.

Table 10.3 Substrate specificities of screened enzymes

	Mesophile (30 °C)		Thermophile (45 °C)	
	KNK 712	KNK 1415	KNK 003A	KNK 505
c-D-Ala	++++	+++	++++	+ +
c-D-Val	++++	+++	+++	+ +
c-D-Leu	++++	++++	++++	+++
c-D-Phe	++++	+++	+++	+ +
c-D-PEG	+++	+++	++	+
c-D-PG	+++	+++	+ −	+ −
c-D-HPG	+++	+++	++	+ −

Isolation of the Enzymes

The DCases from *Agrobacterium* sp. KNK712 and *Pseudomonas* sp. KNK003A were purified after large-scale incubation by ammonium sulfate precipitation and chromatography such as ion exchange and gel filtration. DCase activity was assayed by measurement of D-*p*-hydroxyphenyl glycine (D-HPG) production from *N*-carbamoyl-D-HPG (c-D-HPG). The reaction was started by the addition of 100 *μL* of an enzyme solution diluted with 100 mM potassium phosphate buffer, pH 7.0 (hereinafter abbreviated as KP buffer), containing 5 mM dithiothreitol (DTT) to the assay mixture containing 47.6 μmol c-D-HPG and 100 *μ*mol KP buffer in a total volume of 1 mL. After 20 min incubation at 40°C, the reaction was stopped by the addition of 0.25 *μL* of 20% (w/v) trichloroacetic acid. The D-HPG produced was analyzed by means of the following high-performance liquid chromatography (HPLC) method. *N*-Carbamoyl-amino acids and amino acids were detected and quantified by HPLC at 210 nm on a reverse- phase HPLC column, using 36.7 mM KH_2PO_4 containing 15% methanol adjusted to pH 2.5 with phosphoric acid.

One unit of the enzyme was defined as the amount of enzyme that catalyzed the formation of D-HPG at the rate of 1 *μ*mol/min under the assay conditions already mentioned. The reactivity of the DCase of *Agrobacterium* sp. KNK712 is about 20 times higher than that of *Pseudomonas* sp. KNK003A with c-D-HPG as the substrate, but as concerns heat stability, in contrast, the *Pseudomonas* sp. KNK003A enzyme is more stable, with an about 10°C increase in view of its denaturation temperature. The molecular weights of these DCases were found to be about 37,000 and 38,000 Da, respectively, on SDS-PAGE. The properties of these two enzymes are shown in Table 10.4.

ALKALINE D-PEPTIDASE

D-Aminopeptidase from *Ochrobactrum anthropi* was discovered, and its primary structure has been found to have similarity to the β-lactamases and penicillin-binding proteins. The enzyme acts mostly on peptides with D-Ala at the N-terminus to yield D-amino acids; it does not act on D-amino acid derivatives with bulkier substituents. We proposed that D-aminopeptidase is a new penicillin-recognizing enzyme based on its primary structure, inhibition by β-lactam compounds, and the ability to catalyze peptide bond formation in organic solvents, although the enzyme does not show β-lactamase activity. Here, we describe the screening of soil microorganisms for D-stereospecific endopeptidases using a synthetic peptide $(D\text{-}Phe)_4$, characterization of the new enzyme alkaline D-peptidase (ADP), as well as cloning and sequencing of the *adp* gene from *B. cereus* strain DF4-B.

Table 10.4 Properties of DCases from *Agrobacterium* sp. KNK712 and *Pseudomonas* sp. KNK003A

	KNK712	KNK003A
Molecular weight (monomer)		
SDS-PAGE	37,000	38,000
DNA sequence	34,300	35,400
Thermal stability (60°C, 20 min)		
(residual activity)	5%	95%
Optimal temperature	67°C	
Optimal pH	6.8–7.3	7.0–7.5
Number of amino acids (deduced)		
Total	304	312
Hydrophobic	162	154
Neutral	50	51
Hydrophilic	92	107
Cystein	5	5
G + C Content	60.3%	62.9%
Homology		
DNA	62%	
Amino acid	60%	

Extraction of the Alkaline D-Peptidase

An enrichment culture in LB medium containing a synthetic substrate $(D\text{-}Phe)_4$ led to the isolation of a bacterial strain *Bacillus cereus* DF4-B. The extracellular enzyme hydrolyzing $(D\text{-}Phe)_4$ was purified and characterized. The M_r of the subunit calculated was about 36,000 by gel electrophoresis (SDS-PAGE), 37,000 by HPLC, and 37,952 by mass spectrophotometry. The absorption of the purified enzyme in 0.01 M potassium phosphate buffer, pH 7.0, was maximal at 281 nm. The enzyme is an endopeptidase that acts D-stereospecifically on peptides composed of aromatic D-amino acids, recognizing the D-configuration of the amino acid whose carboxy-terminal peptide bond is hydrolyzed. The enzyme had an optimum pH at around 10.3. Thus, the enzyme was named alkaline D-peptidase (D-stereospecific peptide hydrolase). Each substrate (2 mM) was incubated under standard enzyme assay conditions. The enzyme was active toward $(D\text{-}Phe)_3$ and $(D\text{-}Phe)_4$, forming $(D\text{-}Phe)_2$ and D-Phe. The enzyme is also active toward tripeptides with D-Tyr at the C- or N-terminus and on Boc-

(D-Phe)*n* (*n* = 2–4), forming Boc-D-Phe, (D-Phe)$_2$, and D-Phe. The enzyme had esterase activity toward D-Phe methyl ester and (D-Phe)$_2$ methyl ester. The products from Boc-(D-Phe)$_3$ *tert*-butyl ester were Boc-D-Phe, D-Phe, and D-Phe *tert*-butyl ester.

Table 10.5 Substrate specificity of alkaline D-peptidase from *Bacillus cereus* DF4-B

Substrate	*Relative activity (%)*	K_m *(mM)*	V_{max} *(U/mg)*	V_{max}/K_m *(U/mg/mM)*
(D-Phe)$_6$	1.8			
(D-Phe)$_4$	100	0.398	199	500
(D-Phe)$_3$	90	0.127	130	1020
(D-Phe)$_2$	0.2	50.1	13.7	0.270
D-Phe-L-Phe	<0.1			
(D-Phe)$_2$-L-Phe	14.9	0.522	30.6	59.0
L-Phe-(D-Phe)$_2$	119	0.455	154	346
L-Phe-D-Phe-L-Phe	28.1	1.63	66.0	41.0
D-Tyr(D-Phe)$_2$	83.6			
(D-Phe)$_2$-D-Tyr	83.6			
D-Phe-OMe	15 1.8			
D-Phe-NH$_2$	0.1 0.1			
D-Phe-*p*-Nitroanilide	4.2			
Boc-(D-Phe)$_4$	1.8 0.8			
Boc-(D-Phe)$_3$	3.2 1.1			
Boc-(D-Phe)$_2$	7.0			
Boc-(D-Phe)$_3$-O^{t}Bu	1.2 0.3			
Boc-(D-Phe)$_4$-OMe	0.5 0.2			
Boc-(D-Phe)$_3$-OMe	0.7 0.3			
Boc-(D-Phe)$_2$-OMe	1.4			
Ampicillin	8.9	73.1	262	3.58
Penicillin G	9.7	48.9	250	5.11

The enzyme was not active toward L-Phe methyl ester, (L-Phe)$_2$ methyl ester, (L-Phe)$_4$, Boc-(L-Phe)$_4$, Boc-(L-Phe)$_4$ methyl ester, (D-Val)$_3$, (D-Leu)$_2$, and (D-Ala)*n* (*n* = 2–5). These properties indicated that the enzyme is an endopeptidase that acts D-stereospecifically on peptides composed of aromatic D-amino acids. On the other hand, a dimer was formed when D-Phe methyl ester and D-Phe amide were the substrates. Since the enzyme is found to be a serine peptidase as described it is anticipated that later, the enzyme will be useful in the kinetically controlled synthesis of peptides. Eight stereoisomers of phenylalanine trimer were synthesized, and their effectiveness as substrates for the enzyme was tested. The enzyme recognized the configuration of the second D-Phe of tripeptides and catalyzes the hydrolysis of the second peptide bond from the N-terminus. The calculated V_{max}/K_m values for the peptides containing L-Phe were lower than that for $(D\text{-}Phe)_3$, affected by the neighboring L-Phe. The enzyme also showed β-lactamase activity toward ampicillin and penicillin G.

The calculated V_{max} values of the enzyme for β-lactam compounds were about the same as those for $(D\text{-}Phe)_3$ and $(D\text{-}Phe)_4$, whereas the K_m values were several hundred times larger. On the other hand, carboxypeptidase DD and D-aminopeptidase activities were undetectable. The time course of the $(D\text{-}Phe)_4$ degradation was measured. As shown, $(D\text{-}Phe)_4$ was hydrolyzed to $(D\text{-}Phe)_2$ and D-Phe. No $(D\text{-}Phe)_3$ was detected.

These results coincide with the kinetic properties of the enzyme described earlier. The mode of action of the enzyme was examined with the synthetic substrates D-Tyr-$(D\text{-}Phe)_2$ and $(D\text{-}Phe)_2$-D-Tyr. When D-Tyr-$(D\text{-}Phe)_2$ was the substrate, D-Phe was released first, then D-Tyr was slowly formed. When $(D\text{-}Phe)_2$-D-Tyr was used as a substrate, D-Tyr was released first, then D-Phe was slowly formed. In both reactions, the second peptide bond from the N-terminus of the substrate was hydrolyzed first. These results showed that the enzyme acts as a D-stereo-specific dipeptidylendopeptidase.

The enzyme activity was maximal at 45°C. About 60% activity remained after an incubation at 43°C in 0.1 M potassium phosphate buffer, pH 8.0, for 10 min. No activity was lost between pH 5.0 and 10.0 after an incubation at 30°C for 1 h in 0.05 M buffers at various pH values. The enzyme activity was enhanced by Mg^{2+} (138%), Mo^{3+} (130%), and Ba^{2+} (123%). When measured after incubating at 30°C for 30 min, the activity was inhibited to 94% by phenylmethylsulfonyl fluoride (PMSF), 76% by Ag^+, 74% by Fe^{2+}, and 32% by Hg^{2+} at 5 mM. These results, together with the information about its primary structure as described later, indicated that the enzyme is a serine peptidase.

DNA Technology

The gene coding for alkaline D-peptidase (*adp*) was cloned into plasmid pUC118, and a 1,164 bp open reading frame consisting of 388 codons with an M_r of 42,033 was identified as the *adp* gene. The enzyme would be synthesized with a signal peptide. The deduced primary structure of the enzyme is similar to carboxypeptidase DD from *Streptomyces* R61 (35.0% identical over 346 amino acids), penicillin-binding proteins from *Streptomyces* (*Nocardia*) *lactamdurans* (28.1% over 263 a.a.) and that of *B. subtilis* (28.5% over 309 a.a.), and class C β-lactamases of *Serratia marcescens* (24.9% over 217 a.a.), class C β-lactamases of *Enterobacter cloacae* (25.1% over 191 a.a.), fimbrial protein D from *Dichelobacternodosus* (24.1% over 261 a.a.), D-aminopeptidase from *O. anthropi* (27.5% over 182 a.a.), and esterase from *Pseudomonas* sp. (30.5% over 154 a.a.).

The sequence of Ser-X_{aa}-X_{aa}-Lys is perfectly conserved among this class of enzymes and the consensus sequence is located around 60 residues from the N-termini of most of the enzymes. Thus, we propose that alkaline D-peptidase from *B. cereus* be categorized as a new member of the penicillin-recognizing enzymes, which include penicillin-binding proteins, β-lactamases, and D-aminopeptidase.

Microbial Synthesis of Enzymes

The alkaline D-peptidase not only hydrolyzed $(\text{D-Phe})_3$ and $(\text{D-Phe})_4$ to form $(\text{D-Phe})_2$ and D-phenylalanine but also acted on D-Phe methyl ester and D-Phe amide to form $(\text{D-Phe})_2$. This finding gave us the opportunity to use the enzyme to further investigate the synthesis of D-Phe oligopeptides from D-Phe methyl ester. Because the alkaline D-peptidase was found to be a serine peptidase, we attempted to use it for kinetically controlled peptide synthesis. An expression plasmid pKADP was constructed by placing the alkaline D-peptidase gene (*adp*), amplified by means of the polymerase chain reaction, under the *tac* promoter of pKK223-3. Oligomerization of D-phenylalanine methyl ester by means of the purified enzyme from the transformant *E. coli* was investigated under several conditions. D-Phe dimer, $(\text{D-Phe})_2$, and trimer, $(\text{D-Phe})_3$, were produced in 25.4% and 8.6% yield, respectively, when 50 mM of the substrate was incubated for 8 h with ADP (2.0 and 0.4 U/mL, respectively) in 100 mM triethylamine HCl (pH 11.5). Addition of dimethyl sulfoxide to the reaction mixture resulted in the production of tetramer $(\text{D-Phe})_4$ in 6.7% yield with the decrease of the $(\text{D-Phe})_2$ and $(\text{D-Phe})_3$ production. This is the first example of the synthesis of D-phenylalanine oligomers by means of a D-stereospecific endopeptidase.

ENZYMES INVOLVED IN THE BIOSYNTHESIS

DNA Technology

The DCase genes from *Agrobacterium* sp. KNK712 and *Pseudomonas* sp. KNK003A were isolated and their DNA sequences analyzed as follows. Chromosome DNA was partially digested with *Sau*3AI, and the fractionated DNA fragment of 4-9 kb was inserted into pUC18 and then transformed into *E. coli* JM109. Recombinant colonies were collected, inoculated into enrichment culture medium containing *N*-carbamoyl-D-amino acids or 5-substituted hydantoins, and then incubated at 37°C. After repeating the enrichment cultures, the recombinant clones were isolated. Plasmid DNAs were prepared from these clones and analyzed with several restriction endonucleases. The DCase gene of *Agrobacterium* sp. KNK712 was located in a 1.8-kb *Sal*I–*Eco*RI fragment, and the gene of *Pseudomonas* sp. KNK003A was found in a 1.8-kb *Sph*I–*Acc*I fragment. These fragments were inserted into pUC19, the resultant plasmids being designated as pAD108 and pPD304, respectively. The recombinant strains containing these plasmids showed increased DCase expression compared with the native strains.

Detecting Sequences

The DNA sequences of the genes in the 1.8-kb fragments were analyzed, there being one open reading frame in each case, consisting of 912 and 936 bases with a starting triplet, ATG, and predicted to encode polypeptides of 304 and 312 amino acids with calculated molecular weights of 34,285 and 35,438, respectively. Upstream of the open reading frame, sequences similar to the –35 and –10 consensus sequences for *E. coli* promoters and a putative ribosomal-binding site were found. The G+C contents of the open reading frames were 60.3% and 62.9%, respectively. The homologies of the DNA sequences and the deduced amino acid sequences between these two DCases were 62% and 60%, respectively. An 8-amino acid length of the C-terminal region was different in these two enzymes. The homologies among those of *Agrobacterium* sp. KNK712 and those of two other reported *Agrobacterium* strains which had completely the same sequences, were 97.0% (amino acid sequences) and 93.0% (DNA sequences).

The homologies of the N-terminal regions of 30 amino acids in five DCases were about 50%, the homology between the enzyme of *Pseudomonas* sp. KNK003A and the *Comamonas* sp. E222c DCase being the highest (56.7%), besides the homology among *Agrobacterium* strains. Similarity research using a database (PIR release, 45.0) showed that the aliphatic amidase of *Pseudomonas aeruginosa* showed the highest similarity (30%) to DCase, but a closely related enzyme was not found. Multiple alignment with Clustal W (1.4) software of the DCase and 10 known related enzymes showed that 6 amino acids, that is the 119th amino acid of DCase, glycine (hereinafter abbreviated as 119th Gly), 126th Arg, 127th Lys, 146th Glu, 153rd Gly, and 172nd Cys, were all at corresponding positions.

FUNCTIONAL IMPORTANCE

For its use in a bioreactor, DCase was immobilized on a resin and used repeatedly. Although a DCase requires both high reactivity and high enzyme stability for such use, the known DCases do not have both properties. Thus, the DCase gene of *Agrobacterium* sp. KNK712, which exhibits high reactivity, was mutagenized to increase the enzyme's thermostability as an index of enzyme stability.

Mutagenesis

The DCase gene of *Agrobacterium* sp. KNK 712 was separated from recombinant plasmid pAD108 and then inserted into M13 mp 18. Random mutation was performed using two mutagens, hydroxylamine hydrochloride and $NaNO_2$. For the mutation with hydroxylamine, recombinant phage particles containing the DCase gene were treated with 0.25 M mutagen (pH 6.0) for 1-8 h at 37°C. For the mutation with $NaNO_2$, single-strand recombinant phage DNA was treated with 0.9 M mutagen (pH 4.3) for 30 min at 25°C. After preparation of double-strand phage DNA, the 1.8-kb *Eco*RI–*Hind*III fragment corresponding to the DCase gene was prepared, inserted into pUC19, and then transformed into *E. coli* JM109. The recombinant clones were screened by means of a newly developed colorimetric enzyme assay to select thermostabilized enzyme-producing colonies. The method is as follows. For first screening for colony assaying, recombinant *E. coli* colonies on the plate medium (the plates being called the master plates) were transferred to sterilized filter paper and then lysed for 30 min at 37°C by adding a lysis solution containing lysozyme and Triton X-100. After inactivation of the heat-sensitive enzymes (65°C, 5 min), a color reagent (1 mL), a mixture of two solutions (Sol. A, 30 mM potassium phosphate buffer, pH 7.4, 3 mg/mL *N*-carbamoyl-D-phenylglycine, 10 mM phenol, and 0.8 u/mL D-amino acid oxidase; Sol. B, 70 u/mL horseradish peroxidase and 50 mM 4-aminoantipyrine: Sol. A:Sol. B = 100:1, mixed just before use) was added to the filter paper, followed by incubation at 37°C.

The candidate colonies, which were colored red, were separated from the master plates. These clones were then subjected to the following second screening for cell-free assaying to confirm the mutant clones. Recombinant *E. coli* cells were disrupted by sonication and the heat-sensitive enzymes were then inactivated (65°C, 10 min). The crude enzyme sample was dispensed in 150-μL aliquots onto a 96-well microtiter plate; these aliquots were serially diluted twofold and 50 μL of the modified color reagent (Sol.A : Sol.B = 19:1) was added. The plate was incubated at 37°C, and then the enzyme activity was measured as absorbance at 505 nm. The candidate clones that exhibited high absorbance were separated and the enzyme genes were analyzed.

DNA Sequences

The DNA sequences of the 1.8-kb *Eco*RI-*Hin*dIII fragments that contained the mutagenized DCase genes from these representative clones were analyzed. One or some nucleotides of the DCase gene were changed. The amino acid changes related to the thermostability increase proved that the 57th amino acid of DCase, histidine, was changed to tyrosine, the 203rd proline was changed to serine or leucine, and the 236th valine was changed to alanine.

Table 10.6 Nucleotide analysis of mutagenized DCase genes

Mutant	*Locations of nucleotide changes*	*Amino acid substitutions*
401M	840 C → T	203 Pro → Leu
402M	401 C → T	57 His → Tyr
	715 C → T	(161 Val → Val)
403M	177 C → T	(noncoding)
	499 C → T	(89 Ile → Ile)
	839 C → T	203 Pro → Ser
404M	840 C → T	203 Pro → Leu
406M	839 C → T	203 Pro → Ser
414M	939 T → C	236 Val → Ala

Thermostability

The three thermostability-related sites of the DCase were substituted through site-directed random amino acid changes by use of the PCR technique according to Ito et al., and then thermostabilized DCase-producing strains were screened by means of the colorimetric colony assay method. As a result of DNA analysis of the mutant thermostabilized DCases, it was proved that the following amino acid changes at the sites increased the thermostability in addition to the known amino acid changes; the 57th amino acid of DCase was changed to leucine; the 203rd amino acid to alanine, asparagine, glutamate, histidine, isoleucine, or threonine; and the 236th amino acid to serine or threonine.

Mutations

To produce multiple mutants with combinations of two or three thermostability-related amino acid mutations, restriction fragments containing the mutations were used to replace the corresponding mutation-free DNA fragments. In the DCase of *Agrobacterium* sp. KNK712, the mutation of the 57th amino acid is located in a *NdeI–SacI* DNA fragment of about 190 bp; the 203rd amino acid mutation is in a SalI–ClaI DNA fragment of about 170 bp (hereinafter this fragment is referred to as fragment A); and the 236th amino acid mutation is in a ClaI–SphI DNA fragment of about 75 bp. These fragments were replaced and the thermostabilities of the enzymes were then measured.

Table 10.7 Thermostability analysis of mutant DCases

Location of mutation	*Substituted amino acid*	*Thermostable temperature (°C)*
57 His	Tyr	67.3
	Leu	67.5
203 Pro	Leu	68.0
	Ser	66.5
	Ala	67.7
	Asn	67.0
	Glu	70.0
	His	65.2
	Ile	67.2
	Thr	67.5
236 Val	Ala	71.4
	Ser	72.0
	Thr	69.5
Native type	—	61.8

The thermostabilities of the enzymes increased cumulatively with the accumulation of the individual mutations. Among the mutant enzymes with two amino acid changes, a mutant with mutations of the 203rd amino acid, proline, to glutamate, and the 236th, valine, to alanine, showed a thermostability increase of about 17°C. Among the mutant enzymes with three amino acid changes, a mutant enzyme from KNK455M that had mutations of the 57th amino acid to tyrosine, the 203rd to glutamate, and the 236th to alanine, showed about 19°C increase in thermostability.

APPLICATION OF THE THERMOSTABILIZED ENZYMES

To evaluate the thermostability of the DCases, a cell-free extract (100 μL) was incubated at various temperatures (55–85°C) for 10 min, the denatured protein was removed by centrifugation, and then the residual activity was measured by the standard HPLC method. The heat denaturation profiles of the four mutant enzymes from KNK402M, KNK404M, KNK416M, and KNK455M and the native enzyme were investigated. The relative activities in comparison with the activities of the non-heat-treated enzymes were plotted against temperature.

The thermostabilities of the mutant enzymes increased about 5–20°C under these conditions compared to that of the native enzyme. To examine the effect of pH on DCase stability, a cell free extract (200 μL) was mixed with 800 μL of a pH adjusted buffer (pH 6-10, 0.1 M each), followed by incubation at 40°C for 4 h, and then the residual activity was measured under standard conditions. The relative activity compared with the activity of each nontreated sample was plotted against the incubation pH. The stability of the mutant enzyme from KNK455M was increased in the lower- and higher-pH regions in comparison with those of the native enzyme.

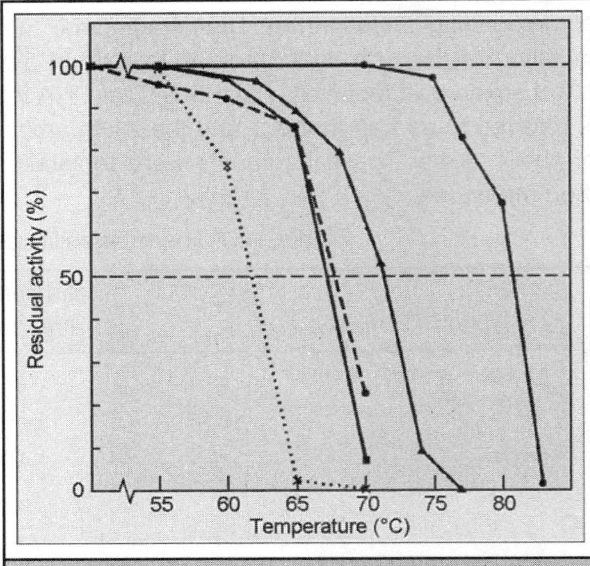

Figure 10.5 Effect of temperature on the stability of the mutagenized enzymes.

Substrate Specificities

The substrate specificity of the enzyme from KNK455M was investigated using a broad range of *N*-carbamoyl-D-amino acids and related compounds. Like the native enzyme, the enzyme showed strict specificity toward *N*-carbamoyl-D-amino acids, *N*-carbamoyl-L-amino acids not being hydrolyzed. *N*-Carbamoyl-D-amino acids having a hydrophobic side chain, as well as *N*-carbamoyl-amino acids having a polar group or short-chain alkyl group, were hydrolyzed. The substrate specificity of the mutant enzyme was almost the same as that of the native enzyme. It seemed that these three thermostability- related amino acid changes little affected the substrate specificity. The other known DCases from *Agrobacterium* showed relatively similar tendencies as to substrate specificity.

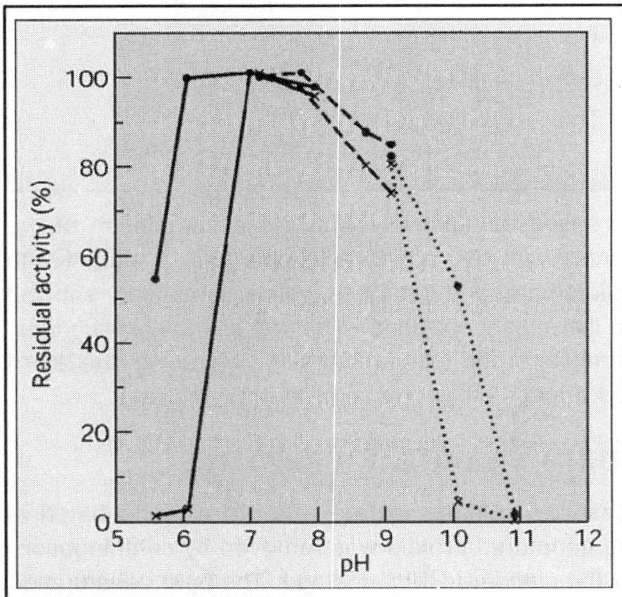

Figure 10.6 Effect of pH on the enzymes stability of the mutagenized DCases.

Kinetic Properties of the Enzymes

The kinetic properties of the enzymes from *Agrobacterium* sp. KNK712 and mutant KNK455M were measured using c-D-HPG as a substrate. As the affinity toward substrates and the reactivity were slightly decreased by the mutations, the mutant enzyme was thought to be utilized under industrial conditions.

Table 10.8 Comparison of substrate specificities of native and mutant DCases

Substrate	Relative activity (%)	
	Native DCase	*DCase of KNK455M*
c-D-HPG	100	100
c-D-Phenylglycine	110	130
c-D-Ala	86	75
c-D-Val	42	47
c-D-Leu	120	110
c-D-Met	180	160
c-DL-Ile	79	76
c-DL-Phe	55	46
c-DL-Ser	58	39
c-DL-Asp	7.2	3.4
c-DL-Norvaline	82	69
c-DL-Norleucine	120	99
c-DL-α-Aminobutyrate	100	87

Table 10.9 Comparison of the properties of native and mutant DCases

	Native DCase	*DCase of KNK455M*
Specific activity (U/mg)	6.8	6.6
Optimum pH	7.0	6.4
pH stability	6.5–7.5	6.0–8.0
Optimum temperature (°C)	65	75
Thermostable temperature (°C)	61.8	80.8
Substrate specificity	c-D-amino acid	c-D-amino acid
K_m (mM, pH 7.0 c-D-HPG)	0.89	1.3
*V*max (μmol/min/mg, c-D-HPG)	9.6	7.6

BIOSYNTHESIS OF AN IMPROVED ENZYME

Expression Vector

To express the DCase gene effectively, the mutant DCase gene was inserted into a newly constructed vehicle, pUCNT. This vehicle was constructed as follows. To insert the improved DCase gene just after the lac promoter of pUC19, a *Nde*I cleavage site was generated in the initiation codon of the *lacZ* gene, through PCR amplification of the *Hin*dIII–*Cfr*10I 1.3-kb fragment. The amplified fragment was used to replace the corresponding fragment of pUC19, and the plasmid constructed was designated as pUCNde.

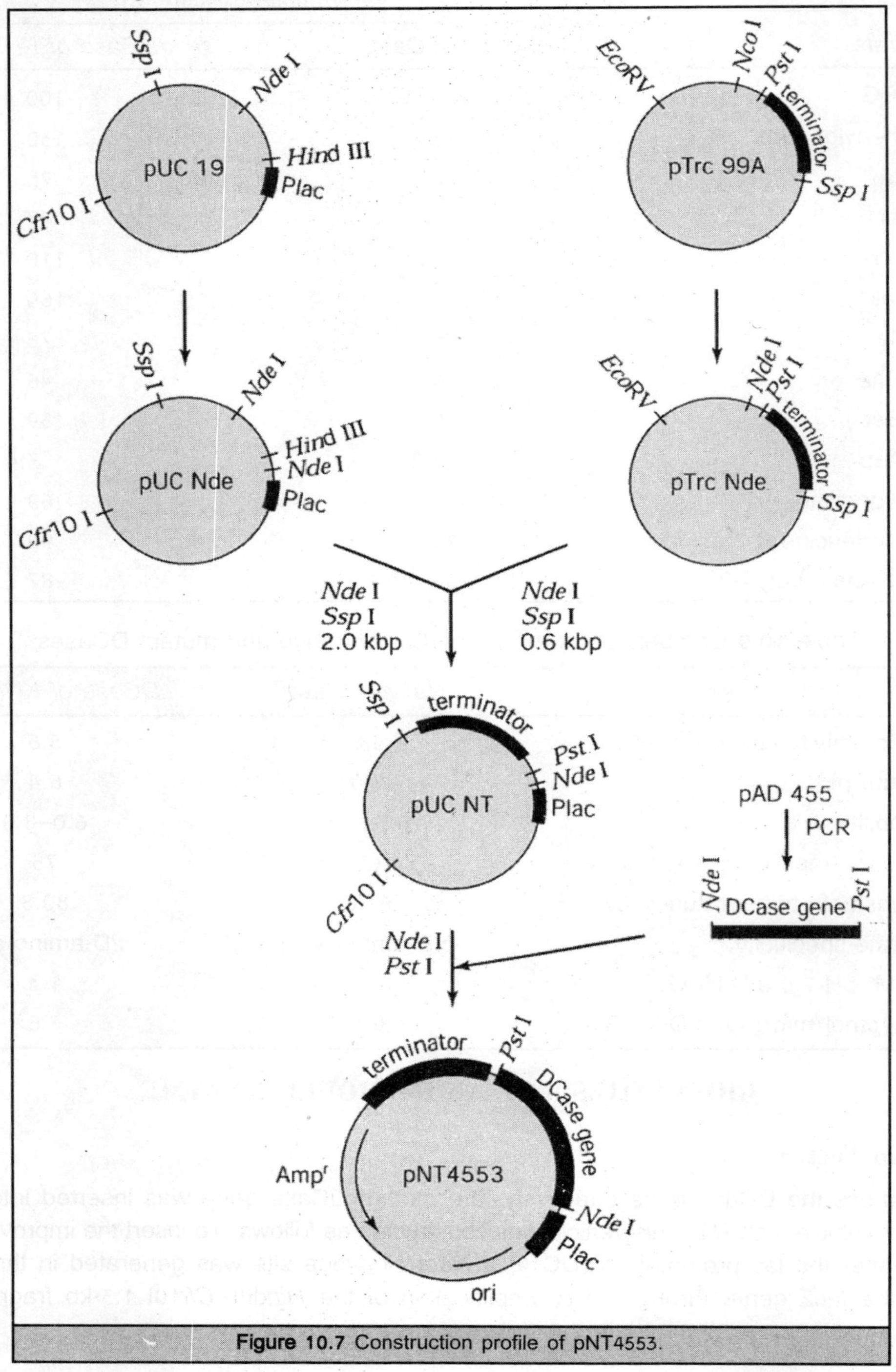

Figure 10.7 Construction profile of pNT4553.

After generating a *Nde*I site instead of the *Nco*I site of pTrc99A in the same manner, a 0.6-kb *Nde*I–*Ssp*I fragment was ligated into the 2.0-kb *Nde*I-*Ssp*I fragment, pUCNT being constructed. The 455M mutant DCase gene was inserted into pUCNT between the *Nde*I and *Pst*I cleavage sites, the recombinant plasmid pNT4553 being constructed.

Thermostabilized Enzyme

The expression plasmid was transformed into *E. coli* HB101, and then the recombinant clone was cultivated overnight in 2 x YT medium at 37°C. In the recombinant *E. coli,* the plasmid was stably maintained without the addition of ampicillin and was expressed efficiently without IPTG, which revealed that the DCase accounted for nearly 50% of all the soluble protein in a cell.

LABORATORY PRODUCTION

Escherichia coli JM109 (pAD108) and JM109 (pAD404) cells were harvested by centrifugation, suspended in 50 mL of 0.1 M KP buffer, pH 7.0, containing 5 mM DTT, and then disrupted by sonication. The cell debris was removed and the supernatant was obtained, as was the enzyme solution. Duolite A-568 was washed with 1 M NaCl, water, and then 0.1 M KP buffer, pH 7.0, and then equilibrated with the same buffer for 18 h at room temperature. The wet resin was subjected to filtration, and then DTT (final concentration of 5 mM) was added to the enzyme solution, followed by stirring at 4°C for 20 h under nitrogen sealing. For adsorption, the weight of the wet resin and that of protein in the enzyme solution were in the ratio of 1 to 0.04. After adsorption, the resin was washed three times with a fivefold volume of 0.1 M KP buffer, pH 7.0, containing 10 mM DTT, cross-linked with a fivefold volume of 0.2% (w/w) of glutaraldehyde at 4°C for 10 min, washed three times with a fivefold volume of the same buffer at 4°C, and then filtered.

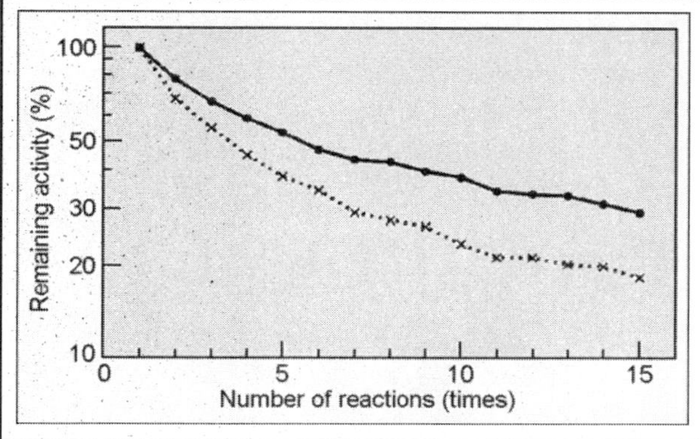

Figure 10.8 Stability of the immobilized DCase on repeated batch reactions. Repeated batch reactions were carried out using the immobilized mutant DCase from KNF404M (circles) and the native enzyme (asterisks).

Stability

For repeated batch reactions with the immobilized DCase, a substrate solution (3%) of c-D-HPG (pH 7.0) was prepared, with nitrogen gas bubbling at 40°C. The DCase immobilized on Duolite A-568 and the prepared substrate solution were stirred at 40°C under a stream of nitrogen gas while the pH was controlled at pH 7.0 with 2 M HCl. Samples were taken for activity measurement at 10 and 60 min, the reaction being continued for 23.5 h in total. The reaction mixture was removed by suction, after which a fresh substrate solution was introduced and allowed to react in the same manner as already described. The residual activity was measured as the relative activity compared with the initial activity and plotted against the reaction times. The stability of the mutant enzyme increased in comparison with that of the native enzyme.

CONCLUDING REMARKS

An enzymatic conversion process involving the improved DCase in an immobilized form was introduced for D-HPG production instead of chemical diazonation. As a result of the introduction of the new process, the efficiency of the reaction was greatly increased, the purification process became simpler because of the decrease in the undesirable color formation of the product, and the energy required for production was reduced. In addition to these advantages, by-products such as inorganic salts were greatly reduced by the change in the reaction mechanism. By means of these advantages, the new process decreases the burden on the environment.

11

BIOPHARMACEUTICAL FILTRATION

Among the most capable and attractive means of micro-particulate (greater than 0.1 μ) separation used in biopharmaceutical processing today is the use of highly porous powdered media in dynamic systems. Powdered media, often called *filter aids,* provide versatility, high solids-loading capacity, high product recovery, low cost, and ease of scale-up in any filtration process. Moreover, recent technical advances have stimulated a much greater breadth of applications that use this media filtration in biopharmaceutical processes, with a concomitant exponential increase in product innovation.

BIOPHARMACEUTICAL PROCESS

The first unit operation in downstream processing is clarification. It involves the removal of cells, cell debris, or precipitated components from a fermentation broth or process supernatant. Four types of solid–liquid separations are common to the industry: (i) The removal of whole cells in fermentation broths where the product is expressed into the supernatant; (ii) The removal of cell debris from process broths after cell disruption and extraction to release the product of interest; (iii) The removal of precipitated contaminants (usually proteins) from a process fluid, for example, in the removal of mis-folded forms of recombinant proteins; and (iv) The selective precipitation of the target protein and the capture of this precipitate from the process supernatant. Most processes use either centrifugation or filtration for solid-liquid separations. Centrifugation, although widely accepted, has some disadvantages in biopharmaceutical applications.

The challenge in many applications is to remove both cellular and subcellular debris without applying excessive shear forces on the solids. Centrifuges depend on differences in density and on the centrifugal force applied for solid-liquid separation. However, centrifugation becomes less practical when the product and waste possess similar densities. In addition, centrifuges can be difficult to maintain, clean, and sterilize. Finally, it is important to avoid the generation of aerosols when dealing with biological fluids, and this is often difficult to avoid when using centrifuges for this unit operation. Depth filters containing diatomite, perlite, and cellulose are suited to low solids applications, because these products tend to blind quickly when subjected to moderate to high levels of solids. Their capacity can be increased by the addition of filter aid (as body feed) into the unfiltered broth, thus extending the lifetime of the pad. The use of powdered media offers many advantages in biopharmaceutical applications. These systems are much more flexible and dynamic because powdered media addition can be adjusted based on changes in incoming solids level. In addition, they are extremely efficient at removing suspended solids and are particularly effective at clarifying supernatants heavily contaminated with colloids and small particles.

The media can be used in all four applications described above, including the selective precipitation of the target protein and the capture of this precipitate from the process supernatant. Once captured on the filter media, the protein or product of interest can be redissolved by adjusting the pH or salt concentration in the wash buffer and recirculating this solution through the filter media bed. After resolubilization, the product can be eluted from the media with approximately two bed volumes of buffer. Capital costs for powdered-media-based systems are relatively low when compared with centrifugation or other filtration systems.

Because the media are disposable, media cleaning and lifetime studies do not need to be performed, hence the cost of validating the filtration process is substantially reduced. In addition, the ability to post-wash the filter cake helps to maximize product recovery, a process that is extremely expensive or impractical with centrifuges. Finally, with improvements in both powder containment and filter equipment design, complete containment is now easily achieved both in the media preparation area and in GMP suites. The combination of these improvements in powder containment, filter design, and powdered media technology position filter aids as a very cost-effective and economical approach to clarification.

POROUS MEDIA

Diatomite, perlite, and cellulose are the most widely used porous media in dynamic process filtrations, with a high percentage of applications using diatomite.

Diatomite is often used in combination with cellulose; however, diatomite is often the key ingredient of static or fixed-bed filters such as filter sheets and filter pads. Diatomite products are especially characterized by an inherently intricate and highly porous structure composed primarily of silica. These products are obtained from diatomaceous earth, a sediment greatly enriched in biogenic silica in the form of siliceous frustules of diatoms, a diverse array of microscopic, single-celled golden-brown algae of the class Bacillariophyceae. Surprisingly, these frustules are sufficiently durable to retain much of their ultrastructure nearly intact through long periods of geologic time when preserved in conditions that maintain chemical equilibrium.

Perlite is a naturally occurring volcanic glass that thermally expands upon processing. After milling, porous, complicated structures result that are also useful in dynamic process filtrations of biopharmaceuticals. Because its structure is not as intricate as that of diatomite, perlite is better suited to the separation of coarse microparticulates from liquids having high solids loading. Like diatomite, perlite is also useful as a functional filtration component of filter sheets and filter pads.

Cellulose, like perlite, possesses a less intricate structure than diatomite. As a result, the use of cellulose is generally limited to coarse filtrations or in specialized applications where a fibrous precoat on the septum is required.

MEMBRANE CARTRIDGE FILTRATION

Bioprocess technology operations have included membrane cartridge filtration for more than 20 years. Cartridge filters have been used for the removal of submicron particles, such as bacteria, and for the removal of particulate material, usually greater than 1 μ. Many applications involve the removal of bacteria in sterile filtration applications; particulate removal cartridge filters can be used as prefilters to protect the sterilizing grade filters. Some examples of applications for liquid filtration include fermentation media, deionized water, sera, and sanitizing chemicals. The purpose of this article is to provide information on operational considerations for cartridge filters and to explain the usage of cartridge filters in bioprocessing. Cartridge filters used in bioprocessing applications are used for the removal of bacteria (sterilizing grade filtration) or for the removal of particulate material (prefiltration). The mechanisms at work during filtration are provided, as well as, a general description of sterilizing grade filters and prefilters. Prefilters are often used upstream of the final (sterilizing) filters to protect the final filter from premature plugging, thereby prolonging the life of the final filter.

Mechanisms of Filtration

There are four mechanisms that can govern the particle removal efficiency of membrane filters: (i) direct interception, (ii) inertial impaction, (iii) Brownian motion or diffusional interception, and (iv) electrostatic attraction between the membrane and particles. Direct interception involves a sieving action that mechanically retains the particles on the filter surface. The filter acts as a screen that stops particles that are larger than the pores (openings) in the membrane. Direct interception is independent of face velocity and mostly involves particles with relatively large diameters. Inertial impaction refers to the deviation of a particle from its streamline during flow because of inertia, resulting in the retention of the particle by the membrane. As the face velocity increases, the probability of inertial impaction increases. Brownian motion or diffusional interception applies to small particles at low face velocities in a gas stream. When air molecules are in a state of random motion, small particles suspended in the air can be struck by moving air molecules and displaced.

The movement of particles resulting from molecular collisions is known as Brownian motion. This phenomena can increase the probability of capture of particles by diffusional interception within the membrane. An electrostatic attraction can exist or be induced between the membrane material and particles in the fluid stream. This attraction can enhance the removal of particles and their retention by the filter membrane. The pores (open area) in most membrane filter materials are typically not straight through the membrane. In membrane filters, the membrane can be considered to be constructed of multiple screens that provide a tortuous path for the fluid through the membrane. In bioprocessing applications, there are many processes that require bacterial removal (sterile filtration). The filters used to provide bacterial removal are generally 0.2-μ sterilizing grade filters.

Removal of bacteria by a 0.2-μ sterilizing grade filter is not only by a sieving action on the surface, but also by size- exclusion entrapment in the membrane structure. The ability of such a membrane to remove bacterial contaminants is related to its thickness as well as to the size of its pores. For example, it has been demonstrated that whereas a single 150-μ thick layer of 0.8-μ cellulose ester membrane can provide only a titer reduction of 70 for *Brevundimonas (Pseudomonas) diminuta*, 10 layers can provide a titer reduction of 2.4 x 10^8. Because membrane filters are constructed of porous materials, they will have a certain amount of open area. The open area allows for fluid flow through the membrane; this area, expressed as a percent, is known as the porosity of the membrane.

Sterile Liquid Filtration

Membrane filters used for sterile liquid filtration are typically constructed of polymeric microporous hydrophilic materials, such as nylon, cellulose acetate, modified polyvinylidenefluoride (PVDF), or other polymers. Most liquids that require sterile filtration are water based, and hydrophilic membranes will spontaneously become wet with the fluid. Absolute rated hydrophilic membrane filter cartridges have been used for sterile filtration of parenterals, diagnostic reagents, purified water and water-for-injection, dry gases, organic solvents, buffers, and biological fluids such as serum, plasma, tissue culture media, nondilute protein solutions, and fermentation harvest fluids. Membrane filters used for the sterile filtration of gas streams, such as nitrogen blankets, storage tank sterile vents, formulation tank sterile vents, sterile air for aseptic packaging, sterile filtration of fermenter air, or vent gases, typically contain a membrane made of hydrophobic materials such as PVDF or PTFE (polytetrafluoroethylene).

Hydrophobic membrane filters are desired in these sterile gas filtration applications because hydrophobic filters do not spontaneously wet with water. When a hydrophilic filter is wetted with water, it will not pass air until the water-wet bubble point of the filter is exceeded. This water-wet bubble point can be greater than 50 psi. Membrane materials can be modified to optimize their performance for certain types of applications. Positive-charged hydrophilic membranes contain cationic (positive charged) functional groups that impart a positive ζ-potential when immersed in an aqueous solution and provide enhanced retention of particles smaller than the absolute rating. Most particles have a negative charge, and the positive charge provides a mechanism, in addition to the tortuous path ofthe membrane, by which the particles are retained. Membrane surfaces can also be modified to expose hydroxyl groups, which tend to prevent the retention of proteins. This is a benefit for dilute protein solutions (typically less than 1 mg/mL), in which the protein is the desired product.

Such fluids include serum-free tissue culture media and protein additives, protein-based therapeutics and diagnostics, dilute protein containing diluents and buffers, recombinant proteins, hormones and growth factors, protein chromatography feeds and eluates, vaccines, and other dilute biologicals. The membranes used for sterile filtration bioprocessing applications are usually pleated and formed into a cylinder.

The filter membrane is typically cast onto a support material (e.g., nonwoven polyester or polypropylene substrate) that provides high tensile strength while retaining flexibility. Layers of the membrane are corrugated, or pleated, along with upstream and downstream layers of courser material (e.g., nonwoven polyester or polypropylene). These layers provide support and drainage for the membrane. The sides of the corrugated membrane pack are sealed, often by a heat seal. The corrugated filter membrane pack is fitted around the core. The purpose of the rigid, inner core is to provide support for the filter element against pressure in the forward direction. An external cage is provided for additional support and protection during handling. The flow path for the cartridge filter is from the outside (cage) to the inside (core).

The material used for the core, cage, and support layers will depend on the intended application for the filter. Polypropylene, polyester, and PTFE are examples of materials that can be used for the cage, core, and support material in a membrane filter. The membrane pack ends are sealed, usually by heat, with end caps. The end caps are constructed of the same types of materials as the core and cage.

The final step in the construction of a membrane filter is to attach the proper end closure and O-ring adaptor by an appropriate welding technique. The end closure is used for placement of the filter in a filter housing. A finned or flat blind end cap on one end is generally attached on one end and an open outlet at the other end. Sealing into the housing is accomplished with a double O-ring seal at the open end to ensure no fluid bypass. Filter elements can have locking tabs at the double O-ring adaptor base for bayonet locking of the element to ensure positive sealing in the housing base. Most single filter elements are 10 inches in length. For the construction of a multiple length cartridge (20, 30, or 40 inches), the sections are attached by welding 10-in. subassemblies end to end.

Material and method

For sterile filtration processes, the biological safety of the membrane filter or filter cartridge should be demonstrated by the performance of the USP Class VI (121°C) *plastics test for biological reactivity*. Another typical qualification used to select filter materials of construction is a listing for food contact in CFR. For sterile aseptic processes, a typical requirement is sterilization by passage through a 0.2-μm sterilizing grade membrane filter, as defined by ASTM Standard F838-83. A membrane filter as well as all system components can contribute extractables to the process stream. Many filter manufacturers document the extractables level of a particular filter in an appropriate solvent as a nonvolatile residue (NVR). These extractables are typically composed of the oligimers or additives of the plastic materials present in the filter element.

It is important to note that the amount of extractables from a filter element as well as any extractable from the rest of the process system can be reduced by flushing the system before filtration. Many filter manufacturers address the issue of effluent quality requirements by the performance of appropriate tests on filter samples from manufacturing lots. These include tests for

1. *Cleanliness*: Per current USP limits under *Particulate Matter in Injections* and conformance with requirements for a non-fiber-releasing filter
2. *Oxidizable substances*: Per current USP requirements under *Purified Water* after flushing
3. *pH*: Per current USP requirements under *Purified Water* after flushing
4. *Pyrogens*: Per current USP requirements under *Bacterial Endotoxins Test* as determined using the limulus ameobocyte lysate (LAL) reagent with an aliquot from a soak solution

Pleated and Nonpleated Filters

Cartridge filters used for prefiltration can be either pleated or nonpleated (depth) filters. Standard design depth filter cartridges are open cylinders of thick filter material (double open-ended). Prefiltration filter cartridges are typically available in 2-1/2 or 2-3 /4-in. o.d., in multiple lengths of 10 in. up to 40 in. long. Prefiltration encompasses a wide variety of fluid types and removal requirements. The following is a discussion of some important characteristic of prefilter construction and a sampling of some of the types available.

The particles or fibers in the prefilter membrane should not become dislodged, slough off, shed, or in some other way contaminate the filtrate (media migration). The filter medium should also not unload or release contaminants retained by the filter medium into the effluent as the differential pressure across the membrane increases. Data that indicate filtration efficiency as a function of increasing differential pressure should be available from the vendor to support the claim of nonunloading.

Polypropylene depth filters

Polypropylene depth filters are made using nonmigrating continuous filaments of polypropylene filament medium without resin binders. Absolute rated polypropylene depth filter cartridges are applicable as prefilters for serum, vaccines, diagnostics, tissue culture products, deionized water, container washing, and final product. These filters are used in fermentation for liquid feeds, makeup water, solvents, and antifoam. In downstream processing, they are used for cell and cell debris removal, buffers, cleaning agents, and sanitizing solutions.

Positive charged versions of these filters are possible and have the added ability to remove organisms and particles such as bacterial endotoxins (pyrogens) that are much smaller than the absolute pore size rating. The membrane for these depth filtration cartridges has an inner (downstream) section in which the pore diameter is constant (this section provides absolute rated filtration) and an outer section in which the pore diameter varies continuously from that of the absolute rated section to as much as 90 μm or more. Pore size variation of the filter cartridge is achieved by varying the fiber diameter while maintaining constant void volume throughout the medium. The constant void volume provides increased dirt capacity and a low clean pressure drop of the filter cartridge.

Polypropylene pleated filter

Polypropylene pleated filters are applicable as prefilters for makeup and rinse water, for serum for culture media, and in downstream processing for cell debris removal and prefiltration of solvents and buffers. These process filter cartridges can be constructed of the same type of membrane material as depth filters. The thin sheet of polypropylene media is pleated and formed into a cylinder with a longitudinal side seal of melt seal polypropylene. The cylinder is then melt sealed to injection molded polypropylene end caps to ensure no fluid bypass. Polypropylene hardware components consisting of an inner support core and an external protective outer cage are incorporated. The process filter cartridge with polypropylene hardware should be rated to withstand differential pressure of 80 psi up to 50°C (122°F) and 60 psi up to 80°C (176°F).

Resin-free cellulose pleated filter

Resin-free pleated cellulose filter cartridges are applicable as prefilters for makeup and rinse water, reverse osmosis membranes, deionized water, and inlet air for fermenters and bioreactors. These filter cartridges are constructed of pure cellulose medium without resin binders, which is pleated into a high area cylinder. The longitudinal side seal of the process filter cartridge should be a polypropylene (approved for food contact usage). Cellulose media cartridges are assembled with hardware components consisting of a perforated inner support core, an outer support cage, and end caps melt sealed to imbed the medium in the plastic. All hardware components should be of pure polypropylene, without filler or reinforcement, to ensure a high quality filtrate with a minimum of soluble extractables. The process filter cartridge with polypropylene hardware should be rated to withstand differential pressure of 80 psi up to 50°C (122°F) and 60 psi up to 80°C (176°F).

Resin-impregnated fiberglass pleated filter

Resin-impregnated fiberglass filters are applicable as pre-filters for deionized water, solvents, serums, diagnostic reagents, and cell debris removal from harvest fluids. The resin impregnation makes available a choice of filters with either a positive or negative ζ-potential when used in aqueous service. Filter cartridges with negative ζ-potential are constructed of glass fiber that has a natural negative ζ potential and reinforced with a resin binder that is also naturally negative when immersed in water.

Positive *f* potential filter cartridges are constructed using a resin binder that coats the glass fibers and imparts a positive ζ potential to the medium. The filter cartridges are available with hardware components consisting of end caps and an internal perforated support core made of either polypropylene or stainless steel and in an external polypropylene outer cage or protective net. Stainless-steel end caps are attached to the filter by an inert synthetic resin. Polypropylene end caps are melt sealed to imbed the medium in the plastic. Filter cartridges with polypropylene hardware should be rated to withstand differential pressures of 80 psi up to 50°C (122°F) and 60 psi up to 80°C (176°F). With stainless-steel hardware, the filter can be rated to 75 psi up to 135°C (275°F).

Housing

During filtration, the membrane filter cartridges are placed in housings that provide a means of fluid contact with the filter cartridge. The housings are typically composed of 304, 316, and 316L stainless steel; plastic housings are also available. The housing design is typically composed of a head and a bowl. The head is the portion of the housing in which the filter is attached. The bowl is clamped to the head.

Membrane filters can also be obtained in complete assemblies. These disposable assemblies consist of a membrane filter cartridge welded into a plastic filter housing. Disposable assembles or capsule filters are typically used in small-scale applications (e.g., 100-L batches or less, depending on the fluid) and are particularly useful in applications in which operator contact with the fluid is not desirable, because the entire assembly can be removed from the system and discarded without operator contact with the product-wet cartridge.

BASIS OF DIATOMITE FILTRATION

A discussion on solid-liquid separations can cover a wide range of techniques and applications, but each method is often defined by the relative size of solids being removed or collected from a given process stream. This chapter will focus on microparticulate filtration, where the solids are typically greater than 0.1 μ. The key objective is to remove unwanted solids from a process stream, so that the refined solution (filtrate) is adequately clarified and qualifies for further downstream processing.

Rigid or Compressible Solids

In general, solids can either be rigid or compressible in nature. Solids that form rigid cakes have some degree of inherent permeability, dictated by particle size distribution and packing arrangement, but most solids in biopharmaceutical applications are typically compressible. These often gelatinous, highly compressible solids retain a degree of permeability if low flow rates and low differential pressures are used for filtration. Unfortunately, these characteristics are contrary to typical processing requirements where this unit operation needs to be completed quickly and economically.

Filtration Behavior

The problem in static fixed-bed filtration is that unwanted solids have very limited permeability when they collect or accumulate onto the filter septum (pad, paper, or fabric or metallic or plastic woven wire screen). If these solids build up on a septum, then eventually these solids will lack sufficient permeability for fluid drainage and filtration terminates. This behavior highlights a mistake frequently applied in filtration practice; often, higher differential pressure does not guarantee faster filtration rates precisely because these solids collapse to form an impermeable cake. Introduction of a filter medium changes the composition of the accumulated cake, and therefore its filtration behavior. Powdered media (filter aids) are essential to this technology because they provide two functions: (i) as a precoat applied before the start of a filtration cycle and (ii) as body feed added to the unfiltered feed throughout the filtration cycle.

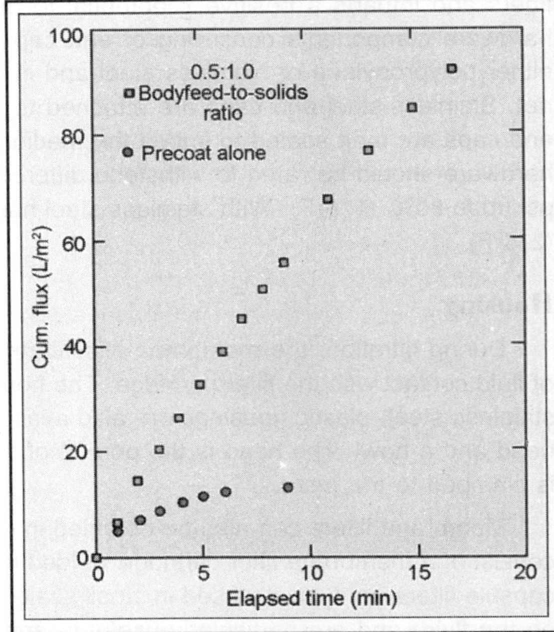

Figure 11.1 Addition of body feed to a filtration process extends the filtration cycle length.

The precoat layer protects the filter septum, preventing penetration and blinding by the unwanted solids. This layer also facilitates septum cleaning and provides for a high-quality filtrate from the beginning of the filtration cycle. The addition of powdered media (as body feed) to the process feed increases permeability in the accumulating filter cake, restricts solids movement, provides channels for filtrate recovery, and extends cycle length. The extended permeability afforded by these media slows the rise in differential pressure for constant flow processes and retards the drop in flow for constant pressure operations. At the conclusion of a filtration cycle, product recovery can be maximized by rinsing or purging the accumulated cake in place. It is extremely important to maximize the body feed in a process. By adding too little, the solids quickly blind the filter cake and the filtration is terminated. Adding too much is not only wasteful but leads to a rapid increase in filter-cake thickness, with a corresponding increase in resistance to flow and a subsequent reduction in cycle length.

Theory of Filtration

The following section is not intended to be a comprehensive treatment of filtration theory, but rather an overview to describe how the permeability of an accumulated filter cake (solids and filter aid) ultimately affects filtration performance (rate). Any discussion on porous media filtration can start with a simplified form of the Darcy equation (equation 1), which is used to describe laminar (nonturbulent) fluid flow through a homogeneous, porous medium.

$$k = \frac{Q \times \Delta x \times \mu}{A \times \Delta P} \quad \ldots(1)$$

The relative permeability (k) is measured in darcy units, and it is described in terms of the instantaneous volumetric flow rate (Q) and viscosity (μ) of the fluid, the thickness (Δx) and cross-sectional area (A) of the porous medium (accumulating filter cake), and the pressure drop or differential pressure (ΔP) associated with flow. The flow rate (Q) can also be represented by the differential change in volume over time (dV/dt). A value of 1 darcy corresponds to the permeability through a filter media 1-cm thick that allows 1 cm^3 of fluid with a viscosity of 1 cP to pass through an area of 1 cm^2 in 1 s under a pressure differential of 1 atm (i.e., 10 1.325 kPa). Although fluid viscosity varies with temperature and shear forces, values for it and the other variables can be obtained from tables and experimental observations. Consequently, the permeability (k) is an empirical value (constant) used to characterize the filter cake composition and not a property that can be readily altered by changing the process flow rate or differential pressure. In practice, permeability may appear to be influenced by flow rate or pressure, but those responses are strictly consequences of composition, and they are often inconsistent with the behavior described by equation 1.

The type and relative usage rate of filter aid coupled with the nature and concentration of solids define composition, which in turn controls permeability and filtration performance. The constancy of k explains why the differential pressure increases as filter cake accumulates (Δx) for constant flow (Q_0) processes (equation 2)

$$k = \frac{Q_0 \times \mu}{A}(\text{constant}) \times \frac{\Delta x}{\Delta P} \quad \ldots(2)$$

and flow decreases as cake accumulates for constant pressure (ΔP_0) processes (equation 3)

$$k = \frac{\mu}{A \times \Delta P_0}(\text{constant}) \times Q \times \Delta x \quad \ldots(3)$$

Permeability measurements are not essential in the application of dynamic filtration technology. In fact, once the unfiltered feed is introduced to the filter, the overall permeability (as measured in darcies) of the filter cake can be reduced by orders of magnitude. Ultimately, the manipulation of permeability (k), through filter aid selection and usage, describes the optimization of this solid–liquid separation process.

As an example, initial (laboratory scale) testing develops enough information to determine the permeability (k_{app}) in a constant pressure application. In fact, a key laboratory test is to determine the elapsed time (t_s) it takes to produce the maximum cake thickness (Δx_f; limited by the physical capacity of the filtration equipment, typically 3 cm) at optimum filter aid usage and terminal pressure (ΔP_f). Hence, the Darcy equation can be put to better use for process scale-up (equation 4):

$$[k_{app}] = \frac{Q_{0\,lab} \times \Delta x_f \times \mu}{A_{lab} \times \Delta P_f} = \frac{Q_{0\,lab} \times \Delta x_f \times \mu}{A_{plant} \times \Delta P_f} \qquad \ldots(4)$$

Notice that differential pressure (ΔP_f) and filter cake thickness (Δx_f) need not be affected by the scale of operations. Differences in allowable cake thickness (Δx_f) must be accommodated by adjusting filtration area (A_{plant}) requirements. A physical capacity test also determines the total volume of process feed that can be filtered on the given area. In the simplest case of equal cake thickness, scale-up becomes a matter of sizing filter area to accommodate production rate (equation 5).

$$\frac{Q_{0\,lab}}{A_{lab}} = \frac{Q_{0\,plant}}{A_{plant}} \qquad \ldots(5)$$

RATING THE MEMBRANE FILTERS

Many filter manufacturers use a nominal micron (μm) rating for particle removal efficiency. This is defined by the American National Standards Institute (ANSI) as an "arbitrary micrometer value indicated by the filter manufacturer. Due to lack of reproducibility this rating is deprecated." Further, nominal rating standards are arbitrary and a comparison of nominally rated filters is imprecise. In addition, nominal ratings can be misleading because the filter can allow passage of particles larger than the rating indicates. In order to establish a more meaningful filter rating, a test is performed on the filter using an appropriate contaminant to meet a specific claim for retention of the contaminant. The test procedure must be sensitive enough to detect the passage of contaminants of interest. The testing should be performed under carefully controlled conditions using industry-accepted reference standards. These reference standards include silica suspensions, latex beads, or microorganisms.

Removal Rating

The removal ratings for particulate removal filters (typically used as prefilters in sterile filtration processes) can be established using the Oklahoma State University (OSU) F-2 Test. This rating method (ISO 4572, ANSI B93.31) has received wide acceptance for use on lubricating and hydraulic fluids. The technique has been used for oils extensively and has been adapted for use in water with contaminants ranging from 0.5 to 25 μm. The test is based on continuous on-line particle counts of different particle sizes, both in the influent and the effluent stream. The β-ratio at a specific particle size is defined as β_x: the number of particles of a given size (*X*) and larger in the influent, divided by the number of particles of the same size (*X*) and larger in the effluent, where *X* is the particle size in micrometers. The percent removal efficiency can be calculated from the β-value. The percent removal efficiency is $[(\beta_x - 1)/\beta_x]\ 100$.

Hydrophilic Membrane Filters

Sterilizing-grade hydrophilic membrane filters require testing using a liquid bacterial suspension challenge that is sensitive enough to detect the passage of any microorganisms, for the establishment of the micron rating of the filter. A sterilizing-grade filter is defined by the FDA's *Guideline on Sterile Drug Products by Aseptic Processing* as one that will produce sterile effluent when challenged with the test organism *Brevundimons (Pseudomonas) diminuta* (ATCC 19146) to the level of greater than 10^7 CFU/cm^2 filter area. Such microorganism retention tests are conducted with *P. diminuta* (ATCC 19146), measuring 0.3 x 0.6 to 0.8 µm, to verify that membrane filters produce sterile fluid filtrates. This is a standard test for the validation of sterilizing-grade filters (0.2-µm pore size rating) in the bioprocessing industry. In practical terms, there is a limit to how many bacteria a filter element can be challenged with before it becomes plugged.

When a filter has been challenged at the level of 5×10^9 CFU/cm^2 it becomes effectively clogged. This challenge level corresponds to about 2×10^{13} CFU for a 5-ft^2 membrane filter. The liquid test involves challenging a test filter with a known quantity of *P. diminuta,* no less than 1×10^7 organisms per cm^2 of filter area. The challenge sample is suspended, for example, in sterile water and then filtered at a defined flow rate and time through a test filter, followed by an analysis membrane. After the challenge is complete, the system is flushed for 5 to 10 min with sterile water. The analysis membranes are removed and placed on Mueller Hinton agar and incubated at 32°C for 48 h. After

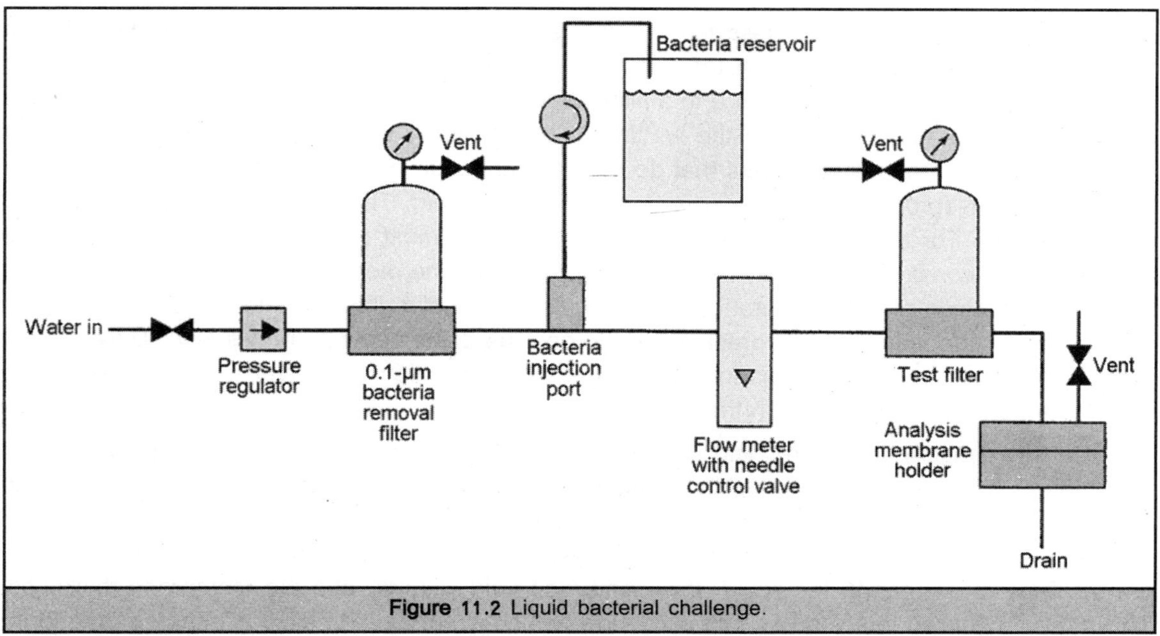

Figure 11.2 Liquid bacterial challenge.

incubation, the plate is examined for the presence or absence of microbial colonies. For some applications (e.g., mycoplasma control or removal) a membrane with a more stringent rating (0.1 μm) is required.

These applications typically involve sera and tissue culture media. One organism that has been used extensively in validating 0.1-μm rated filters is *Acholeplasma laidlawii.* A liquid bacterial challenge is a destructive test; the filter cannot be used after the challenge. Therefore, a correlation is made to a nondestructive test. It is the correlation, not the nondestructive test alone, that provides assurance that the filter performs as required. During the production of a sterile product, the filter should be tested before and after the filtration to ensure that the filter meets the specification, is properly installed and intact, and confirms the filter rating. The three major tests used to determine the integrity of a membrane filter are the bubble point, forward flow, and pressure-hold integrity tests.

These tests are based on the flow of a gas (air or nitrogen gas) through a liquid wetted membrane under applied gas pressure. A fourth test, the water intrusion test, is sometimes used to determine the integrity of hydrophobic membrane filters. For a nondestructive test to be useful, the results ofthe test must be able to predict the ability of the filter to remove bacteria. Such a test should also be easy to perform and highly reproducible. The manufacturing quality control ofthe membrane filters tested to obtain this prediction must be used to establish quality control specifications that are consistently maintained in filter manufacturing. It is the combination of the nondestructive integrity test, the correlated bacteria challenge test, and manufacturing quality control that allows membrane filters to be reliably used for sterile filtration of bioprocessing products.

Nondestructive tests

The industry-accepted, nondestructive tests used to verify sterilizing grade filter integrity are the forward flow, pressure hold, and the bubble point tests. These tests are performed by applying a preset air (or nitrogen gas) pressure to a filter wet with an appropriate fluid. For hydrophilic filters (filters that wet readily with water) the wetting fluid is typically water or a water-based fluid; for hydrophobic membrane filters (filters that do not spontaneously wet with water and are used in air service) 60/40 isopropyl alcohol/water is a typical wetting fluid. The wetting fluid fills the voids in the membrane. For this to be occur uniformly, the wetting must be complete. The result of the wetting can be considered as a layer of water supported by the membrane. When gas pressure is applied to one side of the membrane, the test gas will dissolve into the liquid layer to an extent determined by the solubility and pressure of the gas (as described by Henry's law). Downstream of the membrane, the pressure is lower, and the gas in the liquid is driven out of solution. The result is a net flow of gas through the membrane.

The forward flow, pressure hold, and the bubble point tests all involve the observation or measurement of a gas flow through a wetted membrane. The measured flow is a combination ofthe diffusion ofthe gas through the liquid layer and bulk gas flow through any open passageways. The rate of diffusive flow is a function of the test pressure, the diffusivity of the gas in the liquid, the solubility of the gas in the liquid, the membrane void volume, and the membrane thickness. This diffusive flow is described by Fick's first law, which is presented below in an integrated form:

$$N = \frac{DH\,\Delta P_{\rho}}{L} \quad ...(6)$$

where *N* is the permeation rate, *D* is the diffusivity of the gas in the liquid, *H* is the solubility coefficient of the gas, *L* is the thickness ofthe liquid in the membrane, ΔP is the differential pressure, and ρ is the void volume.

As the applied upstream gas pressure is increased, the diffusive flow increases proportionally. At some point, the pressure becomes great enough to expel the fluid from a passageway and establish a path for bulk flow of gas. As a result, the gas flow through the wetted filter begins to increase in a nonlinear manner due to the initiation of bulk gas flow through these open paths. The pressure at which this occurs can be modeled by a modified capillary rise equation:

$$P = F\,\frac{4\gamma\cos\theta}{D} \quad ...(7)$$

where *P* is the pressure (dynes/cm^2), γ is the surface tension (dynes/cm), θ is the contact or wetting angle, *D* is the diameter (cm), and *F* is the shape factor (to account for the structure of the membrane).

As the pressure is further increased, more passages are opened and the proportion of the total gas flow due to the bulk air flow increases. If the pore size distribution is narrow, then the possible passages through the membrane will provide similar resistance to flow. For a wetted high-area pleated membrane filter, a plot of differential gas flow pressure versus gas flow will show a gradual transition from diffusive to bulk flow. An alternate plot of differential pressure versus gas flow or differential pressure will reveal a rapid transition from diffusive flow to bulk flow. This graph is referred to as a diffusive flow spectrum or a K_L (knee location) curve. The K_L pressure represents the transition from the diffusive flow region to the bulk gas flow region. The bubble point test is a test designed to determine the pressure at which a continuous stream of bubbles is initially seen downstream of a wetted filter under gas pressure. There are several ways of performing this test. The first of these is the visual bubble point test.

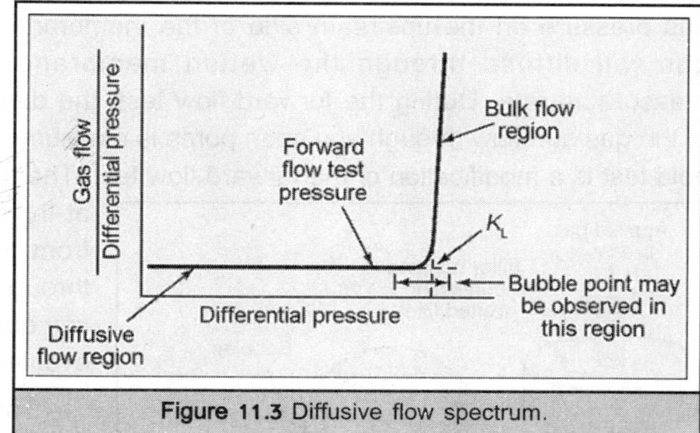

Figure 11.3 Diffusive flow spectrum.

Generally, the visual bubble point test is not used except on small area disks where the whole membrane surface may be examined. To perform a visual bubble point test, gas pressure is applied to one side of a wetted membrane, with a layer of the wetting fluid maintained on the other side. The gas pressure is slowly increased. The rate of increase will affect the measured bubble point value. At some point, the pressure becomes high enough to displace the liquid from one or more passageways. This establishes the beginning of bulk flow of air through the membrane, and a thin stream of bubbles can be observed on the downstream side of the membrane. The pressure at which this stream of bubbles is observed is referred to as the first bubble point. A manual bubble point is measured with a filter installed in a housing, with tubing running from the housing outlet to a beaker of water. The geometry and position ofthe filter housing, and the diameter and length of the tubing will affect the measured bubble point value. In this case, the bubble point is considered to be the pressure at which a steady stream of bubbles is seen coming from the tubing; however, the determination of a steady stream is often subjective.

The manual bubble point will generally occur at a higher pressure than the first bubble point determined by a visual examination of the filter surface, because greater flow is necessary to obtain a clear stream of bubbles from the volume of the tubing. When determining the manual bubble point on a high area cartridge, the diffusion of air through the wetted filter at pressures below the bubble point can be significant. The complications resulting from the length and diameter of the tubing, the housing, and operator interpretation render bubble points determined in this manner not easily reproducible or comparable to bubble points determined by other techniques. The forward flow test and the related pressure hold test were first introduced by Pall Corporation in 1973. During these tests, a wetted membrane is subjected to a predetermined gas pressure on the upstream side of the membrane. The gas will diffuse through the wetted membrane at a measurable rate. During the forward flow test, the diffusion of the gas and flow through any open pores is measured at the specified test pressure. The pressure hold test is a modification of the forward flow test. The upstream side of filter housing is pressurized at the test pressure and then the filter is isolated from the pressure source. The diffusion of gas through the wetted membrane and flow through any open pore is measured as a decay in pressure over a specified period of time.

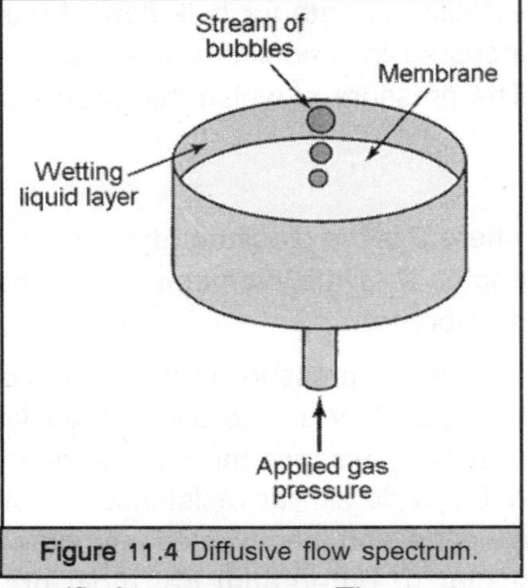

Figure 11.4 Diffusive flow spectrum.

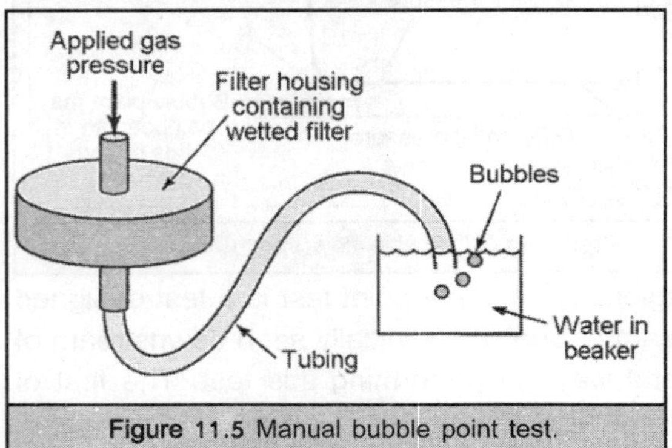

Figure 11.5 Manual bubble point test.

The pressure chosen as the test pressure for the forward flow or pressure hold test must be close enough to the K_L to allow differentiation between different grades of membrane and high enough to present a measurable flow. It is also necessary that the test pressure chosen be such so that the background diffusive flow does not interfere with the sensitive detection of defects in the filter. The exact pressure chosen for the forward flow test must be below the K_L. The test pressure is set by filter manufacturers and should be used during tests performed to establish the correlation with bacterial retention. Integrity tests can be performed with automated equipment. Automated filter integrity test instruments have been developed to provide accurate and reproducible filter integrity test values. The instruments traditionally were designed to perform a pressure hold test. The forward flow value can be calculated from the pressure hold value. Newer equipment designs use direct upstream air flow measurements. The tests are controlled and monitored by a built-in microprocessor, and the equipment often can be used with a programmable logic controller (PLC) for full system automation. Automated devices cannot perform a true bubble point test.

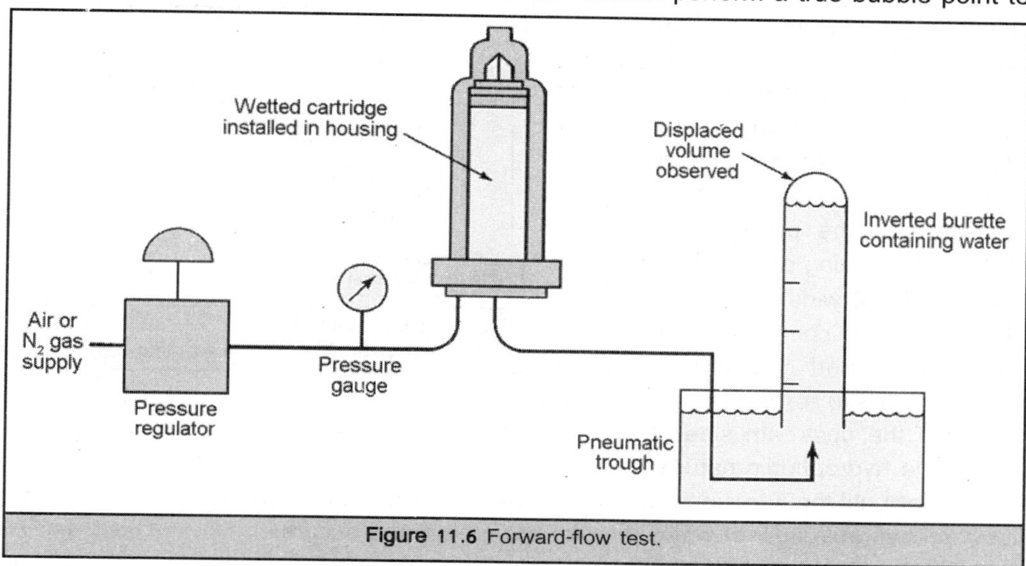

Figure 11.6 Forward-flow test.

Instead, a modified bubble point test is performed, which essentially consists of a series of pressure hold tests. In the automated bubble point test, upstream air pressure is increased in steps. Between each pressure increase, the instrument isolates the upstream volume and monitors the pressure decay upstream of the filter. As the pressure decay exceeds a certain value, the software uses a specific algorithm for calculating the bubble point from relative values of this series pressure decays. The instrument uses a series of pressure hold tests as the basis for the calculation of the bubble point rather than measuring the bubble point directly. The actual algorithms used in the automated test devices may differ between automated filter integrity test instrument manufacturers.

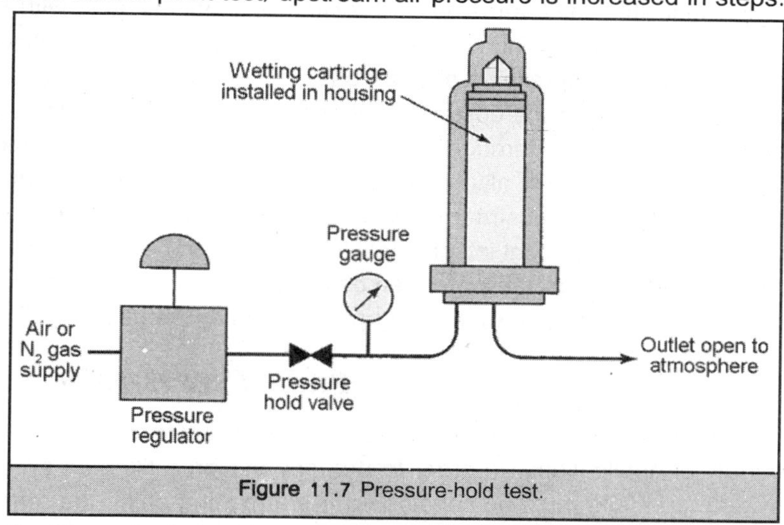

Figure 11.7 Pressure-hold test.

Hydrophobic filters

The water intrusion test can be used to test hydrophobic membrane filters. To perform a forward flow, pressure hold, or bubble point test on a hydrophobic filter, the membrane must be wet with an alcohol solution (e.g., 60/40 IPA/water). For an in situ integrity test, the wetting agent has to be added to the system to perform the test and then removed to restore air flow through the filter. The potential flammability ofthe alcohol is a consideration, and for some applications it can be necessary to validate the complete removal of the solvent to avoid contamination of the product. For these reasons, integrity testing of hydrophobic filters in situ has not been widely used, and off-line testing has been more common. The increasing industry demand and growing regulatory requirement for in situ testing of sterilizing-grade air filters has led to the development of water-based tests. If the upstream side of a dry hydrophobic filter assembly is filled with water and pressurized, the hydrophobic nature of the porous membrane will prevent the bulk flow of water through the filter until the intrusion pressure is reached. At pressure below the intrusion pressure, a small but measurable flow of water through the membrane occurs.

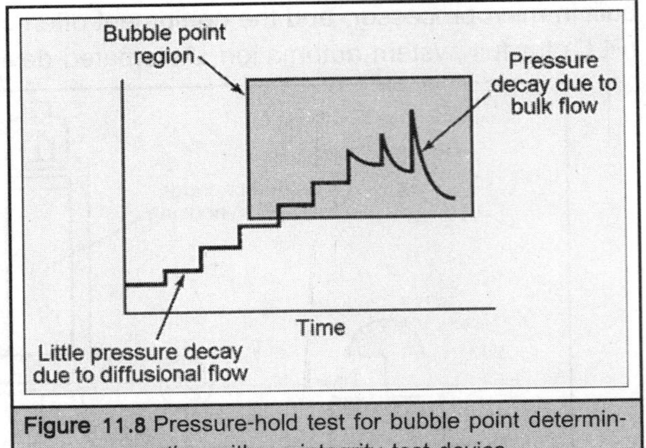

Figure 11.8 Pressure-hold test for bubble point determination with an integrity test device.

The presence of larger pores in the filter will be detected by an increased flow due to bulk water flow through these pores. This principle is the basis of the water intrusion test. At the start of the test, the upstream side of the filter assembly is filled with water and a predetermined gas pressure is supplied. This causes a fall in the water level within the filter housing because of factors such as pleat compression and the passage of trapped air through the membrane. It is important that the changes in the upstream volume caused by these effects have stabilized so that the very small flows of water through the membrane can be accurately measured. Adequate stabilization time (e.g., 10 mins) is also required to ensure thermal equilibrium. Water flow can be measured by determining the upstream flow of gas required to keep the pressure constant. The use of a direct flow measurement integrity test instrument can be used for water intrusion test measurements. After the test is over, the water can be drained and in many cases the fiiter is ready for use. Where dry air or gas is required, a drying procedure can be used.

FILTER SELECTION

As detailed earlier, there are a number of different types of cartridge filter types for a wide variety of applications. In order to properly design a filtration system, all the requirements for the system and fluid must be considered. As a starting point, the removal efficiency must be defined. Other considerations can include process temperature, flow rate, pressure drop restrictions, integrity test regimen, and sterilization. All cartridge filter materials (membrane, cage, core, support material and housing) should be compatible with the process fluid over the temperature and pressure range indicated for a given process application. In general, most bioprocessing fluids are water based and are compatible with most membrane materials, such as polyamides (nylon), PVDF, polysulfone, and cellulose acetate. For many filtration processes, it is desirable to have a staged filtration system.

A staged system will have one or more prefilters of higher micron rating protecting the final filter. The final filter in the stage is the filter that provides the desired effluent quality.

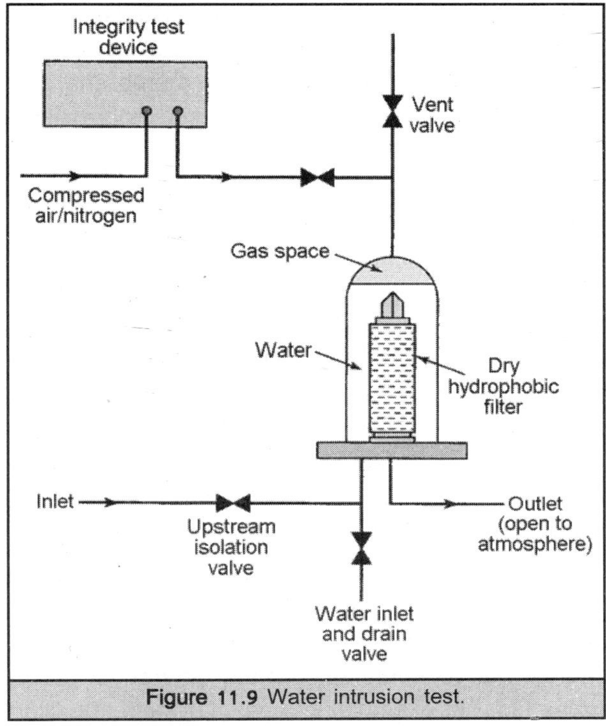

Figure 11.9 Water intrusion test.

The prefilters are typically courser filters that remove larger contaminants in the fluid and reduce the loading of the final filter. For a sterile filtration process, the final filter is typically a 0.2-μm membrane filter. In a staged system, the life of the final filter is enhanced by the use of the prefilters. This will usually reduce the cost of the system, because the cost of the final filter is generally greater than the cost of the prefilters. The size of the filtration system will depend upon the volume of the fluid being filtered and the ease with which the fluid is filtered or its filterability. For production applications, the membrane filters are typically provided as 10-in. segments or filter elements. These segments can be welded together to form 20, 30, and 40-in. filter elements. The diameter of the cartridge is approximately 3 in. Filter housings can hold a single cartridge (10- to 40-in. elements) or multiple filter cartridges. For applications that involve relatively small volumes (for example 10 L), disposable filter assemblies (capsule filters) with smaller membrane areas are available. If prefiltration is required, then the prefilters will also be a part of the filter train. For many applications, there is a one-to-one correspondence of prefilters to final filters. Optimization for a particular process may indicate that fewer, or more, prefilters are needed. In some production environments, it is desirable to have a dual filtration system (filters arranged in parallel) to allow for a change-over from one system to the other.

While one system is being used, the other system can be prepared for use (e.g., it can be cleaned and tested and the filters can be changed out). The decision to use a dual filtration system requires a balancing of the ease of operation with the additional space required for the additional system. For some sterile applications, redundant final 0.2-μm filters (filters in series) are used to provide additional assurance of sterility. If one of the filters fails, because of processing conditions, then the other filter is in place as a backup. If the backup filter is integral, then the fluid has been sterile filtered. The decision point is whether further assurance of maintaining sterility throughout the entire filtration processing time can justify a redundant system.

Resistance of Flow

All components of a system will cause a resistance to flow, which results in a pressure drop (ΔP). Pressure drops, or pressure losses in a system, can be caused by piping, connections, valves, and filling heads and by the filter and its assembly. The pressure drop through a clean filter assembly is due to the sum of component pressure drops, including the filter housing, the filter hardware, and the filter membrane. The filter membrane will cause a resistance to flow, or a pressure drop, in a filtration system. As the filtration progresses and the filter membrane becomes plugged with the particulate contaminants in the fluid, an increasing pressure drop will occur.

The pressure drop will be different, depending on the fluid hydrodynamics (e.g., viscosity) and the filter membrane material characteristics (such as porosity). An adequate pressure source is needed to overcome the pressure drop across the system, including the filter assembly and to allow for the maintenance of a constant flow rate. If the initial clean pressure drop across a filtration system is close to the maximum available pressure, the flow rate will not be maintained because as the filter plugs the pressure drop will increase and there will not be enough pressure to overcome the increased pressure drop throughout the filtration. This will cause a decrease in flow. If possible, where constant flow is required, pumping capacity should be increased. An alternative remedy would be to increase the membrane filtration area so that the initial pressure drop is reduced. Ultimately, the filter will become plugged with contaminants. This will depend on the amount ofthe contaminant in the fluid that is being filtered.

For most pharmaceutical sterile filtration applications the load (percent solids) is typically low. Filters upstream of the sterile filter are more likely to become plugged in a properly designed filter train. As plugging occurs, the differential pressure will increase. System design should include a change-out schedule so that the differential pressure constraints for the filtration system are not exceeded. For example, in a batch system, the filter change-out should occur after the batch filtration has been completed. If multiple in-line stages of filtration are needed, then the differential pressure for each stage must be added, and the sum must be less than the maximum differential pressure allowed for the filtration process to allow for increased pressure caused by membrane fouling. At no time during a process should the pressure drop exceed the recommendations for pressure drop (usually specified for a temperature range of the individual filter unit). At elevated temperatures the recommended pressure drop is typically lower than the recommended pressure drop for ambient conditions. A combination of elevated temperature and a high pressure drop can lead to structural filter damage.

Structural damage to the filter is possible under these conditions because at elevated temperatures the polymeric components of the filter can become softer and are more likely to be deformed if the pressure drop is too high. For example, under extreme conditions, the core of the filter as well as the cage can be crushed. Required flow rate, maximum pressure drop, and available pressure for the filtration system must be considered when the system is designed. Flow rate is typically reported as either gallons per minute (GPM) or liters per minute (LPM). The flow rate achievable through a filtration system is directly related to the applied pressure and inversely related to the resistance to flow. Thus, if the applied pressure is increased, the flow rate will increase. If the resistance to flow is increased, such as by membrane plugging, then the flow rate will decrease. It is important to note that an initial higher pressure (and flow rate) can lead to premature membrane plugging with many products. The nature of microporous membrane filters is that they will tend to plug rapidly if they are subjected to a relatively high flow rate (directly related to a high applied pressure) during the start-up of a filtration. For fluids with a significant particle loading, under conditions of initial high flow rate, the microporous membrane can become rapidly plugged, or fouled, and the pressure drop will increase and the throughput, or filter life, will be reduced. This is especially true for products that may contain gels, such as biological products. Filter life, or throughput, can be increased if the initial flow rate is reduced.

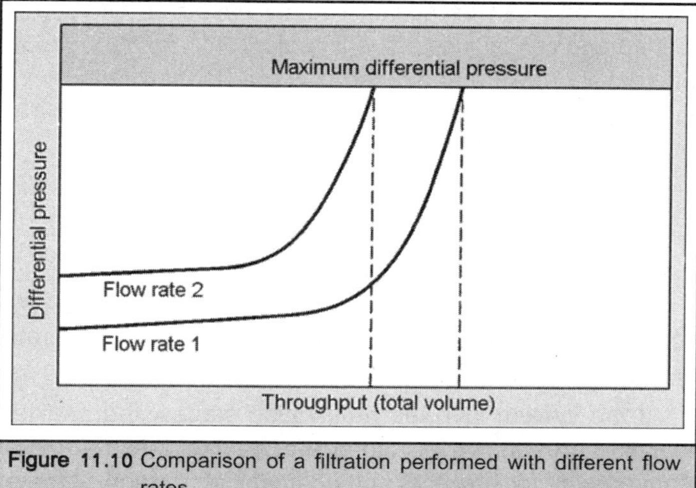

Figure 11.10 Comparison of a filtration performed with different flow rates.

If the initial flow rate is decreased, the pressure drop will increase at a slower rate and the throughput will be increased. Another approach is to increase filter surface area for the given flow rate. Additionally, the pressure drop across a filter assembly can be reduced by increasing the filtration area. Increasing filtration area can often be an economical approach because the increase in throughput is often greater than linear for an increase in filtration area. Fluids with higher viscosities will be more resistant to flow and the pressure drop across the filter will be greater. For fluids with high viscosities, a greater applied pressure will be required to maintain the process flow rate. The smaller the pore size rating of a filter, for a fluid of a given viscosity, the greater the resistance to flow.

Evaluating Particle Size

Although a particle size distribution evaluation can often provide useful information about the contaminant material in a process stream, the filtration system usually cannot be selected without an actual scaled-down filterability test. Actual testing is required because the particle size distribution fails to provide information on the quantitative particle load, its effect on membrane life, and how the membrane filters will perform with the process fluid under process conditions. In addition, other factors such as gels and high molecular weight biopolymers present in some products will also affect filter life and can only be evaluated by actual test. The scaled-down, or filterability, test will require a relatively small representative sample of the process fluid (typically 2 to 4 L). Membrane filter disks with a 47-mm diameter are typically used. If a process stream is subject to variable composition, then the worst case fluid should be defined and tested. Only membrane materials that are compatible with the fluid are tested during the filterability test, and test conditions (e.g., temperature and flow rate) should match the process conditions as nearly as feasible.

Several filtration schemes, including staged filtration schemes, are usually evaluated to optimize the filtration. The filterability tests can be performed at either a constant pressure or a constant flow rate. The use of constant pressure or constant flow rate must be considered because the system size, system life, process time, and throughput can depend on these process parameters.

The flow rate can be scaled linearly based on filtration area. Optimization parameters must be defined and can include effluent quality, time (or throughput) to reach terminal (maximum allowable) differential pressure at a given flow rate, or flow rate for a given applied pressure. System scaling can be based on the following ratios for the final filter in the system, where it is necessary to solve for full scale system area:

$$\frac{\text{Full scale system area}}{\text{Scaled down system area}} = \frac{\text{Full scale system flow rate}}{\text{Scaled down system flow rate}}$$

$$\frac{\text{Full scale system area}}{\text{Scaled down system area}} = \frac{\text{Full scale system throughput}}{\text{Scaled down system throughput}}$$

Throughput (total fluid filtration volume required), flow rate, and process time requirements must be met by the filtration system.

If the system involves prefiltration stages, the prefilter area will usually be equivalent to the final filter area; however, the size of each stage in a multistage system will depend on the overall requirements for practical and economical filtration of the product. Further optimization is possible by performing a filterability test with the prefilters alone. The issues associated with the sizing of a filter system with pre- and final filters are described next. The differential pressure (ΔP) is plotted as a function of volume filtered. The limiting or maximum differential pressure and the total batch volume are indicated on the graph. Ideally, the batch volume (throughput requirement) is reached before the limiting ΔP is reached. This allows a safety factor in the filtration and permits complete processing of the batch without change-out of the filter. The prefilter is unable to adequately protect the final filter. Representative curves are provided for the prefilter, the final filter, and for the combined system. The combined system curve is a sum of the differential pressures for the final filter and the prefilter.

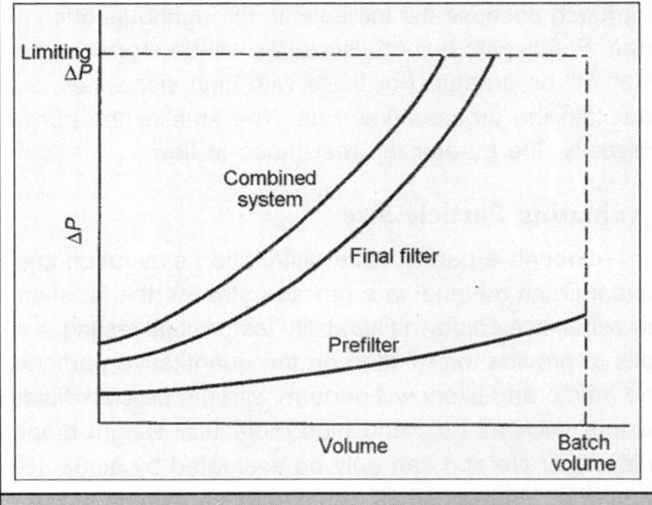

Figure 11.11 Example of filtration system with the prefilter too course.

The combined system reaches the limiting differential pressure (plugging) before the complete batch is filtered. Thus, a finer prefilter or additional final filter surface area would be required for this batch volume to be filtered at differential pressure less than the limiting differential pressure. In this case, the final filter did not reach the limiting ΔP, but the prefilter did. Additional prefilter area would allow the filtration to be optimized. In the optimized system, the combined system is able to completely process the batch at a differential pressure lower that the limiting (plugging) differential pressure. The differential pressures across the final filter are slightly higher that the differential pressure ofthe prefilter. This is the desired situation for a system that requires prefiltration. Depending on the knowledge ofthe filtration process for a particular fluid, it may be desirable to perform a side-stream or a pilot-scale test under process conditions before installing the system at full scale. Further optimization of the system may be possible based on the results ofthe side-stream test. A side-stream test is recommended when the scaled-down test was limited in its ability to match the process conditions for any stage of the filter system.

Specified System Components

Once the filters have been specified, the system components must be considered. A filtration system will typically contain the following filters, filter housings, gaskets or O-rings, pressure gauges, thermocouples, pipe or tubing, connections for the piping, and a pump (or gas pressure source). Some pharmaceutical systems may require an inert gas for product stability as well as pressure. The system components must be compatible with the process fluid, the full-scale process fluid temperature, the process pressure, and the sterilization method. Most sterile filtration applications require a sanitary design. In a sanitary design the components (e.g., sanitary valves) and materials of construction are selected for their ability to prevent the buildup of contaminants in the system. For a sanitary process, 316L stainless steel is typically used for portions ofthe system that will have product contact, provided it is compatible with the process fluid. Internal surfaces should be finished to an appropriate microinch R_s specification (e.g., a minimum of 20 to 25 μin.); external surfaces should be mechanically polished to a high quality sanitary finish. Both internal and external surfaces should be electropolished for a smooth adhesion and corrosion resistant surface.

System welds should be constructed in such a way that the weld porosity is minimized and the joints are of high quality and are clean. Any internal welds that have product contact should be ground smooth and flush to reduce the potential for the buildup of material from the process stream. Proper weld procedures need to be followed. Only sanitary fittings should be used in the system to ensure sterilizibility ofthe seals and cleanliness. Any ports in the system, such as those required for vents, drains, pressure gauges, and thermocouples, should have a sanitary stem or sanitary valve design. Absolutely no threaded connections should be used in the portions of a sanitary system where there is a potential for contact with the process fluid. Threaded connections can lead to system contamination because of a potential buildup of contaminants. There are a number of considerations for the filter housing. Housings are essentially pressure vessels, and as such, may be subject to the appropriate ASME boiler and pressure code.

A housing should be rated for the appropriate pressure and temperature with a safety margin for the process. A full vacuum rating may also need to be considered, especially for housings used in vent service or lyophilizers. The design of the housing as well as the physical placement of the housing in the production facility, must facilitate filter change-out. The size of the housing must meet the requirements for the flow rate and differential pressure process specifications. Inlet and outlet ports should not lead to excessive pressure losses. The system and housing design should not contain any portions through which there is no flow (dead legs). All product contact areas need to be accessible for cleaning purposes. Drains should be designed and positioned to minimize holdup, or retention, of the process fluid in the system. The filter assembly should be optimized for use in a sanitary system, especially where aseptic processing is required. For example, the filter should have an O-ring sealing mechanism in the filter housing instead of a gasket sealing mechanism. For some small batch applications, disposable filter assemblies (in effect filter plus plastic housing) can be used. These assemblies should have proper sanitary fittings.

For most sterile filtration systems a appropriate pump (e.g., a positive displacement pump or a centrifugal pump) is used to provide pumping capacity. The components of the pump head should be of a sanitary design and compatible with the process fluid. Pressure gauges should be installed upstream and downstream of each filter assembly in the system so that the differential pressure across the filter can be monitored. The ability to monitor ΔP can be used to indicate filter life and that filter change-out is required. For some applications it is desirable to have temperature probes or thermocouples installed in the system. During an in situ steam sterilization of a system, the ability to monitor the temperature is needed to ensure that the temperature conditions for steam sterilization are met. For a process that requires a temperature other than ambient, the temperature of the fluid should be monitored.

Temperature measurement should be performed immediately downstream ofthe filter installation to ensure that the filter installation is at the correct temperature. Throughout the system, sample ports may be required to allow for sample collection during a process. It is important to ensure that the collection of a sample does not cause contamination in the system. This can be attained by using sample ports that are fitted with septa. For systems that are not hard piped (plumbed with permanent connections), the ability to make sterile (or aseptic) connections must be available.

Sterile Filtration Process

In sterile filtration processes, the downstream side of the filter must be sterilized and must remain sterile during the process. The sterilization process must be properly validated to ensure that the sterile condition is met for a given system. Filters can be sterilized by a number of techniques, including in situ steam sterilization, autoclave sterilization, irradiation, and chemical treatment (hydrogen peroxide vapors or ethylene oxide). For any method, it is important to ensure that the sterilization is adequate for the system and that the technique does not damage the filter. The Parenteral Drug Association (PDA) published a technical monograph on the validation of steam sterilization cycles.

The majority of sterile processes employ in situ steam sterilization or autoclave sterilization and will be the subject of this discussion. Methods that involve steam are validated through the use of thermocouples or biological indicators to ensure that the system has been properly sterilized. The maximum steam temperature for most sterilizing grade filter cartridges is 140°C; the maximum conditions for temperature and pressure should be obtained from the filter manufacturer.

Typically, the steam pressure is held at a minimum of 15 psi at 121°C for a minimum of 30 min. Although steam temperatures up to 140°C can be used, the service life of the filter may be shortened, especially where multiple uses require repeat sterilization. A decision must be made on the appropriate sterilization technique for the process. Larger systems tend to be in situ steam sterilized, while small volume systems tend to use autoclave sterilization. A system that utilizes in situ steam sterilization can be automated, which is becoming a common practice for a new system.

Integrity Testing

The issues that must be considered for integrity testing of a filtration assembly include the technique (forward flow, pressure hold, bubble point, or water intrusion for hydrophobic filters), the wetting fluid (water or product wet), and the test time (after installation, after sterilization, after filtration) and the test location (in situ or off line). Once a plan for integrity testing is established, the system must be designed so that the integrity test regimen can be properly carried out. Field experience with the integrity testing of sterilizing grade filters shows that various combinations of integrity test procedures are in use and that different testing schedules, both pre- and postuse, are followed.

In some sterile processes, redundant filters are used for convenience in integrity testing or for increasing assurance of sterility. For others, filter usage is based on filter life studies. A common approach is to perform the test or combination of tests that meet regulatory guidelines and that provide the highest degree of accuracy commensurate with the economics and the practicalities of the process. The highest degree of confidence in the maintenance of filter integrity can be attained by routinely integrity testing the filter system, either by the forward-flow or pressure-hold integrity test, before sterilization, after sterilization, and after filter use.

Pre- and poststerilization integrity tests are recommended, because both tests can provide information about the process conditions. However, if only one preuse integrity test is to be performed, then an integrity test after sterilization will provide more useful information, because improper sterilization procedures can lead to filter damage. Regardless of any preuse testing performed, a postuse test is always recommended.

The assurance of bacterial retention during the filtration of critical fluids, such as parenterals, biological liquids, and media for fermentations, is extremely important and, at the very least, filters are integrity tested after use. It is often desirable for a filter user to perform a product–wet integrity test after the filter has been used to filter the product. Product–wet testing allows the filter user to perform the integrity test under actual process conditions and does not require a flush step to remove the product from the filter membrane. Flushing out the product and testing with water can lead to a false failure if there is incomplete removal of the product that may have a lower surface tension than water and thus reduce the bubble point to below the test pressure required for a water-wet integrity test.

Appropriate integrity test parameters for specific applications should be obtained from the filter manufacturer in writing because the filter manufacturer can provide the appropriate test parameters for a fluid and can ensure the proper relationship to the claims for the specific filter. The filter user should confirm performance of the tests before incorporating the test parameters into their standard operating procedures. If any anomalies are noted during confirmation, then these must be resolved. Integrity test parameters can be provided with either air or nitrogen (or other inert gas) pressurization. Carbon dioxide cannot be used as a test gas, because the diffusive flow through the filter will be extremely high and the gas is reactive. Pure oxygen is typically not acceptable as a test gas, because it is a strong oxidizer. Process conditions, including temperature and gas used for pressurization, must be considered for the issuance of integrity test values.

OPTIMIZED FILTRATION PROCESS

An optimized filtration process maximizes throughput and product recovery and minimizes pressure drop while maintaining the desired filtrate clarity, all in a reasonable time. There are five main parameters one needs to consider when selecting an appropriate grade of powder media for a given process.

Prevention of Contamination

Careful attention should be paid to soluble metals when selecting a grade. High concentrations of soluble metals in reagents can lead to contamination of biopharmaceutical products, for example, aluminum contamination of albumin. Also, high concentrations of soluble metals are known to oxidize proteins and enzymes and can activate proteases in fermentation broths. Therefore, the use of high-purity reagents for filtration will help reduce or eliminate product contamination and degradation issues further downstream.

The industry demand for high purity and high performance reagents has led to the development of a new generation of powdered media. These products, sold under the trade name Celpure offer greater filtration capacity with a corresponding reduction in powdered media consumption and reduction in disposal costs.

Table 11.1 Typical celpure properties (applicable to all grades)

Property	*Value*
Soluble aluminum (mg Al/kg)	
Extraction in fermentation broth, pH 4.5.	<3
Soluble iron (mg Fe/kg)	
Extraction in fermentation broth, pH 4.5	<3
Conductivity in H_2O (μS-cm)	<12.5

Filtration processes with Celpure grades typically use less media because of a combination of the higher solids loading capacity and improved flow properties of the media. This results in a reduction in overall processing times compared with conventional grades of diatomaceous earth, such as the widely used Celite grades. Finally, the products are extremely pure and have very low levels of extractable metals such as aluminum and iron and correspondingly low electrical conductivities.

Clarity Specification

Most processes have a clarity specification that needs to be met or exceeded. Achieving stringent clarity specifications can extend the life of downstream filters and protect chromatography columns. The filtrate clarity achieved is dictated by the grade selected and the nature of the turbidity removed. Once the grade is chosen, its level of usage (body feed addition) combined with the available differential pressure to induce flow will control the volume of unfiltered feed that can be processed by a given filtration area.

Grade Selection

Product throughput and filtrate clarity are tightly linked when it comes to grade selection. The goal is to select a grade that achieves the desired clarity and maximizes throughput. By selecting a grade that is too fine, the clarity specification can be exceeded, but the throughput rate may be extremely low with correspondingly high differential pressures.

Operating Pressure

Actual operating pressure can be limited by shear sensitivity of the product or feed-stream by equipment constraints or by overall plant design. Many processes have mechanical pressure limits of approximately 30 to 45 psi or 2 to 3 bar.

Recovery of Products of Interest

This can be sometimes overlooked in initial filtration studies. At the conclusion of any filtration cycle, the accumulated solids should be washed with a product-compatible buffer to maximize product recovery. Upon completion of the filtration cycle (including rinse), it is important to confirm that the product of interest does not interact with the filtration media. Product recovery issues can often be improved by selecting a more permeable grade of filter aid. As you increase the porosity of a filter aid, the available surface area decreases and any nonspecific interaction decreases. All these factors are tightly linked; therefore, we recommend a systematic approach to grade selection and optimization.

DEVELOPMENTAL APPROACH

As stated above, the overall goal is to achieve optimum clarity and maximum throughput for a process while minimizing pressure and product losses. The following is a systematic approach to grade selection. To perform constant pressure filtrations, a pressurized vessel is required to deliver unfiltered feed to the filtration housing.

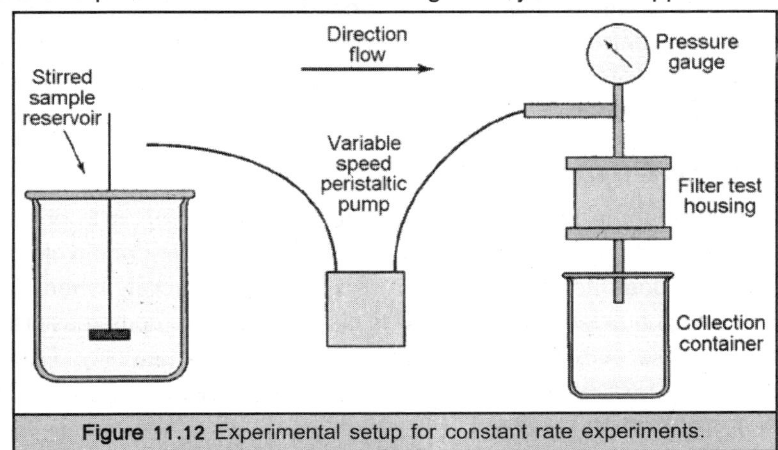

Figure 11.12 Experimental setup for constant rate experiments.

Diversity

Any information on distribution and size of feed solids is useful in grade selection. If this information is not available, a recommended procedure is outlined below. Determine percent solids by weight in your fermentation broth. Filter a small aliquot onto a preweighed filter disk, dry in an oven, and reweigh to determine percent solids in unfiltered feed.

1. Use a Whatman 934-AH filter paper as a septum with the filtration housing. This is an open-pore septum and enables you to precoat without having to custom make a wire screen.

2. Resuspend 1.5 g of media in 100 to 150 mL of buffer (similar in pH to your sample solution; 1.5 g of most grades gives a precoat of approximately 2 mm on 20 cm^2 of surface area). Start with the coarsest grade, Celpure 3000 or Celite 545 AW. By starting with the coarsest grade, you will always achieve maximum throughput and optimal clarity for a given process.
3. Recirculate the solution until filtrate is clear (typically 2 to 3 mins). The recommended flux rate is 40 L/m^2/min. On a 20-cm^2 surface area, this is equivalent to 80 mL/min.
4. Once the precoat has been formed, reduce the volume of buffer in the chamber until 2 to 3 mm above surface of precoat.
5. Add media to the unfiltered feed solution. As a starting point match the percent of solids in your feed with powdered media addition (2% solids by weight, add 2% by weight of powdered media). Add product/ media solution to filter chamber at a flux rate of 1 to 5 L/m^2/min. On a 20-cm^2 surface area, this is equivalent to 2 to 10 mL/min.
6. Measure initial flow rate and monitor both flow rate and pressure throughout the coarse of the filtration. Allow the pressure to rise to 30 psi or 2 bar (or maximum system pressure), and maintain this pressure for the duration of the filtration by reducing flow or until flow rate reaches 20% of initial flow.
7. At the end of the filtration cycle, wash the filter bed with two bed volumes of compatible wash buffer. This will maximize the product recovery for a process. Finally, measure the total volume processed, the time of filtration, filtrate clarity, and the bed height in the filter housing. Save a sample for product recovery analysis.

If clarity is not acceptable, then repeat steps 1 to 7 with a less permeable grade of filter aid. Powdered media are rated based on their Darcy permeability. Celpure 3000 and Celite 545 AW are considered the coarsest grades and Celpure 65 and Celite Filter-Cel the finest. For constant pressure filtrations, follow above procedure through step 4. Then switch to a pressurized vessel containing unfiltered broth mixed with appropriate grade of powdered media. Ramp the pressure up to the desired output and monitor decay in flow rate. Repeat experiments with different grades as outlined above.

Body Feed Optimization

Once grade selection is completed, the next step is to optimize the body feed addition. The level of body feed addition controls the permeability and volume of accumulating cake. Higher pressure does not mean greater through-put, especially when body feed usage coupled with the handling characteristics of the solids dictate filter cake compressibility. The dilatant nature of diatomite resists cake compression and thereby retains permeability. Feed properties (pH, viscosity, etc.) also affect filtration performance; therefore, these factors must be considered in process development and optimization studies.

Past work with biological solutions at high solids levels (greater than 5% by weight) suggests that constant flux (filtration) rates above 10 L/m^2/min exhibited steep rises in differential pressure because of compression and subsequent loss of cake permeability. Modest flux rates (5.0 L/m^2/min) extended filtration cycles, often producing better economy. Although dilution using a compatible solution can reduce shear and also make filtration easier, the goal to expedite processing implies that it will be used only after others options have failed. As described earlier, an excellent starting point is to match the percent solids in the process stream on a weight basis with percent media also on a weight basis. If this ratio of 1:1 is not sufficient, i.e., the filtration terminates quickly because of a rapid rise in pressure and subsequent blinding of filter cake with solids, then consider increasing your body feed addition. Repeat the above steps using body feed additions of 1.5:1 or 2:1. If a body feed addition of 1:1 allows you to process your batch with little or no rise in differential pressure, then you can optimize the body feed by reducing the ratio of powdered media to solids. Reduce the body feed until an acceptable throughput rate is achieved. This approach will allow you to size an appropriate filter and minimize the size of filter needed to process a given batch.

Applications

Candle elements, pressure filters, and filter presses are all used in biopharmaceutical applications. Your choice of filter will depend on a number of factors, including plant design, equipment compatibility with existing systems, clean-in-place and steam-in-place considerations, scale of operation, and overall capital costs. All these factors need to be considered in selecting an appropriate filter. Once the filter design has been selected, the data generated in the methods development stage can be used to predict filter size. A smaller filter can be considered if the cycle time is short, and multiple cycles (including time for cleaning and setup) can be performed, in a typical 8 hour shift, to clarify the batch. The use of advanced filtration media in dynamic filtration systems is a flexible, efficient, and attractive means for the solid–liquid separation of microparticulates from various process streams. Powdered media provide versatility, high solids-loading capacity, high product recovery, low cost, and ease of scale-up in any filtration process.

These systems are cost effective and extremely flexible and efficient at removing suspended solids and clarifying supernatants rich in colloids and small particles. Newer products such as Celpure demonstrate improvements in purity and performance and further address the needs of the biopharmaceutical industry. In addition to the advances that have been made in powdered media, improved systems for media handling and containment are now available from vendors that readily meet the needs of the industry. Finally, a wide variety of economical filter housings of various designs, including candle elements, pressure filters, and filter presses, are manufactured to biopharmaceutical specifications.

12

GAS FILTRATION

In the bioprocessing industry there exist a number of applications for gas (or air) filtration. The type of filter used for the application will depend on the specific removal requirements. The filtration required can be to the 0.2-μ (sterilizing) level or to a courser filtration level for particulate removal. Some air applications for which sterile filtration can be a requirement are fermenter inlet air, fermenter vent gas, vents on water-for-injection tanks, and vacuum break filters during lyophilization. Courser filtration can 90 be used for particulate removal. In many cases, the course filter acts as a prefilter to the sterilizing grade filter. Other applications use coalescers as prefilters for the removal of liquid droplets, such as oil or water.

TYPES OF GAS FILTERS

Packed towers were the first gas filters used by the industry for air sterilization. Packed towers consist of beds or pads of a fibrous material, such as paper, cotton wool, glass wool, or mineral slag wool. Membrane filters typically are constructed of a porous, hydrophobic membrane material, such as PVDF (polyvinylidienefluoride) or PTFE (polytetrafluoroethylene). When filters are used to produce sterile gas, the filters can be referred to as sterilizing grade, 0.2-μ filters. Microbial retention tests can be used to verify that the filters produce sterile gas. The standard test organism is *Brevundimonas (Pseudomonas) diminuta* (ATCC 19146). The removal efficiency of filtration media used in packed towers or membrane filters can be dependent on the following mechanisms:

1. Direct interception by the fibers
2. Inertial impaction
3. Brownian motion or diffusional interception
4. Electrostatic attraction between the fibers and particles

Brownian motion (also referred to as diffusional interception) applies to small particles at low face velocities. When air molecules are in a state of random motion, small particles suspended in the particles can be struck by moving air molecules and displaced. The movement of particles resulting from molecular collisions is known as Brownian motion. This phenomenon can increase the probability of capture of the particle by the fibers in a packed tower or by the diffusional interception in a membrane. The filtration mechanism for *direct interception* is a sieving action that mechanically retains the particles.

The filter material acts as a screen that stops particles that are larger than the pores. Direct interception is independent of face velocity and mostly involves particles with large diameters. *Inertial impaction* refers to the deviation of a particle from its streamline during flow because of inertia. This deviation can result in the retention of the particle by the fibers in the packed tower or by the membrane in a membrane cartridge filter. As the face velocity increases, the probability of inertial impaction increases. An *electrostatic attraction* can exist between the fibrous material in the depth filter, or the membrane material, and the particles in the air stream. This attraction can enhance the ability of the filter to remove particles.

Packed Towers

Packed towers for air filtration are comprised of beds of fibrous material. The diameter of the fibrous material is between 0.5 and 15 μ, and the space between the fibers can be many times this range, depending on how the tower is packed. The filter consists of a steel structure filled with loose fibrous packing. The air inlet is on the bottom and the outlet is on the top of the filter. The packing is supported by a grid or perforated plate. For proper operation, it is necessary to ensure that the appropriate packing density for the application is obtained. If the tower has been packed properly, there will not be movement of the fibers during use. Fiber movement during use can lead to channeling of air, which will lead to an inefficient packed tower, because only a portion of the bed will be acting as a filter. Fiber repositioning can also lead to the dislodging of trapped microorganisms. Once the filter has been packed with the appropriate fibrous material, a support grid or plate is fitted to the top of the bed to ensure that the bed remains compressed.

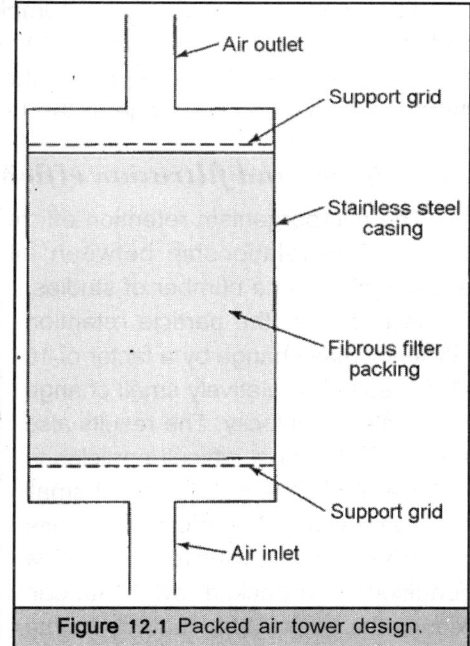

Figure 12.1 Packed air tower design.

After the first steam sterilization of a packed tower air filter, the packing will tend to settle further. Additional fibers can be added to the bed after the first steam sterilization to maintain the correct packing density. Bonded fiber mats have been developed to use in place of loose fibers. Resins are used to bond the fibers, and the mats must be tested to determine if they are compatible with steam sterilization. When mats are used, it is necessary to have a good seal between the mat and the tower wall, so that channeling does not occur. Fiber pads often contain mineral wool with an average fiber diameter of 4 to 5 μ. Thin sheets of small diameter fibers are another option for packed towers. The sheets are placed on top of each other with a mesh or grid in between each sheet for support. The edges of the sheets are sealed between flanges.

Steam sterilization of air filters

Before using a packed tower for the sterilization of fermenter inlet air, the filter itself must be sterilized. There are two techniques that can be used: steam sterilization and dry heat sterilization. Passing steam at a pressure of 15 psig through a packed tower for 2 h is typically adequate for sterilization. Actual conditions required for a specific installation may vary. The presence of air in the packed tower during the sterilization can prevent complete sterilization of the packing. A drain at the bottom of the packed tower is used to purge the steam and to drain any residual condensate in the filter.

It is necessary to remove condensate from the bed because wetted fibers are less efficient for microorganism removal and may also decrease the retention efficiency of the packed tower well below its design value, especially if channeling through the wetted media occurs. Some fiber material as well as material used to bond fibers can be degraded by steam sterilization. An alternative to steam sterilization is dry heat sterilization, which will avoid the possibility of steam degradation and fiber wetting. This can be accomplished by using a heating device at the inlet of the tower and passing air at a temperature of 160 to 200°C through the bed for 2 h. During dry heat sterilization, the filter is isolated from the rest of the process. Fiber material can be stable up to 800°C, thus there is no chance of damage to the bed from this sterilization technique.

Air velocity and filtration efficiency

The microorganism retention efficiency of packed towers is dependent on the inlet velocity of the air. The relationship between air velocity and filtration efficiency has been determined experimentally in a number of studies. In the example, the particle retention efficiency may change by a factor of 10 as a result of a relatively small change in the inlet air velocity. The results also show that the most difficult particles to remove are in the size range of small microorganisms. Therefore, it is quite possible to encounter air flow conditions in a packed tower that can reduce the statistical probability for the complete retention of all microorganisms. The hydrophilic nature of the fibrous material used in packed towers (e.g., glass wool) can contribute to a reduction of the microorganism removal efficiency of the packed tower. Water vapor can enter the system with the air discharge of the compressor. The air and water vapor mixture is initially at

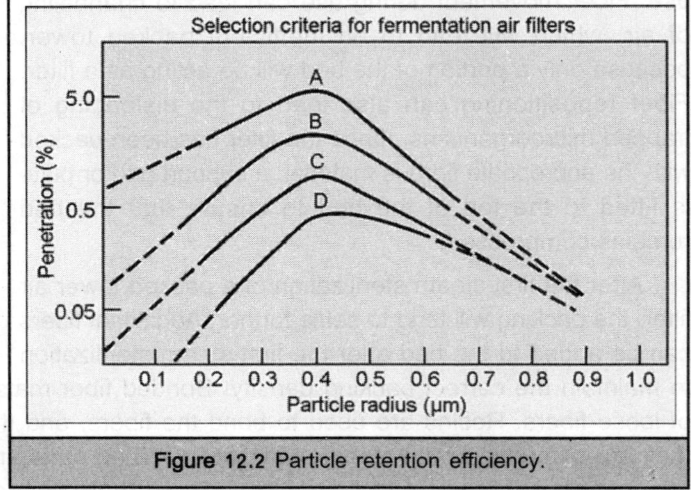

Figure 12.2 Particle retention efficiency.

an elevated temperature. As the gas stream is cooled, water droplets may condense and wet the hydrophilic glass fibers. The wetted fibers are less efficient for microorganism removal and may decrease the retention efficiency of the packed tower well below its design value, if channeling through the wetted media occurs. Also, organic components present in the compressor exhaust can provide a nutrient source for the retained microorganisms and increase the possibility of bacterial growth and eventual penetration through the depth filter medium.

Although a number of approaches have been tried to overcome the wetting problem, such as heat tracing to maintain an elevated temperature, these are expensive to operate and have not consistently resolved the problem. The pressure differential across the tower may also increase significantly with wetting. Although it is obvious that operating deviations such as these do reduce the reliability of a packed tower for air sterilization, there is no quantitative procedure to determine if the filtration efficiency is adequate to ensure a sterile inlet air condition. The lack of such a quantitative procedure adds an element of uncertainty that affects both the operation and maintenance decisions. Skilled operators are necessary to monitor the packed tower to ensure that it is packed and operating efficiently. The actual source of contamination of a production batch is often difficult to identify, and the air filtration system is always suspect during a contamination outbreak. Without a technique for testing the efficiency of a packed tower, it is difficult to ascertain whether the packed tower is the cause of the contamination outbreak.

Fibrous air filters

In order to model a fibrous air filter, several assumptions must be made. It is assumed that (i) Once a particle is trapped by a fiber, it then will remain trapped; (ii) At a particular depth across the filter, the particle concentration does not vary; and (iii) The removal efficiency at a given depth is equivalent across the filter.

The following equation describes how the concentration of particles varies with the depth position in the filter:

$$\frac{dN}{dt} = -Kx \qquad \ldots(1)$$

where N is the particle concentration, x is the depth, and K is a constant.

Solving the equation between a depth of 0 and$x$ and a particle concentration of N_0 particles entering the filter and particle concentration of N particles leaving the filter yields:

$$\ln\left(\frac{N}{N_0}\right) = -Kx \qquad \ldots(2)$$

The relationship between depth and the logarithm of the ratio of particles removed to particles incident is known as the log–penetration relationship. This relationship has been used in sizing depth filters.

The constant K in equation 2 will vary with the type of packing and is dependent on linear velocity through the packed bed. If the relationship between the constant K and linear velocity through the bed is known, equation 2 can be used to size a packed bed for a given log reduction of particles.

Another consideration for the design of a packed bed is the pressure drop across the bed. Typically, the ΔP is linear with the linear air velocity for a given depth. The pressure drop across the packed tower can be dependent on the type of medium, the packing density, the air density, and the linear air velocity through the filter. As an example of an equation for the pressure drop across the packed bed is given in Richards:

$$\Delta P = \frac{2\rho v^2 \alpha x C}{\pi D_f} \quad \ldots(3)$$

where ΔP is the pressure drop, v is the linear air velocity, α is the ratio of filter density over fiber density, x is the filter bed depth, C is the drag coefficient, D_f is the fiber diameter, and ρ is the air density. The relationship in equation 3 indicates that the pressure drop is proportional to the square of the linear air velocity. At relatively low linear air velocities (2 to 3 ft/s), the relationship is linear.

Membrane Filter Cartridges

Membrane filter cartridges are available as either prefilter (particulate contaminant rating) or sterilizing filter (bacterial contaminant rating) configurations. Prefilters in air service can be used for particulate removal or aerosol removal. Prefilters are positioned upstream of the final (sterilizing) filters to protect the final filter from premature plugging, thereby prolonging the life of the final filter. The following are brief descriptions of the membrane filters (sterilizing and prefilter types) that can be used for air filtration.

Sterilization

Membrane filters used for fermenter and bioreactor sterile inlet air and exhaust gas vents, sterile pressure gas, sterile nitrogen blankets, storage tank sterile vents, formulation tank sterile vents, and sterile air for aseptic packaging usually contain a membrane made of hydrophobic materials such as PVDF or PTFE. Hydrophobic membrane filters are desired in these sterile gas filtration applications because hydrophobic filters do not spontaneously wet with water. When a hydrophilic filter is wetted with water, it will not pass air until the water-wet bubble point of the filter is exceeded, and this water-wet bubble point can be greater than 50 psi. The inherent hydrophobicity of membrane filters used for fermenter air sterilization allows these filters to be able to remove bacteria completely from inlet air, even when exposed to moisture. The filter

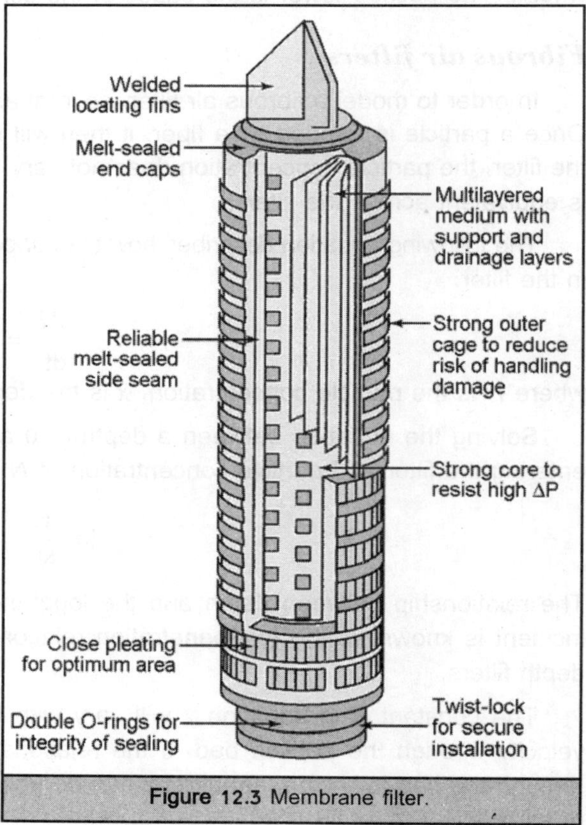

Figure 12.3 Membrane filter.

cartridge is expected to remove bacteria and bacteriophage from air streams with 100% efficiency. The hydrophobic membrane filter material is pleated together with a layer of material (for example, nonwoven polypropylene) on the upstream and downstream side of the membrane. These layers provide mechanical support to the membrane and proper drainage of the fluid.

The pleated membrane pack is formed into a cylinder. The longitudinal side seal of the pleated membrane filter pack should be an integral, homogeneous melt seal without any additional or extraneous materials, glues, or resins. A rigid, perforated inner core is present to provide support against operating pressure. An outer cage placed on the upstream side of the membrane filter pack is provided for additional support and protection during handling. The cage provides retention of structural integrity against accidental reverse pressure (usually up to about 50 psi). Polypropylene and PTFE are examples of materials that can be used for the cage, core, and support material in a membrane filter. End caps are attached by melt sealing to imbed the medium in the plastic. All hardware should be specified as produced from virgin materials. Membrane filters are available in a variety of shapes and sizes. The most typical configuration for pharmaceutical air applications are 10-in. elements. These elements can be flame welded together end to end to form multi-length configurations (typically up to 40 in.). Multiple elements can be used in a filter housing, as required. Membrane filter elements may be integrity tested to ensure that the air is sterile. The integrity tests have been correlated with bacteria removal efficiency by direct microbial challenge testing.

Hydrophobic membrane

Hydrophobic membrane process filter cartridges can be subjected to multiple sterilization cycles and must be designed to be repeatedly steam sterilized in either direction of flow or repeatedly autoclaved. The vendor should provide information on filter cartridge sterilization procedures and operating data on process limitations (time and temperature). These filters are typically capable of withstanding multiple in situ steam sterilizations (for example, at least 165 cumulative hours at up to 142°C). If needed, the membrane filter can be autoclaved and aseptically installed into the process application. Hydrophobic membrane filters can be in situ steam sterilized in 30 min of exposure time (or longer if the process requires a longer time) at a temperature of 125°C. Because the filters are hydrophobic, no additional drying time is required. It is necessary for membrane filters as well as for packed towers to drain the entrained condensate on the inlet side of the filter. A membrane filter housing can be isolated under pressure at the end of an aeration period and therefore does not require sterilization for each cycle.

Removal of aerosols

For some applications, it is necessary to use a prefilter to remove particulate material or liquid aerosols. It is often desirable to use a prefilter to protect the sterilizing grade filter. Typically, the use of a prefilter will reduce the overall filtration economics.

Removal of particulate material

Membrane filters composed of materials such as polypropylene or cellulose can effectively be used to remove particulate material from an air stream. The filter micron ratings range from the order of 1 to the order of 100 μ. The appropriate filter can be selected for the particulate contaminant removal. The following are descriptions of examples of membrane filters that can be used as prefilters in pharmaceutical air filtration, including porous stainless steel filters, cellulose pleated filters, and polypropylene pleated filters.

Porous stainless steel filters cartridges

Porous stainless steel medium is made by sintering small particles of stainless steel or other high alloy powder together to form a porous metal medium. Porous stainless steel can be formed as a flat sheet or, when used in filter elements, as a seamless cylinder. This special manufacturing process produces a high dirt-capacity medium that is temperature and corrosion resistant. The recommended alloy is type 316LB, which has a higher silicon content than type 316L and provides a stronger, more ductile product with better flow properties. Elements have absolute ratings of 0.4 to 11 μ in gas service applications. Porous stainless steel filter cartridges are chemically or mechanically cleanable, offering economy of reuse. Porous stainless steel filters are used for steam filtration and for air sparging.

Cellulose pleated filter cartridges

Pleated cellulose filter cartridges (8 μ rating) are applicable as prefilters for inlet air for fermentors and bioreactors. These filter cartridges can be constructed of pure cellulose medium, without resin binders. The cellulose membrane can be pleated into a high area cylinder and has a longitudinal side seal with an appropriate polypropylene (approved for food contact usage by the FDA). Cellulose media cartridges are assembled with hardware components consisting of a perforated inner support core, an outer support cage, and end caps melt sealed to imbed the medium in the plastic. All hardware components should be of pure polypropylene, without filler or reinforcement, to ensure a minimum of soluble extractables.

Polypropylene pleated filter cartridges

Polypropylene pleated depth filters (8-μ rating) are applicable as prefilters for prefiltration of exhaust gases. These process filter cartridges are constructed using non-migrating continuous strands of nonwoven polypropylene filaments. The medium should have a constant pore size section (downstream) for absolute rated filtration and a continuously graded upstream section for effective prefiltration. The thin sheet of polypropylene media is pleated and formed into a cylinder with a longitudinal side seal of melt seal polypropylene. The cylinder is then melt sealed to injection molded polypropylene end caps to ensure no fluid bypass. Polypropylene hardware components consisting of an inner support core and an external protective outer cage are incorporated.

Liquid aerosol removal

A coalescer can be used for the removal of liquid aerosols containing water or oil droplets. This is desirable as a prefiltration for a sterilizing air filter, because the liquid aerosol could prevent the flow of air through the sterilizing grade filter. Coalesers operate efficiently if they are able to separate the liquid and the gas in the liquid aerosol. The three basic steps that are required are (i) aerosol capture (ii) unloading or draining of the liquid, and (iii) separation of the liquid and gas.

The coalescer has a gravity separator, which allows for the removal of large liquid aerosols (typically >300 μm). The coalescer flow direction is in to out to prevent re-entrainment. The liquid is captured through the coalescence of fine aerosols (0.1 to 300 μm) to large droplets (1 to 2 mm). The large droplets flow downward from a drainage layer. The separation is completed when the liquid is drained from the system. The air leaves the system from the top of the assembly.

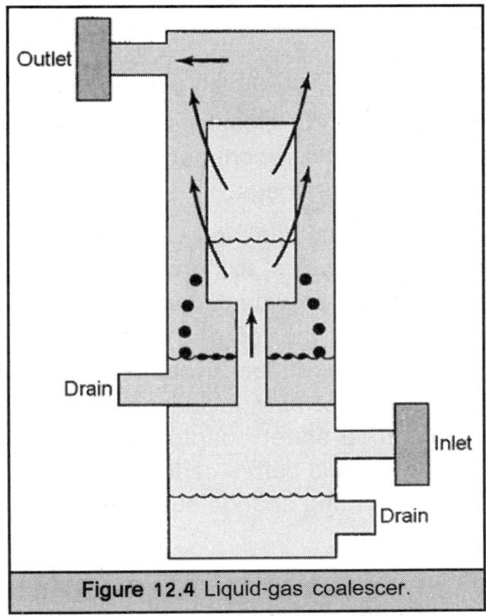

Figure 12.4 Liquid-gas coalescer.

Filter housings

Cartridge filters used for air filtration are placed in sanitary, air or gas service filter housings. Cartridge filter housings usually consists of a head and a bowl. The filter is attached to the head, and the bowl is clamped to the head to provide a complete enclosure of the filter. The size of the housing should be adequate for the flow and differential pressure requirements. Filter housings for bioprocessing applications are typically constructed of stainless steel (e.g., 304, 316, 316L, etc.) or carbon steel, with 316-series stainless steel internal hardware and cartridge seating surfaces. Housings typically have quick-release mechanisms such as V-band clamps or fast-action swing bolts to facilitate filter change-outs. Design operating pressure of all filter housings should be specified as minimum psig and rated for full vacuum service.

Design maximum operating temperature of the housing should also be specified. The housings or pressure vessels that are within the scope of the ASME Boiler and Pressure Vessel Code, Section

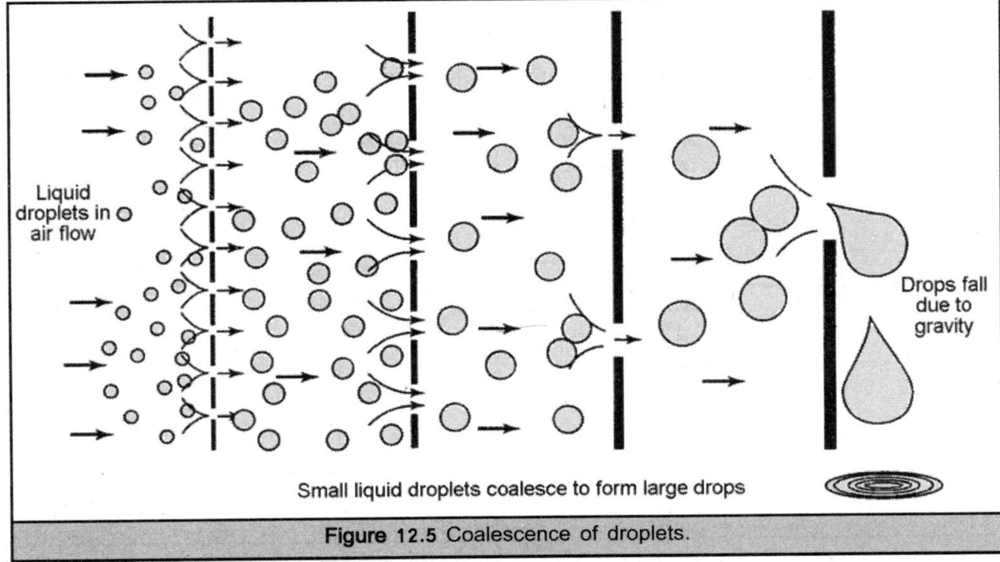

Figure 12.5 Coalescence of droplets.

VIII, Division 1, should be designed and U stamped per the code. TIG weld construction should be used in sanitary-style housings to minimize weld porosity and ensure high-quality, clean joints, with all internal welds ground smooth and flush. All weld procedures and welders should be qualified to ASME/BPVC. Housings should be capable of in situ steam sterilization in accordance with the manufacturer's recommended procedures, and housing or system design should provide for condensate drainage.

Gasket material and O-ring elastomers must also be capable of withstanding repeated steam sterilization cycles, along with being compatible with process fluids. Industrial style housings (used for prefilters) provide cartridge mounting on a tie rod and sealing to the tie rod assembly by use of a seal nut at the top of the assembly. Tube sheet adapters should be seal welded to the tube sheet to prevent fluid bypass. Filter cartridges are thereby sealed in the housing independent of any cover assembly, ensuring positive sealing and no fluid bypass. Filter cartridges should be seated on the tube sheet adaptor assemblies above the tube sheet to ensure complete drainage of nonfiltered fluid before cartridge replacement. This prevents potential contamination of downstream surfaces during change-out of filter elements.

GRADE MEMBRANE FILTERS

Microorganism retention tests can be conducted to verify that membrane filters produce sterile air. Liquid challenge tests with *Brevundimonas (Pseudomonas) diminuta* (ATCC 19146), measuring 0.3 x 0.6 to 0.8 μm, is a standard challenge test for the validation of sterilizing grade filters (0.2 μm) in the pharmaceutical industry. Aerosol challenge tests with *P. diminuta* should approximate extreme air flow conditions. Aerosol challenges with T_1 bacteriophage (0.05 x 0.1 μm), for example, can provide a test of a filter's retention efficiency of extremely small organisms, so small that the bacteriophage in a liquid suspension will penetrate a 0.2-μm sterilizing grade filter. The retention

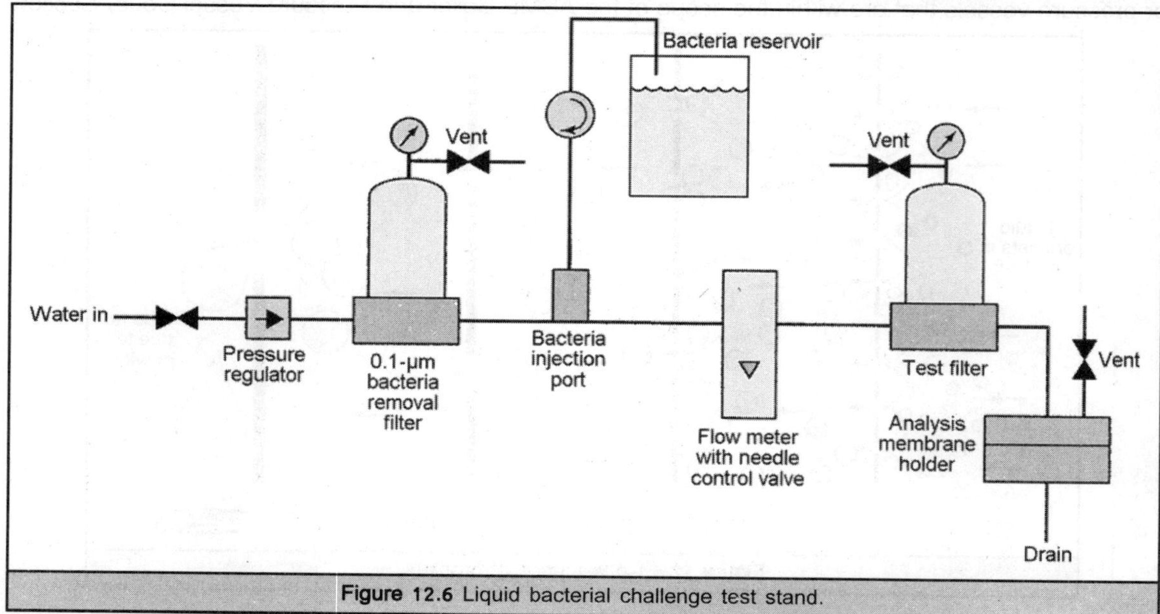

Figure 12.6 Liquid bacterial challenge test stand.

efficiency of a given filter is less when a liquid challenge is used instead of an aerosol challenge. Thus, a liquid challenge test is a more stringent test of a filter's retention capability. A liquid challenge test can also provide retention information for process conditions such as extreme moisture after sterilization or air entrained with water drops.

An example of the technique used to perform a liquid bacterial challenge on sterilizing grade membrane cartridges was published by Pall Corporation. The liquid test involves challenging a test filter with a known quantity of *P. diminuta,* no less than 1×10^7 organisms per 2 square centimeter of filter area. The challenge sample is passed through the filter suspended in sterile water at a defined flow rate and time. All of the effluent from the test filter is passed through an analysis membrane. After the challenge is completed, the system is flushed for 5 to 10 minutes with sterile water. The analysis membranes are removed and placed on Mueller Hinton Agar at 32°C for 48 h. After incubation, the plate is examined for the presence or absence of microbial colonies. The aerosol challenge test system can consist of a nebulizer loaded with the challenge microorganism suspension, a separate line for dry air makeup, and split-stream impingers to sample the aerosol challenge with and without the test filter. During the aerosol challenge, an aerosol is generated with a nebulizer. The aerosol is introduced into the test filter at a given flow rate. The filter effluent is collected in dual liquid impingers. Controls are performed simultaneously via a split stream by using a two-channel timer to direct air flow, on an alternating basis, from the test side filter impingers to the unfiltered control side impingers for recovery.

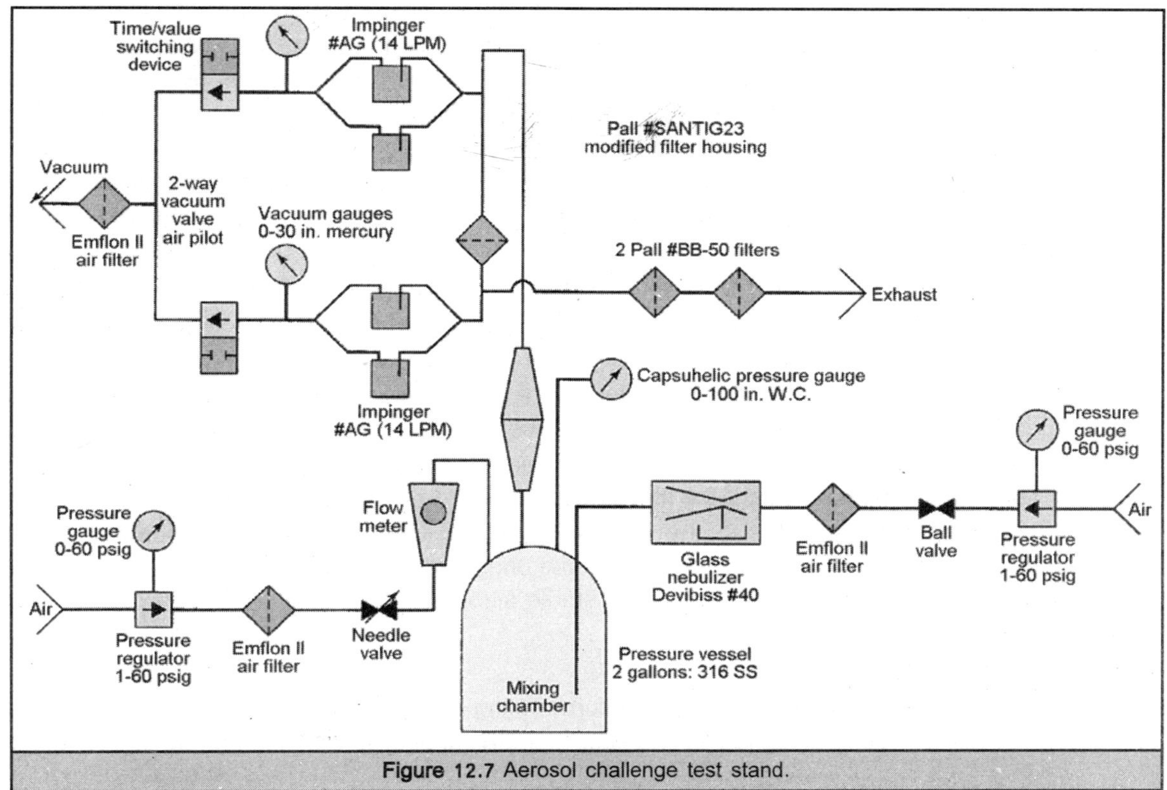

Figure 12.7 Aerosol challenge test stand.

The impingers contain sterile buffer and after the challenge is completed, the buffer can be analyzed for the test organism. If *P. diminuta* is the test organism, then the buffer is analyzed by putting the buffer solutions through an analysis membrane and placing the membrane on Mueller Hinton Agar for 48 h before titering. If T_1 bacteriophage is the test organism, then samples of the buffer are diluted with nutrient broth and mixed with liquid nutrient agar (1.5% agar concentration; 48°C) and *Escherichia coli* in the log phase of growth. After vortex mixing all three components, the mix is poured over nutrient agar plates and incubated for 3 to 5 h at 37°C, so that the plaques can be counted.

Filter

During the sterilization of fermenter air, it is necessary to achieve the highest possible assurance of filter integrity and removal efficiency. The installation of integrity-testable filters and the performance of routine integrity testing by the user are essential to demonstrate that the system is performing to specification. Tests that qualify the retention characteristics of a membrane filter can be defined as destructive or nondestructive tests. Destructive tests are performed using an appropriate contaminant to meet a specific claim for retention of the contaminant. The test procedure must be sensitive enough to detect the passage of contaminants of interest. For sterilizing grade 0.2-μm membrane filters, the industry standard test organism (i.e., contaminant) is *P. diminuta* (ATCC 19146).

The organism and minimum challenge level (10^7 CFU/cm^2 filter area) are specified in the ASTM standard F383,-83 and referenced in the FDA guideline on sterile drug process produced by aseptic processing. Because most filter users would not want to perform a destructive test in a process environment and because the filter is not useable after the destructive test, nondestructive tests related to the retention results of the destructive test are used instead. From the relationship developed between a nondestructive and a destructive test, membrane filter performance can be safely and conveniently verified in the production environment. The relationship between a nondestructive integrity test and the assurance of bacterial retention constitutes a filter validation study and is extremely important for microbial retentive filters used in critical fluid processes.

There are four nondestructive integrity tests that can be used for sterilizing grade hydrophobic membrane filters used for air filtration: (i) bubble point test, (ii) forward flow test, (iii) pressure hold test, and (iv) water intrusion test. These tests are described in Filtration, Cartridge. For a nondestructive test to be useful, the results of the test must be able to predict the ability of the filter to remove bacterial. The manufacturing quality control of the membrane filters tested to obtain this prediction of bacterial removal for the filters must be used to establish quality control procedures that are consistently maintained in the manufacturing of the filter. The combination of the nondestructive integrity test, the correlated bacterial challenge, and manufacturing quality control allows membrane filters to be used reliably for sterile filtration.

Removal Efficiency

Many filter manufacturers use a nominal micron rating for removal efficiency. A nominal rating is an arbitrary micron rating assigned by the filter manufacturer. Such ratings are subject to a lack of reproducibility. An alternative method for rating filters is the Oklahoma State University (OSU) F-2 Test. This rating method has received wide acceptance for use on lubricating and hydraulic fluids. Pall Corporation, for example, uses this method for oils extensively and has adapted it for use in water with contaminants ranging from 0.5 to 25 μm. The test is based on continuous on-line particle counts of different particle sizes, both in the influent and the effluent. The β ratio at a specific particle size is defined as β_x: the number of particles of a given size (X) and larger in the influent, divided by the number of particles ofthe same size (X) and larger in the effluent, where X is the particle size in microns. The percent removal efficiency can be calculated from the β-value. The percent removal efficiency is $[(\beta_x - 1)/\beta_x)]$ 100. As an example, Pall prefilters are given a micron rating that corresponds to a 100% removal efficiency or the value in microns at which the OSU F-2 Test gives a β-value of more than 5000.

USES OF GAS FILTRATION

Fermentation and downstream processing are two major operation categories in bioprocessing. During both fermentation and downstream processing, sterilizing-grade 0.2-μm hydrophobic membrane filters can be required for processing gas (or air) streams. Sterilizing-grade 0.2-μm hydrophobic membrane filters are used during fermentation for the sterilization of fermentation inlet air and for the filtration of fermentation exhaust gas. During downstream processing, sterilizing grade filters can be found in use as sterile tank vents. In a final purification, membrane filters can be used for vacuum breaking in processes such as lyophilization. As discussed above, hydrophobic membrane filters are desired in sterile gas filtration applications because hydrophobic filters do not spontaneously wet with water. Sterile air filtration applications discussed in this section include fermenter inlet air, exhaust gas (vent), high temperature air filtration, and lyophilizer vacuum break filtration. Special attention will be given to the specific requirements for a sterilizing filter when used for these applications. After the discussion ofthe applications is a general discussion of recommendations for sterilizing-grade filter usage.

Fermentation

During a fermentation process a specific cell (yeast, bacteria, or mammalian) is grown to provide a desired product. Products can include cells, antibiotics, amino acids, or recombinant proteins. There can be a variety of sizes for the fermenter, ranging from very small (100 L or less) cell culture reactor to very large scale antibiotic production (100,000 L). In these applications, there is often the need to maintain sterility in both liquid and gas (air or nitrogen) feeds to support growth of the desired cells. Those applications specific to fermentation are described here, and filtration of utilities used in fermentation such as steam, air, and water are discussed in the "Utilities" section.

Compressors

Compressors are often used to generate air flow for the manufacturing facility. There are two types of compressors, oil free and oil lubricated. In older facilities where oil-lubricated air compressors are commonly used, prefiltration of inlet air is necessary for removal of oil droplets. A coalescing filter can provide greater than 99.9% removal of oil and water droplets in the 0.01- to 0.5-μm range and larger. This also acts as an excellent prefilter for the hydrophobic membrane pleated filters that are commonly used for sterilizing the inlet air to the fermenter. The typical gas flow rate per 10-in. filter module is 200 to 400 standard cubic feet per minute (SCFM). For oil-free compressors, a prefilter acts to remove dirt in the air system, extending the service life of the final filter. For use with fermentation air, a cellulose pleated filter with an absolute rating of 8.0 μm is normally the filter of choice. Alternately, polypropylene (2.5-μm rated) pleated filters also serve as excellent prefilters for this application. The typical gas flow rate per 10″ filter module is 75 SCFM.

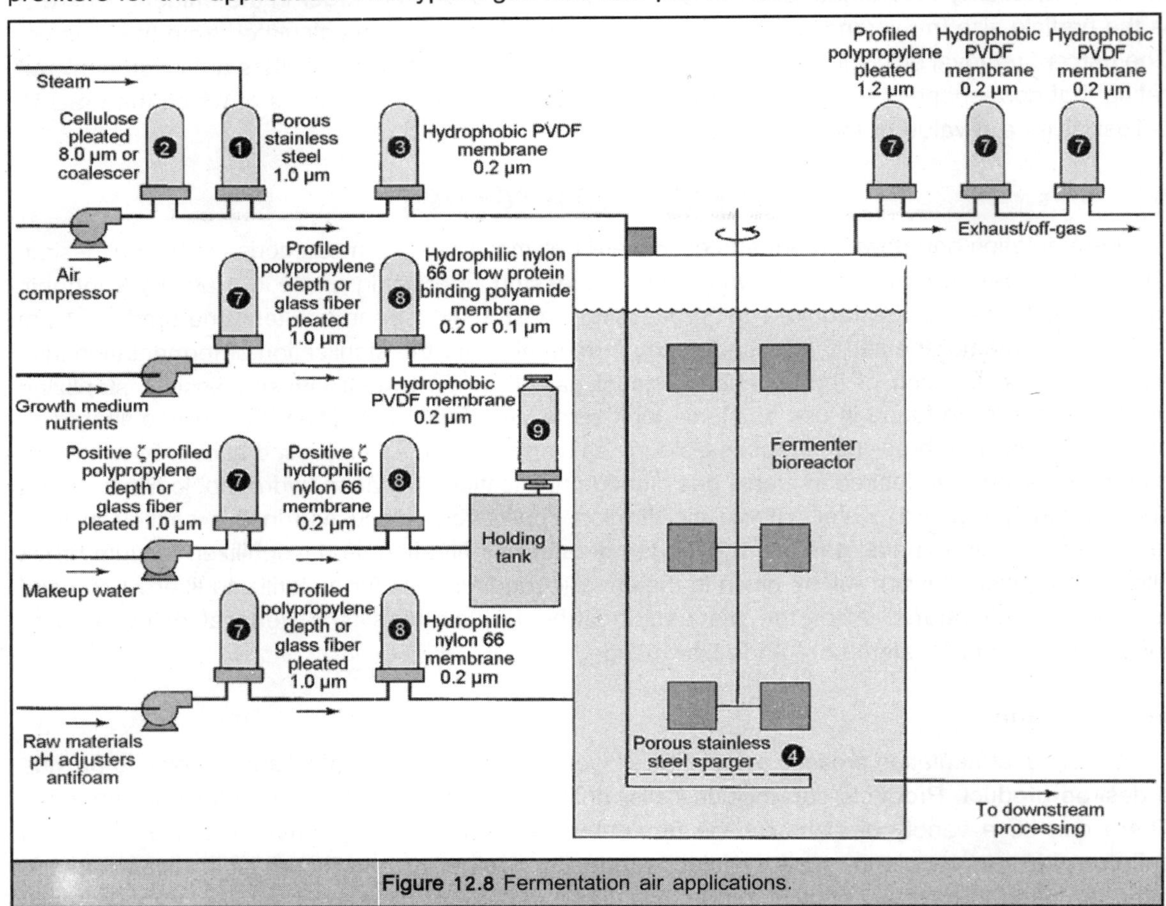

Figure 12.8 Fermentation air applications.

Sterile gas filtration

One of the largest applications for sterile gas filtration is the sterile gas used for an aerobic fermenter during a typical production cycle. Typically, 1 vol of air per volume of broth per minute is used. Thus, for a 100,000 L fermenter on line for 48 h a total of 1.01 x 10^9 cubic feet of gas requires sterilization. The contaminants present in compressed gas can include dust, lubricating oil, hydrocarbons, water, rust, and microorganisms including molds, bacteria, and viruses. Microorganisms in gas are often associated with carrier particles, such as dust. Water and oil can be present as bulk liquid, vapor, or an aerosol. The gas distribution system can give rise to contaminants such as rust and water. The concentration and size distribution of particles in compressed gas are variable. The size range is generally between 0.001 and 30 μ, with a concentration between 10^{-2} to 10^{-4} g/m^3.

Bacteria and bacteriophage, when present in gas feeds, can enter fermentation tanks or bioreactors and contaminate the product. Bacteriophage or other viruses can destroy the producing cells and reduce yields. The process requirements to supply this sterile air can be quite restrictive. The air sterilization process must (i) process a large volume of compressed air, (ii) provide a high degree of reliability, and (iii) operate economically. Several methods have been considered for the sterilization of fermenter inlet air. These include filtration, heat, irradiation, washing with sterilizing chemicals, and electrostatic precipitation. Washing and electrostatic precipitation are not effective for the removal of microorganisms. Heat and irradiation are not economical. Filtration is the only technique that meets all the requirements for sterilizing fermenter inlet air.

Packed towers, an early filtration approach, were used widely in the industry. Since the early 1980s, filtration technology has advanced, and there has been an ongoing trend since the early 1980s to replace depth filters with hydrophobic membrane cartridge filters. The recommended filters for sterilization of air feeds to fermenters and bioreactors are the hydrophobic membrane pleated filters. The hydrophobic (water repelling) nature of these membranes can provide for bacteria and bacteriophage removal with 100% efficiency under moist or dry operating conditions. This is an important benefit over fiberglass towers and cartridges. Filters for sterile air feeds should have a 0.2-μm absolute bacterial rating in liquids and a 0.01-μm particulate rating in air service.

The typical gas flow rate per 10-in. filter module is 75 to 100 SCFM. For some fermentations, the requirement maybe for the filtration of fermenter air at an elevated temperature. If an application involves hot air and a longer service life is desired, then a filter that can withstand the elevated temperature is required. High Temperature Emflon filters manufactured by Pall Corporation can be used in continuous service at a temperature up to 120°C. These filters have a 0.2-μm microbial rating in liquid service and a particulate removal rating of 0.01-µm in gas service. The filter membrane is made of inherently hydrophobic PTFE, and the cage, core, and end caps are specifically designed for high temperature applications.

Air dispersion

Sparging acts to disperse air evenly in the fermenter or bioreactor containing the growth media and product. The product of choice for this application is a porous stainless steel sparging element, which can provide an exceptionally uniform and fine aeration gas dispersion. These elements are fabricated with one face of porous metal and one face of solid metal. If both surfaces of the sparging elements were porous, bubbles from the under surface may coalesce with bubbles from the top surface.

Porous stainless steel sparging elements should be positioned horizontally in the fermentation tank, with the porous stainless steel facing upward. Fine grades of porous stainless steel (e.g., 3.0 μm absolute liquid rated) are ideal even for sheer-sensitive mammalian cell cultures because of their high gas transfer and low shear aeration capability. Elements are typically available in standard and custom designs.

Prevention of contamination

The purpose of a vent filter on a sterile fermentation tank is twofold: to prevent contamination of the tank and to provide containment of the material inside the tank. Prevention of contamination in the tank is desirable for processes that involve long fermentation cycles or require a sensitive fermentation medium (e.g., tissue culture medium). Genetic engineering techniques as well as fermentation of pathogenic organisms (such as organisms used for the manufacture of vaccines) have made it necessary to protect the environment and prevent the escape of microorganisms from the fermentation tank. The exhaust filtration system for a recombinant or mammalian cell fermenter or bioreactor must yield sterile air to the environment and provide a sterile barrier to prevent ingression of contaminants. Additionally, it must be in situ steam sterilizable and typically has a clean differential pressure less than 1 psid. The removal efficiencies for simple depth filters (as described earlier) are typically poor under wet or variable flow conditions. Therefore, membrane filters are recommended for vent filtration applications.

The fermenter or bioreactor exhaust gas line can be contaminated with microorganisms or cells, growth medium components expelled from the fermenter or bioreactor as droplets or as solid particles, and aerosol condensate droplets formed during cooling ofthe gas in the exhaust system. These aerosol droplets, when present, can potentially block the final filter and must be removed before reaching the final filter. Mechanical separation devices, such as cyclones, condensers, and demisters, may not achieve effective aerosol removal below 5 μm. Removal efficiencies and pressure drops also vary significantly with flow rate in such equipment.

Recent studies have shown that aerosols in exhaust lines are predominantly in the very fine 1- to 5-μm range. The contaminants present will depend on the fermentation conditions, the growth medium, and the design of the exhaust gas system. The basic requirements for a vent filter are ability to provide sterility, a low pressure drop, and in situ steam sterilizable. The recommended exhaust filtration system design entails two stages using a polypropylene pleated depth-filter cartridge as a prefilter to a 0.2-μm absolute rated hydrophobic membrane pleated filter cartridge. The purpose of the prefilter (typically 1.2-μm absolute rated) is to remove aerosolized particles and liquid droplets containing cells or growth media from the fermentation off-gas or exhaust air. This serves to extend the service life ofthe final sterilizing filter.

If the medium contains only fully dissolved components, such as with a sterile filtered cell culture medium, and if the fermentation is run at low temperatures (<30°C) and low aeration rate (1 to 1.5 volume of air per volume of media per minute [VVM]), the prefilter may be optional. The typical gas flow rate per 10-in. filter module is 40 SCFM. Like the final sterilizing filter, the pleated polypropylene prefilter should be multiple steam sterilizable. As additional benefit of the prefilter is to retard foam-outs from reaching the final sterilizing filter. Sterilizing-grade 0.2-μm absolute rated hydrophobic membrane pleated final filters, with PVDF or PTFE membranes, can prevent organisms from entering or leaving the controlled reaction zone, even in the presence of water droplets and saturated gas.

Steam sterilizability and integrity test values correlated to microbial retention studies under worst-case liquid challenge conditions provide the highest degree of assurance performance. Redundant systems using a second 0.2-μm rated sterilizing filter in series are recommended for high-risk recombinant organisms. Condensate control is usually the most critical consideration for this application. In cases in which there is condensate accumulation and if the fermenter is operated with overpressure in the fermenter head, the amount of condensate accumulation can be reduced if a pressure control valve is placed at the fermenter exit, upstream of the exhaust gas filter. An alternative technique for the prevention of condensate accumulation is to use a heating section in the exhaust gas pipe upstream of the filter installation. This can be also be accomplished by specifying steam jacketing on exhaust filter housings. In this case, the exhaust gas temperature at the terminal filter must lie above the temperature of the exhaust gas at the fermenter exit. The heater must be properly sized based on the process parameters.

Downstream Processing

Starting with the cells and conditioned broth medium from the fermenter or bioreactor, the objective of downstream processing can be to produce a highly purified, biologically active protein product, free of contaminants such as endotoxins, bacteria, particles or other biologically active molecules. This phase of bioprocessing typically comprises a series of unit operations including cell and cell debris separation, fluid clarification and polishing, concentration and purification, and membrane filtration sterilization of the purified product. Cartridge filters are used in many stages of downstream processing, which involves filtration of both the harvest fluid and product intermediates as well as filtration of air and gases required throughout the process. Air filtration applications include vacuum break filters for lyophilizers, sterile nitrogen blankets, tank vents, and sterile air for container cleaning. Absolute rated cartridge filters eliminate contaminants and impurities from air, nitrogen, and other gases used in downstream processing to prevent contamination of product and further protect concentration and purification equipment.

Vent filtration ensures containment and freedom from product contamination during fluid transfer operations and protects processing equipment during sterilization cycles. In fermentation, cartridge filters are used to maintain the sterility of the makeup water, feeds, additives, media in holding tanks, and in fermenter or bioreactor exhaust. Cartridge filters are typically used in downstream processing for the filtration of air, gases, and venting applications when it is necessary to vent tanks during fluid transfers; pressurize tanks using inert gases such as nitrogen and argon; protect vacuum lines, sterile vent holding tanks, and lyophilizers; and for gas purging, blanketing, drying, and when sterilizing equipment by in situ steaming or autoclaving. The recommended filters for nonsterile particulate removal applications are polypropylene pleated filters. Hydrophobic membrane pleated filters such as PVDF or PTFE are recommended for aseptic processing. The absolute removal rating for the latter filters should be 0.2 μm determined under liquid flow conditions. The typical gas flow rate per 10-in. filter module is 75 to 100 SCFM. In downstream processing applications, Cell and cell debris separation and clarification processes can be broken into a primary separation, secondary separation, and a cell concentrate section.

During some primary separations, a cyclone can be used for particulate removal; a sterilizing air filter can be used as a vent on the cyclone. A variety of holding and receiving tanks can be used during the separation and clarification process; these tanks can be fitted with sterile vent filters. During secondary separation, a nitrogen blanket may be needed; the nitrogen gas can be sterile filtered with a hydrophobic filter. Downstream processing can also involve the concentration and purification of clarified harvest fluid. Applications for air filtration include sterilizing- grade vent filters for solvent or buffer tanks and for holding or buffer tanks needed for ultrafiltration and chromatography. The final pharmaceutical product will often need to be packaged. During filling processes, a sterile nitrogen blanket and thus a sterilizing grade hydrophobic filter may be needed. The final product is can be placed into a container; sterile air or nitrogen may be needed for container cleaning. A vent filter can be required on holding tanks. The typical gas flow rate per 10-in. filter module for tank vent applications is 75 to 100 SCFM.

Grade filters

Sterilizing grade filters are used in freeze dryer installations to filter the gases used to maintain the chamber pressure and to break vacuum during operation and in sterilizers for vacuum break purposes.

Equipments required

Blow-fill-seal equipment can be used for the aseptic filling of pharmaceutical products. The container is formed and sealed aseptically. Air filtration is required to ensure sterility in this unit operation. For buffer-tank air hydrophobic membrane filters are used to supply sterile air to a buffer tank on the blow-fill-seal machine. This blanket air is used to drive the sterile solution through a pneumatically controlled dosing system. The air used in the buffer tank is referred to as the gas cushion or buffer tank air. Hydrophobic membrane filters provide sterile air used to form the hot

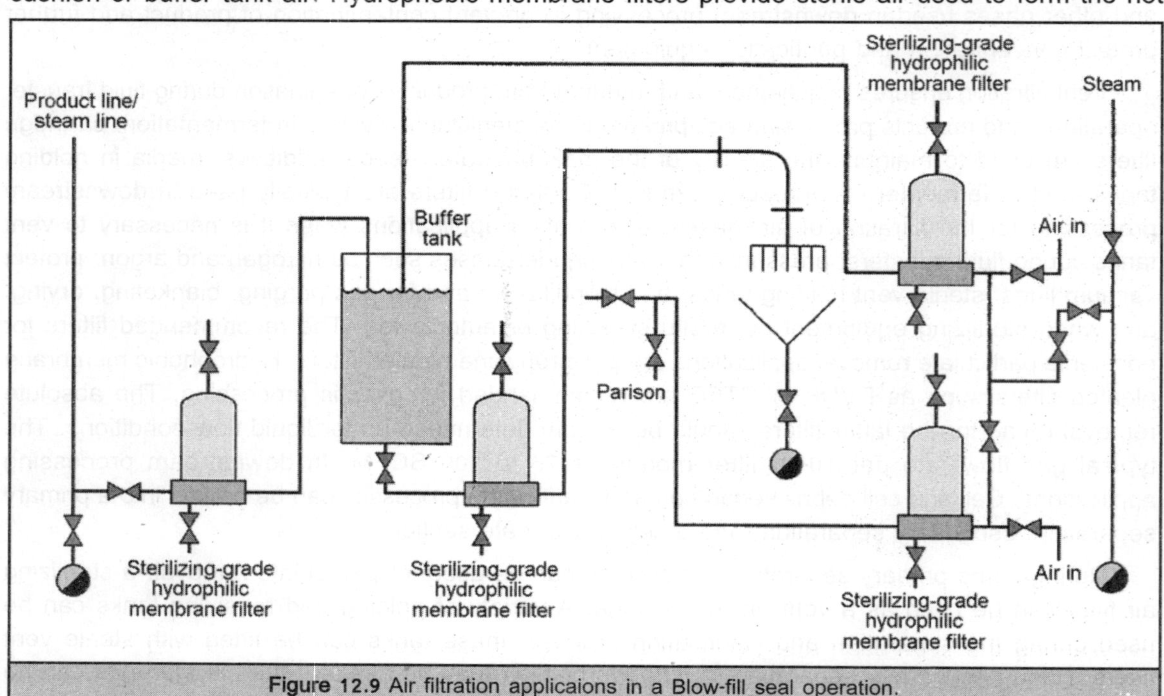

Figure 12.9 Air filtration applicaions in a Blow-fill seal operation.

moldable plastic tube (parison). The air used to form the parison is known as the parison support air. The parison is subsequently blow molded into the shape of a ampule strip or a bottle. Typical requirements for the filters used include that the filters be steam sterilizable, integrity testable, and the proper size to prevent restriction of gas flow. Hydrophobic membranes are used to prevent wetting out and to maintain high flow rates even in moist conditions.

Requirements

There are a number of peripheral unit operations required during a sterile process.

Liquids

Sterilizing grade membrane filters can be used in vent applications in which the fluid in the tank is at an elevated temperature. One such application is the vent used to prevent contamination in a water-for-injection tank. The water is at 80°C or higher. When a sterilizing-grade air filter is used for this type of vent service, a steam-jacketed housing is typically used. It is only necessary to maintain the temperature of the filter cartridge at a temperature *slightly above* the dew point of the vapor. The steam introduced into the jacket should be at ambient pressure. Continuous operation of the jacket at a significantly higher steam pressure and temperature can reduce the service life of the filter caused by accelerated aging of the hardware by oxidation.

Role of steam

Process equipment and final filters are frequently sterilized by direct steam flow in situ, during the normal line sterilizing cycle. This eliminates the need for making aseptic connections and risking recontamination. Filtered steam is required for this sterilizing-in-place of filters, piping, vessels, and filling equipment. Steam is also required for general equipment cleaning and sterilizing. The steam often contains significant amounts of pipe scale and other corrosion products. This particulate material should be removed in the interest of overall cleanliness and to avoid burdening the prefilters and final filters. Particulate contamination in process steam is efficiently retained by porous stainless-steel filters with an absolute gas rating of 1.2 μm. Porous stainless-steel filter assemblies are typically sized at steam flow rates of 30 to 40 ACFM per square foot of filter medium.

GUIDELINES

Field experience with the integrity testing of sterilizing grade filters shows that various combinations of integrity test procedures are in use and that different testing schedules, both pre- and post-use, are followed. Perhaps the best approach is to perform the test or combination of tests that provides the highest degree of accuracy commensurate with the economics and the practicalities of the process.

Since the requirements for the use and testing of sterile gas filters can depend on location and application, several different types of filter tests can be used. Filter users typically use the forward flow or pressure decay tests, which require the use of a low surface-tension solvent, such as 60:40 isopropyl alcohol/water or 25:75 *t*-butyl alcohol/water. This is a well-proved, widely accepted approach directly correlatable to bacterial challenge. Air filters are typically integrity tested when they are installed and retested periodically based on service conditions and operating requirements (e.g., once a month). Change-out schedules based on filter life studies have also been used in conjunction with integrity testing.

Because filters in these applications are often sterilized in situ and can be damaged during in situ steam sterilization by reverse pressurization if the sterilization procedure is not properly controlled, filter life study data or a periodic integrity test regimen can be used only if the filter sterilization procedure is validated and in control. In certain cases (such as in facilities in which the use of a nonflammable fluid is required, or when disposal of organic solvents is a concern), some filter users in aseptic processes have recently considered the use of water-based tests for hydrophobic filters. There are a number of integrity tests possible for a hydrophobic air filter. The selection of the appropriate test and the appropriate test schedule depends on the specific application. After an integrity test has been completed, it is typically desirable to remove the wetting fluid from the filter. This can be accomplished by blowing clean, dry (-40°C dew point) air or nitrogen through the filter. It is necessary to qualify this procedure, because every system is different.

Precautions

Filter cartridge change-out is usually based on actual experience, with a safety factor. Filters should be inspected on a monthly basis for oxidation. This should be supplemented by monitoring the pressure drop across the filters during operation to determine if the filters are plugging and routine integrity testing to confirm filter integrity during the service life ofthe filter. Alternatively, filter life studies, with an appropriate safety factor, could also be used to set a change-out schedule. Actual conditions for each application should be used during filter life studies. For elevated temperature applications, conditions leading to oxidation need to be considered. Aging of the membrane filters by oxidation depends on the status ofthe system.

Oxidation does not occur when the cartridge is being steamed, because there should be no air present in a properly operating steam-in-place system. If a cartridge is exposed to air at an elevated temperature, oxidation of the material in the filter, such as polypropylene hardware, will be accelerated. Oxidation will also occur when the filter is in a stagnant situation, that is, it has no air flow going through it. The flow of air through a filter can moderate the temperature environment, whereas under stagnant conditions the temperature ofthe filter will rise to the temperature of the housing. Stagnant conditions can exist when the tank is not being used or when the tank is empty. To prolong service life, the steam jacket should be turned off when there is no air flow through the filter for extended periods of time, when operating conditions permit.

Guidelines for Membrane Filters

Membrane filters can be sterilized by chemical sterilants (such as ethylene oxide, hydrogen peroxide in vapor form, propylene oxide, formaldehyde, and glutaraldehyde), radiant energy sterilization (such as γ-irradiation) or steam sterilization. The most common method of sterilization is steam sterilization. Steam sterilization of a membrane filter can be accomplished either by an autoclave or by in situ steam sterilization. In situ steam sterilization can be effectively accomplished by a variety of different process arrangements. Steam sterilization is often the most critical portion of the process, and it is important that the procedures followed lead to sterilization of the system and do not impart any damage to the membrane filters. Some general procedural recommendations are included for prevention of either forward or reverse pressurization damage. Reverse pressurization conditions can be prevented by using a noncondensing gas, such as air or nitrogen, at the end of a steam cycle. If this step is not followed, and the steam valve is shut off without the introduction of air or nitrogen, the filter housing will act like an isolated system.

As such, the temperature will be different on the upstream (outside of the housing) and downstream sides ofthe filter. Due to the temperature difference, steam will condense at different rates on the upstream and downstream sides of the membrane. This can lead to pressure differences of 5 to 15 psi in the reverse direction. Even if this reverse pressure condition exists for a short period of time, the filter can suffer permanent damage. Condensation during the steam cycle can lead to long cycle times and filter damage by excessive forward pressurization. If during steam sterilization the filter membrane is wet with condensate and in order to overcome the resistance of the wetted membrane the differential pressure in the forward direction is greater than required maximum (typically 5 psi), then the filter can be damaged by excessive forward pressurization. In this case, the filter will appear crushed. Precautions should be taken to minimize condensate accumulation. Drains for condensate removal should be strategically placed and should be cracked during the steam cycle.

13

L-ASPARTIC ACID

L-Aspartic acid is widely used as a food additive as well as an ingredient of infusion solutions and medicines. In addition, a dipeptide, L-aspartyl-L-phenylalanyl methyl ester, is used as a synthetic low-calorie sweetener in many countries. Moreover, studies for production and application of L-aspartic acid derivatives have been carried out by many companies in Europe, America, and Japan because L-aspartic acid is also chemically reactive. L-Aspartic acid derivatives are hoped to function as a kind of biodegradable precursor for environment friendly products. It is presumed that each of these would be produced at the 1,000-kiloton level in the near future. However, the high price of L-aspartic acid produced using the existing process has been a high hurdle in expansion of its use. Therefore, we tried to develop a novel process to produce L-aspartic acid economically.

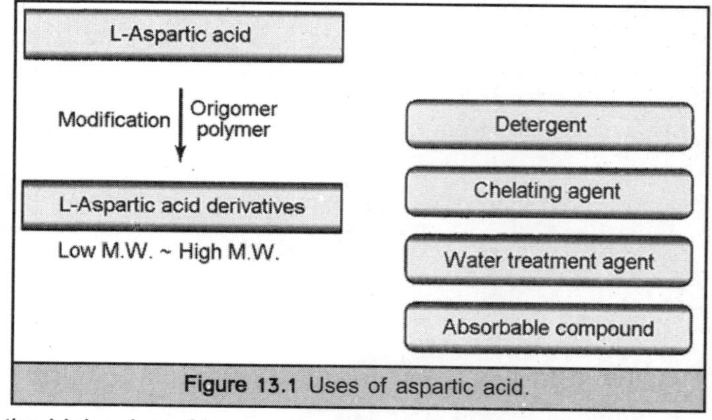

Figure 13.1 Uses of aspartic acid.

INDUSTRIAL PRODUCTION

Since Quastel and Woolf observed the reversible formation of L-aspartic acid from fumaric acid and ammonium ion by a cell suspension of *Escherichia coli*, many reports have been published on aspartase, which catalyzes this reaction. This enzyme degrades L-aspartic acid into fumaric acid and ammonium ion under physiological conditions, but the reverse reaction proceeds in high concentrations of ammonia. High-yield conversion of L-aspartic acid from fumaric acid and ammonia was first reported by Sumiki et al. in 1928. This method employed resting yeast cells to produce L-

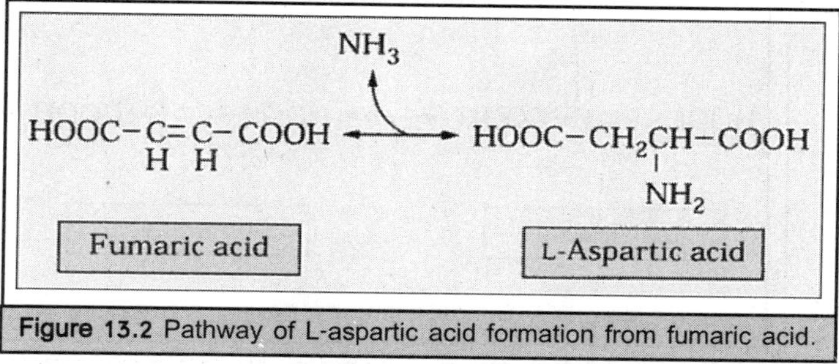

Figure 13.2 Pathway of L-aspartic acid formation from fumaric acid.

aspartic acid. Fermentation methods using medium containing fumaric acid, and enzymatic methods using bacterial aspartase activity have also been reported. A process for industrial production of L-aspartic acid was first established in 1973; bacterial cells having aspartase activity were immobilized in a gel matrix to stabilize the enzyme activity. A continuous operation was also reported. In the immobilized-enzyme process, the enzyme cost was reduced compared with that of a batch process. However, in the immobilized-cell process, cells had to be immobilized and packed into reactor columns under aseptic conditions. Contamination-free operations were needed, which resulted in additional costs. Also, it did not seem to be possible to use the reactor columns for conventional purposes because different specifications were required according to each individual reaction mode or the chemical properties of substrate and product.

CHARACTERISTICS OF MEMBRANE REACTOR

As already mentioned, there are several biological and engineering disadvantages that need to be solved in conventional production methods. To overcome these disadvantages, we have developed a novel bioreactor process, the membrane reactor system, using microbial cells without artificial immobilization for the enzymatic reaction. Coryneform bacterium MJ-233, which is employed in this process, is characterized by nonlytic properties under non- growing conditions. The cost of the enzyme is low because the bacterial cells can be reused repeatedly for enzymatic reactions without immobilization. Intracellular components such as protein and nucleic acid do not leak out, so the purity of the products is very high. From the engineering stand point, the ultrafiltration membrane system is employed for recycling bacterial cells. The advantages of the ultrafiltration process are as follows:

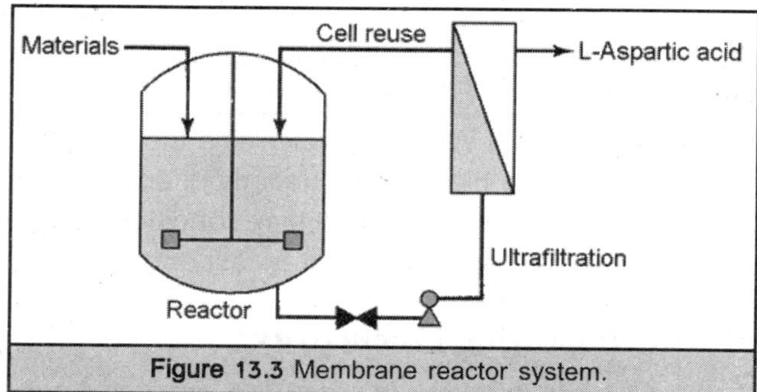

Figure 13.3 Membrane reactor system.

1. Damage of bacterial cells by compression is reduced compared with centrifugation.
2. Capital outlay and running costs are relatively low.
3. Operation and maintenance are easier.

REQUIREMENTS

We tried to develop a new bioprocess for L-aspartic acid production using maleic acid as a starting material. Maleate isomerase, which isomerizes maleic acid to fumaric acid, and aspartase, which converts fumalic acid and ammonium ion to L-aspartic acid, were used for L-aspartic acid

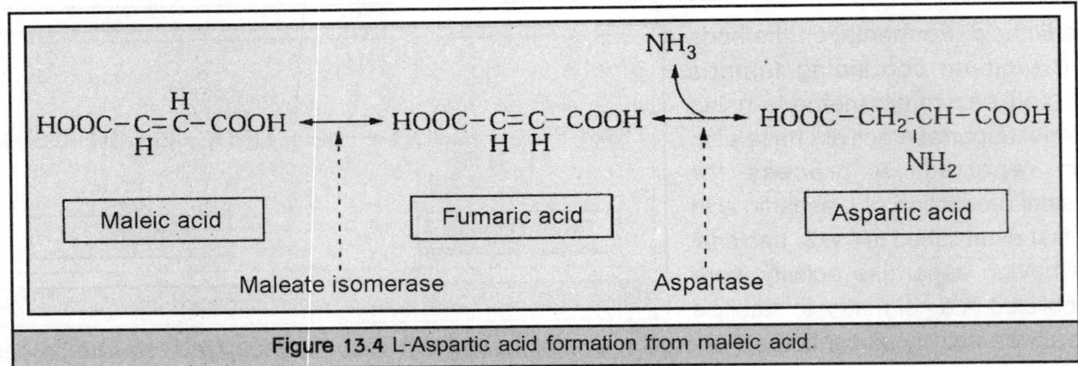

Figure 13.4 L-Aspartic acid formation from maleic acid.

production from maleic acid. The equilibrium of the isomerase reaction strongly favors fumaric acid formation. This results in almost complete conversion of maleic acid to fumaric acid, so it is very useful for L-aspartic acid formation. On the other hand, the equilibrium of the aspartase reaction also favors L-aspartic acid formation under high concentrations of ammonium ion. Therefore, the conversion yield from maleic acid to L-aspartic acid can reach 99%.

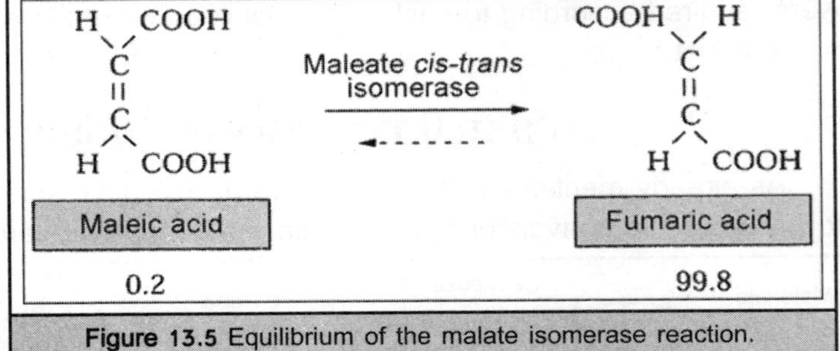

Figure 13.5 Equilibrium of the malate isomerase reaction.

Factors Affecting the Production

In order to establish a one-pot reactor system using two enzymes, maleate isomerase and aspartase, we analyzed the properties of each enzyme, including pH profile. We could determine the potential for coexistence of two enzymes in a one-pot reactor even in the high concentrations of ammonium ion that are an essential condition for L-aspartic acid formation by aspartase.

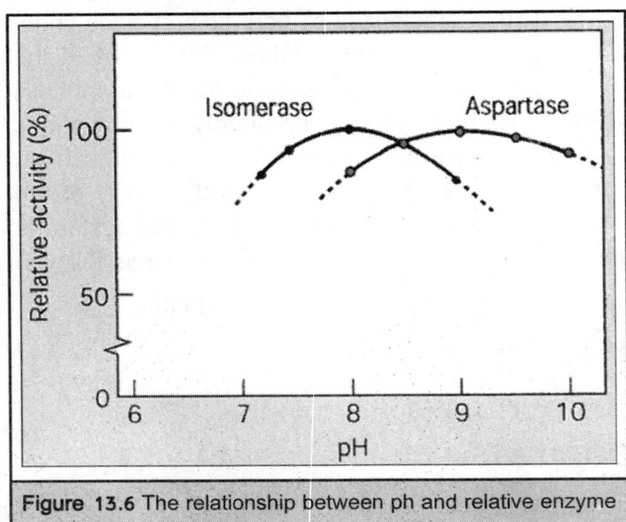

Figure 13.6 The relationship between ph and relative enzyme reactivity.

Improving Productivity

In order to improve the productivity of the process, genetic engineering techniques were applied to coryneform bacterial strain MJ-233. An efficient transformation system in coryneform bacteria using electroporation and other techniques has also been developed. Both maleate isomerase and aspartase can be overproduced in coryneform bacterial strain MJ-233. These recombinant cells can therefore be employed for L-aspartic acid production from maleic acid as a starting raw material.

Host–vector interaction

One of the important factors in the application of genetic engineering techniques for industrial use is the development of a stable host–vector system. For this purpose, we isolated various plasmids from coryneform bacteria and identified a genetic element that is important for stable plasmid maintenance. We then constructed a highly stable host–vector system.

Hereditary control

Isolation of both the maleate isomerase gene and the aspartase gene from a genomic library was achieved by hybridization with DNA probes synthesized on the basis of the N-terminal amino acid sequences of each purified enzyme. Using this methodology, we cloned maleate isomerase and aspartase genes from the *Alcaligenas faecalis* and coryneform bacterial strain MJ-233, respectively, and achieved high-level expression in recombinant cells of coryneform bacterial strain MJ-233.

Physiological Characteristics

Coryneform bacterial strain MJ-233 exhibited nonlytic properties under nongrowing conditions and, therefore, showed no leakage of intracellular macromolecules, as already mentioned. Taking advantage of these physiological characteristics, an ultrafiltration system was adopted for the separation and recycling of cells from the reaction mixture. Employment of the ultrafiltration system greatly simplified the purification step. In this ultrafiltration system, the filtration capacity can be easily modified by varying the number of modules. A hollow-fiber tubular-type membrane made of polysulfon, which can withstand high temperature and alkaline pH, was chosen for our process. We can operate this system at high temperature and alkaline pH, allowing us to carry out a contamination-free process. L-Aspartic acid can be produced at high purity using this membrane reactor system and, therefore, more economically compared with conventional fermentation processes. The host–vector system in coryneform bacterial strain MJ-233 was developed and L-aspartic acid can now be economically produced at high yield using recombinant intact cells in a membrane reactor system. We hope that our method can provide the quantities of L-aspartic acid needed to meet the increasing demands for biodegradable chemicals in an expanding market.

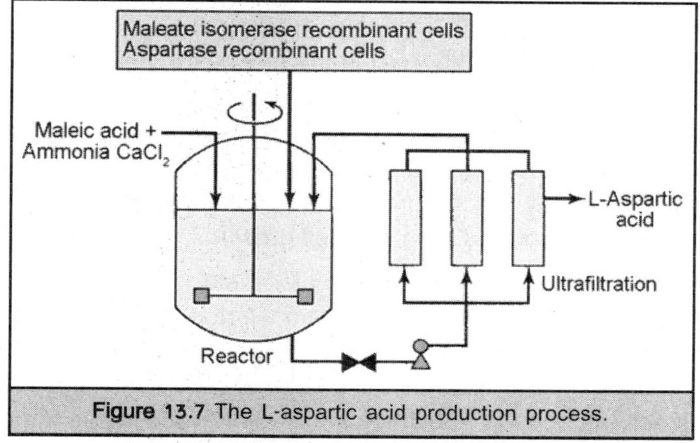

Figure 13.7 The L-aspartic acid production process.

14

NITROGENOUS TOXICANTS

Mammalian cells are used for the production of several important biologicals including interferons, growth factors, vaccines, hormones, and monoclonal antibodies. The process efficiency of cell culture fermentation should be increased in most cases for economically viable production. Maintaining high numbers of viable cells in the bioreactor for extended periods of time results in higher product concentration and higher reactor productivity, thus increasing process efficiency. The cell growth rate and the number of cells attained in the bioreactor are determined, in part, by the concentration of nutrients available to the cells. Even though fed-batch or perfusion systems can eliminate nutrient depletion, they are still prone to inhibitory metabolite accumulation. Several metabolic by-products have been reported to inhibit cell growth.

Ammonia and lactate are major metabolic by-products of glucose and glutamine metabolism, and their presence has been demonstrated to diminish growth of mammalian cells and to influence protein production. Lactate is generated mainly from incomplete oxidation of glucose, and it affects the cells through alteration of culture pH and osmolarity. The effects of ammonia on cells are more direct and occur at much lower concentrations. Thus, ammonia is generally more important as an inhibitory metabolic by-product in cell culture. Ammonia is produced mainly from glutamine, an essential component of cell culture media. Glutamine gives rise to ammonia through chemical degradation and cellular metabolism. Concentration of ammonia in cell culture is influenced by many factors, including the mode of reactor operation, cell concentration, glutamine concentration, and cellular metabolism.

Typically, ammonia concentrations reach 2 to 5 mM in batch cultures. This concentration can be still higher in fed-batch reactors where glutamine is the limiting substrate and therefore supplemented continually, leading to increased ammonia accumulation. Although different cells exhibit different tolerance to ammonia, ammonia concentrations as low as 2 mM have been shown to adversely affect mammalian cells. Numerous studies have reported the effect of ammonia on mammalian cells in different cell culture systems. It has been established that ammonia not only affects cell growth but also alters cell metabolism, product expression, and product quality. The action mechanism for ammonia is complicated and not well characterized. Ammonia was shown to alter intracellular pH, membrane potential, and intracellular composition and to result in futile cycles. Minimization of ammonia inhibition should yield better cell growth and increase the longevity of the cultures.

Several techniques were used to minimize ammonia accumulation in cell culture systems. Media modification involves replacement or elimination of glutamine from cell culture medium. The success of this method depends on the cell line, and in most cases the cells should be adapted or genetically altered. Alternatively, the ammonia generated in the media can be removed by physical and chemical means. In this article, we review the current knowledge on ammonia inhibition in cell culture. The sources of ammonia accumulation; the effects of ammonia on cell growth, metabolism, product secretion, and product quality; the mechanism of ammonia inhibition; and the methods of reducing ammonia concentration in cell culture are discussed.

ACCUMULATION OF NITROGENOUS COMPOUNDS

There are two major sources for accumulation of nitrogenous compounds specially the ammonia and in both cases *glutamine* is involved. Glutamine is one of the important energy sources for the mammalian cells in vitro. Addition of glutamine has been shown to stimulate cell growth and antibody production. Because almost all conventional cell culture media contain glutamine, the problem of ammonia accumulation is of universal importance. However, depending on the cell line and media composition, the effects can be different from one case to another.

Degradation of Glutamine

Glutamine is not stable in aqueous solutions and is degraded to yield a five-member ring structure, pyrrolidone carboxylic acid, and ammonia:

$$\text{Glutamine} \rightarrow \text{Pyrrolidone carboxylic acid} + NH_4^+ \quad \ldots(1)$$

This reaction is irreversible, and the extent of the reaction is affected by the chemical environment of the medium. Glutamine degradation follows first-order kinetics, the half-life of glutamine is reported to be between 6 and 20 days, depending on the conditions. Glutamine degradation is pH and temperature dependent. Alkaline pH increases the degradation rate, and altering pH from 7.2 to 7.6 was reported to elevate the degradation rate by 300%. The presence of serum, especially if not heat inactivated, can elevate degradation of glutamine and subsequent ammonia accumulation caused by glutaminase and asparaginase activity. The media composition, especially the phosphate levels in the media, affects glutamine degradation significantly. Glutamine degradation is an important problem for cell culture. The degradation not only generates inhibitory ammonia, but also decreases the level of glutamine available to the cells. The cell performance is affected negatively because of these two factors.

Metabolic Production of Ammonia

Although ammonia can be produced from other amino acids, glutamine metabolism is established to be the most important pathway. Glutamine metabolism primarily takes place in mitochondria. Glutamine is initially deaminated by glutaminase, with ammonia as a by-product. Ammonia can also be removed from glutamine via biosynthetic pathways, where the ammonia group is used in the formation of a biosynthetic product, particularly for pyrimidine and purine base synthesis.

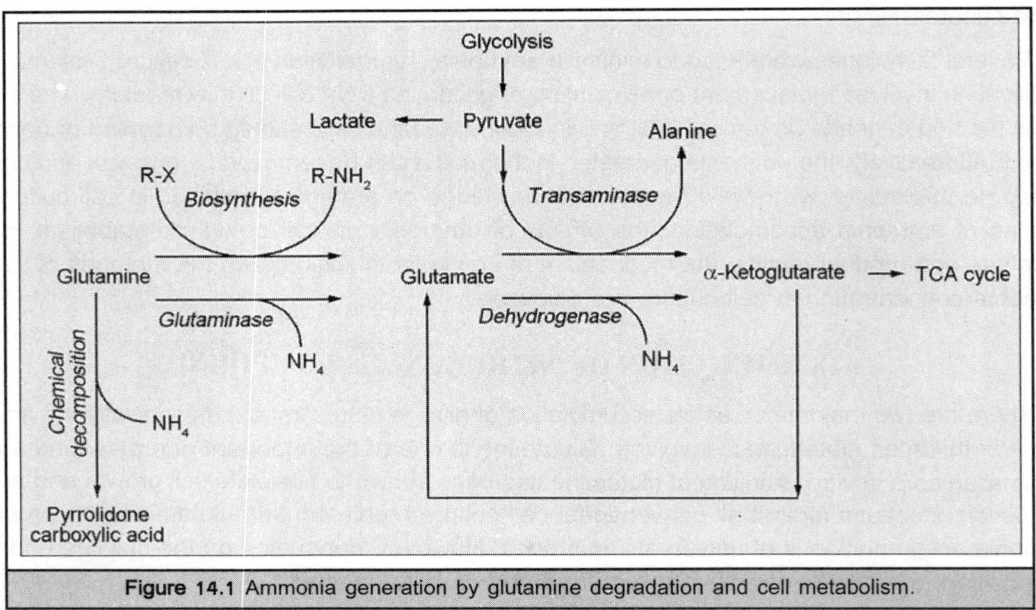

Figure 14.1 Ammonia generation by glutamine degradation and cell metabolism.

The product formed by either route is glutamate. The second step of glutaminolysis is the conversion of glutamate to α-ketoglutarate. Again, this conversion may be accomplished by more than one enzyme. An active transaminase can remove the ammonia group from glutamate. Less-active transaminases appear to deliver the ammonia group to the glycolytic intermediate 3-phosphoglycerate to form serine, which subsequently can be transformed to glycine. Glutamate can also be transformed by glutamate dehydrogenase to form α-ketoglutarate, which liberates a second ammonium ion. α-Ketoglutarate is then metabolized via the tricarboxylic acid (TCA) cycle.

Table 14.1 Amount of ATP produced and ammonia generated in different metabolic pathways

End product	*ATP produced (mol/mol Gln)*	*NH_4 produced (mol/mol Gln)*	*NH_4/ATP*
Alanine	9	1	0.11
Aspartate	9	1	0.11
Acetyl-CoA	15	2	0.13
Pyruvate	12	2	0.17
Lactate	9	2	0.22

The amount of ammonia generated is dependent on the metabolic pathway for glutamine. Depending on the pathway and the fate of glutamine, the ammonia yield on glutamine varies. Based on stoichiometry, Glacken reported possible scenarios for glutamine metabolism and ammonia yield, between 1 and 2 mol of ammonia are generated per mole of glutamine consumed. These yield coefficients, however, are very high and are not typically observed in cell culture. For hybridoma cells, the ammonia yield was reported to be between 0.5 to 1 mol of ammonia generated per mole of glutamine consumed. Glutamine concentrations in typical cell culture media range between 1 and 7 mM. Thus, the ammonia concentrations in cell culture can reach 5 mM. Consequently, the higher glutamine concentrations found in special media formulations yield much more ammonia levels. Glutamine concentrations used in fed-batch cultures, for instance, can vary between 8 and 40 mM, and very high ammonia concentrations are obtained.

INFLUENCE OF AMMONIA ON MAMMALIAN CELL

The influence of ammonia concentration on mammalian cell growth and product expression has been reported for a variety of cell lines and culture conditions. Most of these studies emphasized the influence of ammonia on cell growth. The effects of elevated ammonia concentrations on metabolic rates and cell productivity have been studied less extensively. Although it is not discussed in detail in this article, the effect of ammonia on virus replication and virus production from cell culture was demonstrated clearly in the literature.

Role of Ammonia on Growth

Although ammonia inhibited cell growth in all these studies, the sensitivity of the growth rate to ammonia concentration varied among the cell lines used. The extent of ammonia inhibition seems to be more severe for primary cells, whereas transformed cells tolerate higher ammonia levels. For 3T3 cells, ammonia levels as low as 0.6 mM reduced cell growth by 60%. However, when the cells are transformed, the reduction was reduced to 15%. Addition of 2 mM ammonia reduced growth by 30% for BSC-1 monkey epithelial cells. For established cell lines, inhibitory effects of ammonia were observed at higher ammonia concentrations. Hassell et al. reported reduction of cell growth at 2 mM for a variety of cell lines. Several reports indicate ammonia inhibition for hybridoma cells with varying severity.

Inhibitory concentrations of ammonia were observed to be in the range of 2 to 10 mM in these studies. For BHK cells, even lower concentrations of ammonia resulted in growth inhibition. Wentz and Schugerl reported a 80% reduction at 1 mM ammonia, and Butler and Spier reported a 75% decrease in growth at 3 mM of ammonia in BHK cells. For CHO cells, the inhibitory ammonia levels seems to be higher. Kurano et al. obtained a 50% reduction in growth at 8 mM ammonia. Some cell lines are reported to be more resistant to ammonia. For instance, Schneider observed only a 20% reduction in growth at 6 mM ammonia for HeLa cells, and Hansen and Emborg did not observe any inhibition at up to 8 mM for a CHO clone. Other factors that can influence the inhibitory effects of ammonia on growth are the serum level and cell adaptation. Holley et al. for instance, observed that increasing the level of serum from 0.1 to 10% reduced the ammonia inhibition. Miller et al. observed adaptation of hybridoma cells to 8 mM ammonia. Maiorella et al. used serial passaging in the presence of 5 mM ammonia and demonstrated that the resulting cells survived better in 10 mM ammonia. Adaptation to high ammonia concentrations can be used for selection of ammonia-resistant cell lines. The effects of ammonia on cell growth in most of these studies reported were quantified as the reduction in cell density, and only a few studies evaluated the specific growth rates as a function of ammonia concentration. Ammonia inhibition on the specific growth rate can be described by a second-order inhibition model:

$$\mu = \frac{\mu_0}{1 + [NH_4^+]^2 / K_a} \qquad \ldots(2)$$

where μ and μ_0 are the growth rates in the presence and absence of ammonia, respectively; $[NH_4^+]$ is the initial ammonia concentration; and K_a is the inhibition constant. The square root of K_a corresponds to the ammonia concentration at which the specific growth rate decreases by 50%. Equation 2 describes the cell growth rate accurately with parameters μ_0 = 0.037/h and K_a = 24 mM^2. For comparison, K_a value of 26 mM^2 was reported for the CRL-1606 hybridoma cell line, however, the value reported for the SB-4082 hybridoma cell line was almost one order of magnitude lower (K_a = 3.2 mM^2).

Ammonia and Optosis

Even though ammonia affects the cell growth rate and maximum cell density, effect on cell death is not well documented. Ozturk et al. reported no effect of ammonia on the cell-specific death rate in the range of 0 to 5 mM. When the ammonia concentration was increased from 0 to 3.75 mM, the specific death rate did not change. Ozturk and Palsson also observed that elevated ammonia concentrations did not significantly influence the cell viability. Thus, the reduction in cell density at higher ammonia concentrations was simply the result of lower cell specific growth rates. On the other hand, Goergen et al. (89) used much higher concentrations of ammonia (up to 17 mM) and observed an increase in the death rate. Newland et al. also reported the influence on death rates by ammonia at high concentrations. More quantitative studies are needed to verify the effects of ammonia on specific death rates and the mechanism(s) involved.

Involvement of Ammonia on Metabolism

A number of studies reported the effects of ammonia on cell metabolism. For a murine hybridoma cell line, Ozturk et al. observed a roughly twofold increase in glucose and glutamine consumption rates in cultures exposed to 3.75 mM of ammonia, compared to control conditions. The production rates of lactate and ammonia were also enhanced by the presence of ammonia, indicating an elevated metabolic state of the cells. The yield coefficients of ammonia from glutamine and lactate yield from glucose decreased about 15% with 3.75 mM ammonia. A similar elevation in metabolic rates was reported by Miller et al. for AB2-143.2 hybridoma cells and by Alex et al. for C127 cells. Glacken has also shown an increase in the glutamine consumption rate by added ammonia. McQueen and Bailey indicated an increase in hybridoma cell yield from glucose and glutamine that resulted from a decrease in growth rate and an increase in metabolic rate. Miller et al. and Ozturk et al. studied the effects of ammonia on amino acid metabolism. For this cell line, glutamate, serine, glycine, and alanine were produced and all other amino acids were consumed.

The consumption or production rates of these amino acids were doubled at 3.75 mM ammonia. The increase in the amino acid consumption or production rates were parallel to the glutamine consumption rate because they increased at higher ammonia concentrations by the same magnitude. Only serine, arginine, alanine, and valine showed a relatively different response to ammonia concentration. The consumption rate for serine decreased. The rate of increase for arganine (1.5-fold increase at 3.75 mM ammonia) was lower compared to glutamine (1.8-fold increase). On the other hand, the increase in alanine production (2.1-fold) and valine consumption (2.6-fold) was higher. The yield coefficients of lactate from glucose and ammonia from glutamine decreased, whereas alanine yield from glutamine increased as a result of ammonia addition. Miller et al. observed similar data for ammonia yield from glutamine.

Table 14.2 Amino acid utilization and production rates

	$[NH_4^+]$			
Amino acid	*0 mM*	*1.25 mM*	*2.5 mM*	*3.75 mM*
Aspartate	1.74	2.21	2.58	3.75
Glutamate	(2.92)	(1.69)	(2.03)	(3.25)
Asparagine	2.13	2.26	2.67	4.18
Serine	(2.45)	(2.07)	(2.03)	(0.47)
Glutamine	48.50	58.00	67.50	89.00
Histidine	0.12	0.23	0.35	0.20
Glycine	(0.58)	(0.32)	(0.66)	(0.66)
Threonine	3.10	3.01	3.86	4.14
Arginine	3.18	3.24	3.61	4.89
Alanine	(34.92)	(42.78)	(55.33)	(72.14)
Tyrosine	3.75	4.51	5.08	6.84
Methionine	2.14	2.98	2.54	3.51
Valine	5.68	8.46	9.61	14.70
Phenylalanine	2.21	3.48	3.86	4.57
Isoleucine	7.06	10.84	11.26	14.16
Leucine	8.64	11.14	13.35	18.57
Lysine	5.39	6.94	8.17	11.73

Hansen and Emborg also observed a decrease in ammonia yield from glutamine at high ammonia concentrations for CHO cells. A decrease in the lactate yield coefficient has been reported at elevated ammonia concentrations for other mammalian cells. Similarly, McQueen and Bailey reported a decrease in lactate yield from glucose and in ammonia yield from glutamine. An increase in the alanine yield from glutamine was reported by Miller et al. An increase in amino acid metabolic rates was observed for the C127 cell line. Ozturk et al. showed a 12% increase in the alanine yield from glutamine at elevated ammonia concentrations, whereas the ammonia yield decreases by the same amount. These data indicate that at elevated ammonia concentrations, relatively more of the glutaminolytic flux is via the alanine trans-aminase pathway than that of glutamate dehydrogenase. This shift could be due to the fact that ammonia is a by-product of the dehydrogenase reaction, which is believed to operate nearly at equilibrium; thus, elevated ammonia concentrations would shift the equilibrium toward glutamate. Both Miller et al. and Ozturk et al. showed an increase in the specific glutamine consumption rate at elevated ammonia concentrations. This increase in the consumption rate suggests that inhibition of glutaminase by ammonia is not operative in the hybridoma cell line investigated.

Ammonia in the Production of Energy

Mammalian cells use various pathways to produce energy in the form of adenosine triphosphate (ATP). The production rate of ATP can be estimated using the lactate production and oxygen consumption rates:

$$q_{ATP} = q_{Lac} + 2(P/O) \cdot q_{O_2} = q_{Lac} + 6 \cdot q_{O_2} \qquad ...(3)$$

where q_{ATP}, q_{Lac}, and q_{O_2} are the production rates of ATP, lactate, and oxygen, respectively, and P/O is the phosphorylation ratio. The term q_{Lac} corresponds to the contribution of glycolysis, and the term $6 \cdot q_{O_2}$ corresponds to the contribution of oxidative phosphorylation.

Miller et al. observed a decrease in oxygen uptake at high ammonia concentrations in a continuous reactor operation. On the other hand, Kimura et al. and Ozturk et al. reported no significant effect of ammonia on oxygen consumption rates for a human leukemia line and a hybridoma line, respectively. Miller et al. and Ozturk et al. reported an increase in the ATP production rate at high ammonia concentrations. Although cells generated more ATP at higher ammonia levels, this increased energy production was not, however, used for growth, because the growth rate was inhibited. It appears that the cells produce more energy under stressful conditions and use it for maintenance. The relative contribution of oxidative phosphorylation was observed to decrease, and the contribution of glycolysis increased with elevated ammonia levels.

Ammonia and Product Expression

Because ammonia affects cell growth and metabolism, it is conceivable that it also affects the product expression. Either directly or indirectly, ammonia can affect the product concentration obtained in the culture and thus the process efficiency. The indirect effect of ammonia on production results from the reduced cell growth rate and subsequent cell density achieved in the culture. Even though the specific productivity may not have been altered, the cell density and longevity of the culture can be restricted by high ammonia levels. Thus, cultures with high ammonia levels result in lower product concentrations. Several studies have reported a decrease in monoclonal antibody production at high ammonia concentrations, mainly because of lower cell densities obtained. Although the titer was reduced, the specific antibody production rate remained constant under these conditions. The direct effect of ammonia on specific productivity varies and the mechanism(s) are poorly understood. Dyken and Sambanis observed a slight increase in protein secretion from AtT-20 cells.

Ammonia was reported to alter the secretion rate of interferon (INF) γ and synthesis of INF-β; however, the specific production rates were not provided in these studies. Hansen and Emborg reported a decrease in tissue-type plasminogen activator (t-PA) production from CHO cells at high ammonia concentrations. Specific antibody productivity was evaluated for a number of hybridoma cell lines in the literature. Glacken reported a decrease in antibody productivity at elevated ammonia levels for the C-1606 hybridoma line. Other investigators, however, reported an unaltered specific antibody production rate.

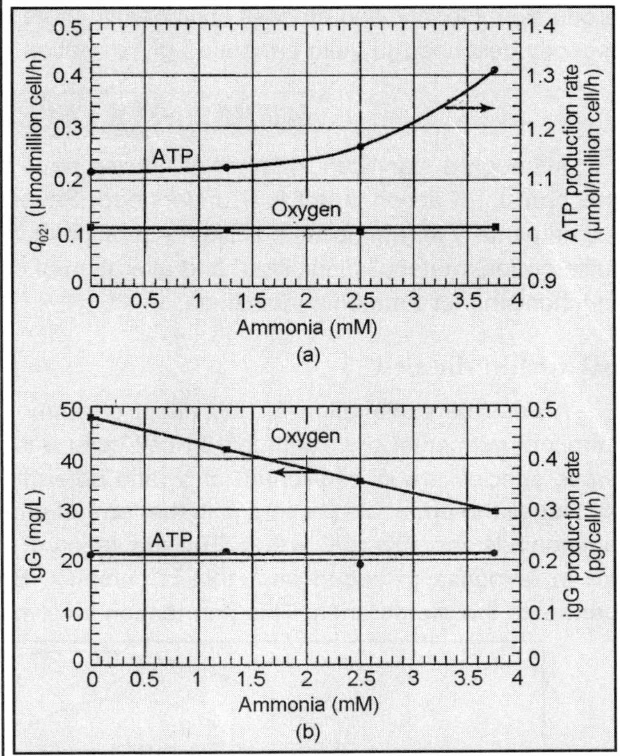

Figure 14.2 Effect of ammonia on murine hybridoma culture. (a) Oxygen consumption and ATP production rate, (b) monoclonal antibody concentration and antibody production rate.

Ammonia and Quality of Products

Ammonia was reported to affect the conditions in intracellular compartments and enzymatic reactions for protein processing. High ammonia concentrations were reported to alter protein after translations, glycosylation, and secretion. Ammonia can alter the intracellular or intracompartmental pH, alter membrane potentials, and directly interact with enzymes.

Thorens and Vassalli reported inhibition of sialic acid transferase for Immunoglobulin (Ig) M at 10 mM ammonia. The results indicate a pH increase in Golgi because of ammonia load. Similarly, Andersen and Goochee reported significant reduction in the terminal sialylation of O-linked glycosylation of recombinant granulocyte colony-stimulating factor. Borys et al. studied the N-linked glycosylation of recombinant mouse placental lactogen-I from CHO cells and observed inhibition of glycosylation by ammonia. The effect of ammonia was also dependent on extracellular pH.

Gawlitzek et al. studied the glycosylation pattern of recombinant proteins expressed by BHK-21 cells. For the production of HuIL-2 variant in BHK cells, ammonia was observed to increase the intracellular uridine diphosphate-*N*-actylglucosamine (UDP-GlcNAc) pool. UDP-GlcNAc is a precursor substrate for the glycosylation process in the cytosol and Golgi. High ammonia levels led to a decrease in terminal sialyation and to an increase in branching. Totest the hypothesis that UDP-GlcNAc is involved, the authors used glucosamine, which is the precursor for UDP-GlcNAc. Kopp et al. studied product consistency and glycosylation patterns in recombinant CHO-expressed glycoproteins. High ammonia levels influenced glycosylation patterns significantly for INF-ω and t-PA. The alteration of product glycosylation by ammonia was described in monoclonal antibody production. To minimize the effects of ammonia, a multilevel pH control was proposed for commercial production. Genetic and process engineering strategies for control of ammonia in cell cultures were recently described to yield enhanced glycosylation.

SUPPRESSIVE EFFECT OF AMMONIA

Although a significant number of articles have been published on the affect of ammonia, the mechanism of action is not fully understood. Several mechanisms were hypothesized and tested experimentally for ammonia inhibition. Alteration of intracompartmental pH and membrane potential, futile cycles, metabolic inhibition, and alteration of critical ribonucleotides are identified as potential mechanisms for ammonia inhibition.

pH within the Cell

A possible explanation for the effect of ammonia is simply the alteration of intracellular and intracompartmental pH. Ammonia in cell culture is present in both gaseous and ionic form, and these species are in equilibrium at a ratio determined by the medium pH. At normal pH values, almost all the ammonia present is in the form of ammonium ion, and the concentration of gaseous ammonia is very low ($pK_a = 9.2$). The regulation of intracellular (cytosolic) and intracompartmental pH in response to added ammonia is complex. Both gaseous ammonia and ammonia ion can permeate the membranes. This permeation involves both passive (diffusion) and active transport.

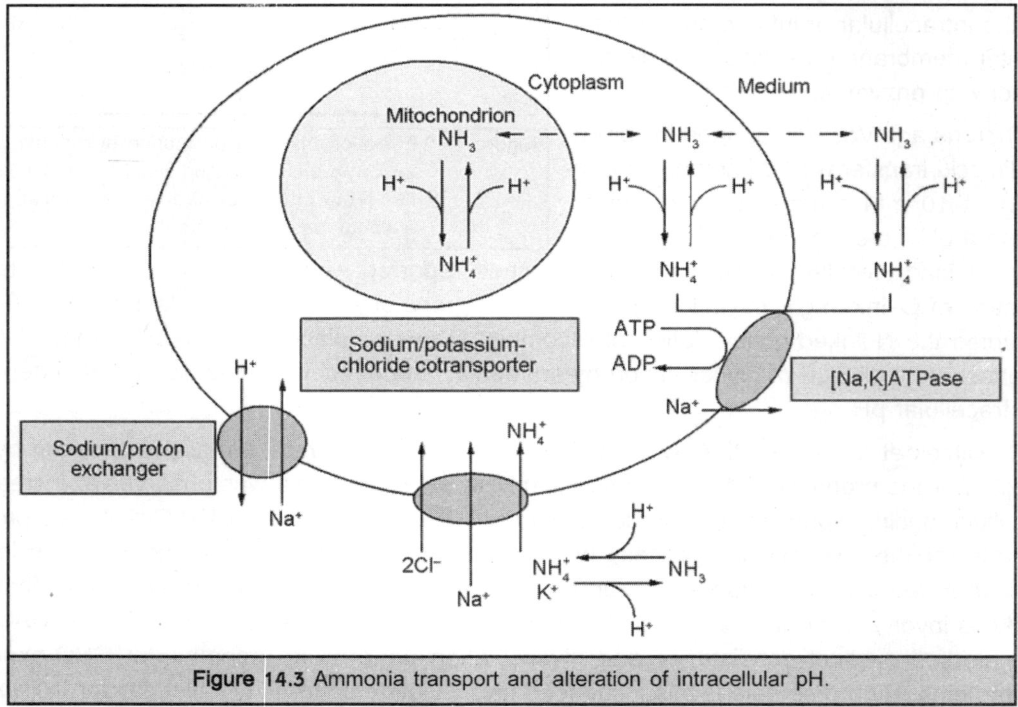

Figure 14.3 Ammonia transport and alteration of intracellular pH.

The permeability of gaseous ammonia is much greater than that of the ammonium ion. Thus, initially, gaseous ammonia permeates rapidly, raising the intracellular pH. Then, the slower penetration of ammonium ion decreases the intracellular pH. McQueen and Bailey and Ozturk et al. studied the response of hybridoma cells to added ammonia and verified this mechanism. After the addition, there is an immediate increase in intracellular pH. This is due to rapid gaseous ammonium diffusion into the cells. The intracellular pH decays to steady-state intracellular pH values that are lower than those observed prior to the exposure to ammonia because of the transport of ammonium ion.

Table 14.3 Effect of ammonia on intracellular pH at external pH of 7.20

[NH_4^+] mM	*Peak pH_i*	*Steady-state pH_i*	*Delta pH*
0	7.23	7.19	0
1.25	7.3	7.14	–0.05
2.5	7.42	7.1	–0.09
3.75	7.43	7.07	–0.13

The rate of diffusion for ammonium ion through the cell membrane is four to five orders of magnitude lower than the rate of diffusion for gaseous ammonia. However, ammonium ion can be actively transported across the cell membrane via [Na, K] ATPase, [Na,K,Cl]-cotransporter, and [Na,H]-exchanger. The dynamics of ammonia transport for mammalian cells were analyzed by mathematical models. The transport of ammonia and ammonium ion to the cells were studied in detail by Martinelle and Haggsrom for the murine myeloma cell (Sp2/O-Ag14). In addition to diffusional transport, these authors identified the active transport of ammonium ion via [Na,K]ATPase and [Na,K,C]-cotransporter in hybridoma cells. The presence of K^+ could inhibit the [Na,K] ATPase active transport and alter the transport of ammonium ion. When K^+ (10 mM) was added, the change in intracellular pH due to ammonia addition was observed to be negligible.

The authors postulate that one of the reasons for ammonia inhibition is the increased energy demand resulting from energy wasted because NH_4^+ is actively transported to the cells via [Na,K]ATPase, and they proposed the use of K^+ for minimization of ammonia inhibition. Use of KOH instead of NaOH for pH control is proposed to increase K^+ concentration in the culture. A change in intracellular or intracompartmental pH can alter the activity of numerous enzymes and, depending on the location, different results can be obtained. The *net* result of ammonium ion transport to the cells is a decrease in intracellular pH. However, the ammonia transported to the cells further diffuses to compartments in the cells. Gaseous ammonia easily permeates intracellular membranes and the pH in mitochondria, Golgi, endoplasmic reticulum, and lysosymes are elevated by its presence.

The pH in vesicles is normally lower than in cytosol, and the pH can be altered by ammonia. The rise in pH by the transport of ammonia gas into the lysosomes has been demonstrated. The generation of ammonia inside the cell caused by cellular metabolism alters intracompartmental pH. Ammonia is generated mainly in mitochondria because of glutamine metabolism. Ammonia gas readily diffuses out of the mitochondria, and the pH decreases. Ammonia diffuses not only to the cytoplasm of cells but also to the compartments. In all cases, ammonia that diffuses from mitochondria increases the pH in other intracellular compartments (i.e., Golgi, endoplasmic reticulum, and lysosymes). McQueen and Bailey have shown that the net result of ammonia addition is a decrease in intracellular pH for the hybridoma line ATCC TIB 131. When internal pH was altered by external pH, McQueen and Bailey associated ammonia effects to the variations in intracellular pH, because both ammonia addition and low external pH resulted in lower cell yields on glucose and glutamine.

On the other hand, Ozturk et al. individually evaluated the cell growth rate and the metabolic rates and could not relate the effects of ammonia to the effects of lowering external pH. There is a decrease in glucose consumption and an increase in glutamine uptake rates when pH was controlled below 7.2. However, a decrease in intracellular pH as a result of ammonia addition increased both rates. ATP production was also influenced differently from ammonia addition (increasing ATP production) and from lowering external pH (decreasing ATP production). Miller et al. observed an increase in glucose consumption rate as a result of ammonia addition, although the decrease in extracellular pH led to a decrease in glucose consumption rate. Thus, it can be concluded that the mechanisms of the ammonia effects may not be as simple as they were originally thought to be. The alteration of intracellular pH by ammonia cannot alone explain the observations.

Transport Across the Membrane

As previously mentioned, ammonium ions can be transported across cellular membranes via [Na,K]ATPase and [Na,K,Cl]-cotransporters. Ammonium ions compete with K^+ ions and disturb the membrane potential. The [Na,K]ATPase transport system has a high energy demand, and the competition of ammonium ions with K ions decreases the efficiency. Martinella and Hagstrom reported futile cycles originating from ammonia diffusion and ammonium ion transport. These futile cycles increase the maintenance energy required by the cells. When ammonia is generated in the mitochondria, it diffuses out. However, ammonia is transported back in the form of ammonium ion, and cells use [Na,K]ATPase for this transfer. Thus, the alteration of membrane potential and elevation of maintenance energy are possible mechanisms for ammonia inhibition.

Enzymatic Pathways

Ammonia can also interact with the enzymes and alter the metabolic pathways. Phosphofructokinase (PFK) can be activated by ammonia. PFK is a highly regulated key enzyme for glucose metabolism, and its activation by ammonia can explain the increase in glucose consumption at high ammonia levels. Ammonia can also affect the metabolic pathways in mitochondria. The glutamine metabolism pathway, for instance, involves glutaminase, transaminase, and glutamate dehydrogenase reactions. The glutaminase enzyme is known to be inhibited by ammonia. However, Ozturk and Palsson did not observe a decrease in the specific glutamine consumption rate at elevated ammonia concentrations. The glutamate dehydrogenase reaction is known to operate near equilibrium, and elevated ammonia levels would shift the equilibrium toward glutamate.

At high ammonia levels, Ozturk and Palsson observed a shift to the glutaminolytic flux via the alanine transaminase reaction from glutamate dehydrogenase, confirming this change. Thus, at high ammonia concentrations, the ammonia production by the cells decreased and alanine yield increased. This finding was also supported by other investigators. The involvement of ammonia in metabolic pathways can explain the effects on cell metabolism. Growth inhibition by ammonia, on the other hand, can be explained by the presence of futile cycles. Futile cycles are cyclic reactions in which the hydrolysis of ATP is the only result; thus, cells have to increase ATP production to maintain the energy required for their survival. Tagler et al. studied the glutamate dehydrogenase and trans-hydrogenase enzymes in isolated mitochondria and liver cells. This enzyme system catalyzes the following reactions:

$$\text{Glutamate} + \text{NAD}^+ \leftrightarrow \alpha\text{-Ketoglutarate} + \text{NH}_4^+ + \text{NADH} \quad \ldots(4)$$

$$\text{NADH} + \text{NADP}^+ + \text{ATP} \leftrightarrow \text{NADPH} + \text{NAD}^+ + \text{P}_i \quad \ldots(5)$$

$$\alpha\text{-Ketoglutarate} + \text{NADPH} + \text{NH}_4^+ \leftrightarrow \text{Glutamate} + \text{NADP}^+ \quad \ldots(6)$$

Ammonia can also drive the glutaminase–glutamine synthetase enzyme system in a futile cycle mode. Glutaminase enzyme catalyzes the deamidation of glutamine to glutamate and ammonia. Glutamine synthetase, on the other hand, uses ATP to convert the glutamate to glutamine:

$$\text{Glutamine} \leftrightarrow \text{Glutamate} + \text{NH}_4^+ \text{ (Glutaminase)} \quad \ldots(7)$$

$$\text{Glutamate} + \text{NH}_4^+ + \text{ATP} \leftrightarrow \text{Glutamine} + \text{ADP (Glutamine synthetase)} \quad \ldots(8)$$

Thus, for both glutamate dehydrogenase–trans-hydrogenase and glutaminase–glutamine synthetase cycles, the ATP is dissipated and the reactions are driven by ammonia. This increases the maintenance energy of the cells and can explain the inhibitory effects of ammonia. Although these futile cycles have been studied for isolated mitochondria, their functionality has not been demonstrated for mammalian cells in culture.

Inhibitory Effect of Ribonucleotides

Another explanation for ammonia inhibition comes from detailed studies on various ribonucleotides by Ryll and Wagner and Ryll et al. The intracellular pools of UDP-*N*-actylglucosamine (UDP-GlcNAc) and UDP-*N*-actylgalactosamine (UDP-GalNAc) have been shown to be responsible for inhibition of protein, DNA, and RNA synthesis. The pool of UDP-GNAc refers to both UDP-GlcNAc and UDP-GalNAc collectively, and it was demonstrated to be elevated in response to increased ammonia levels for a variety of cell lines. Ammonia is incorporated into the glycolytic pathway in cytosol. Fructose-6-phosphate and ammonia combine to form glucosamine-6-phosphate, a direct precursor for the UDP-GNAc pool in the cytosol. In the mitochondria, ammonium leads to formation of carbamoyl phosphate.

When the ammonia concentration is high in the mitochondria, then subsequently up-regulated carbamoyl phosphate can enter the cytoplasm. Carbamoyl phosphate in the cytosol stimulates de novo synthesis of UMP and UTP. Finally, UTP and glucosamine-6-phosphate are combined to supplement the UDP-GNAc pool. Ryll et al. studied the biochemistry of the UDP-GNAc pool formation and regulation of the enzymes involved. The formation of the UDP-GNAc pool was observed to be dependent on both glucose and ammonia. Ammonia increases the size of the UDP-GNAc pool, which has been correlated with growth inhibition. It is also important to note that the UDP-GlcNAc is also known as a precursor substrate for glycosylation processes in the endoplasmic reticulum and Golgi. Thus, the effects of ammonia on glycosylation could also be explained by this mechanism.

INHIBITION REGULATION

Several techniques were used to minimize ammonia accumulation in cell culture. The techniques can be classified into two groups: minimization of ammonia generation and removal of ammonia.

Production of Ammonia

As mentioned before, ammonia is generated mainly by the spontaneous degradation and cellular metabolism of glutamine. Ammonia generation can be minimized by controlling these processes.

Regulating degradation of glutamine

As previously mentioned, the glutamine degradation rate is dependent on several factors, and it can be minimized by several strategies. Temperature is the most obvious factor, and the media should be kept cold when stored. In the reactor, however, the temperature cannot be lowered much. Other factors, such as media pH, can be used to minimize the degradation. The pH optimum for cells varies between pH 7.0 and 7.4. Cultivation at lower pH values should decrease the degradation rate and ammonia generation. Serum, if used, should be inactivated to minimize any glutaminase activity. Finally, glutamine degradation is dependent on several ions, such as phosphate. Various media should be evaluated for a given cell line, and glutamine degradation rates for each should be compared.

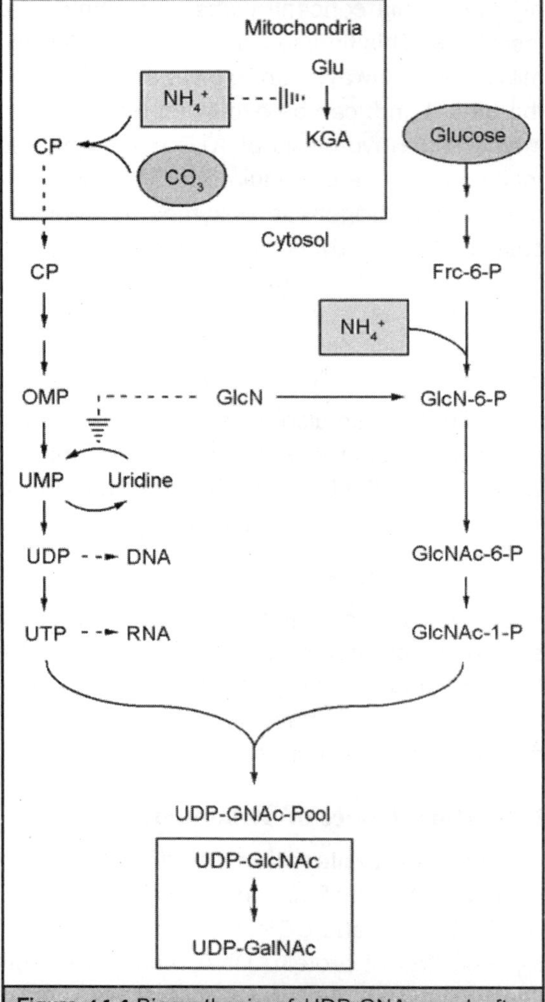

Figure 14.4 Biosynthesis of UDP-GNAc pool after ammonia application.

Attempts have been made to replace glutamine by glutamine-containing peptides. Peptides such as alanyl-glutamine or glycyl-glutamine are more stable, and the media prepared with these peptides can be autoclaved.

These peptides were also shown to generate less ammonia. Although the cells exhibited a considerable lag phase, the cell yields from peptide-substituted media were comparable to those obtained from standard media. Holmlund et al. used glycyl-glutamine and *N*-acetyl-glutamine for CHO cultures. Although cells did not perform well in the presence of N-acetyl-glutamine (low growth and t-PA production), glycyl-glutamine provided good culture performance.

Regulating cellular metabolism

The specific ammonia production rate is affected by the cellular environment, and variables such as pH, dissolved oxygen (DO), temperature, and the level of metabolites can be manipulated to minimize the production rate. There seems to be an optimal pH for reducing the ammonia production rate. For instance, Ozturk and Palsson observed the ammonia production rates to be minimal at pH 7.2. The cell growth rates were optimal, and glutamine consumption was also observed to be minimal at this pH. The ammonia yield from glutamine, on the other hand, was reduced at low pH values. Cells seem to produce less ammonia when they are growing under optimal conditions. At very low (1% air saturation) and very high (100% air saturation) DO levels, the ammonia generation rates increased.

In most cases, the ammonia production follows closely with the glutamine consumption with a relatively constant yield coefficient. Reducing the glutamine concentration ([Gln]) decreases the glutamine consumption (q_{Gln}), and Monod-type saturation kinetics can be used to describe the data:

$$q_{Gln} = q_{Gln}^{0} \frac{[Gln]}{K_m + [Gln]} \qquad \ldots(9)$$

where q_{Gln}^{0} and K_m are constants. The value of K_m is in the order of 0.2 to 1 mM.

Reducing glutamine not only decreases the glutamine consumption and ammonia production rates, but also decreases the yield of ammonia from glutamine. When glutamine was fed in a controlled fashion below 1 mM levels, Glacken et al. observed a substantial decrease in ammonia production in Madin-Darby bovine kidney (MDCK) and human fibroblast cells. A similar strategy was successfully used to reduce ammonia production by 50% in hybridoma cells.

Although a glucose limitation alone did not cause a change in ammonia secretion, Ljunggren and Haggstrom observed an enhancement in ammonia reduction when glucose and glutamine were limited. The replacement of glucose by fructose, mannose, and galactose virtually eliminated the ammonia-induced generation of UDP-GNAc.

Altering the amino acid composition of the media seems to affect ammonia generation. Hiller at al. increased the concentrations of leucine, isoleucine, valine, and lysine and observed a decrease in ammonia secretion in a hybridoma line. In these experiments, alanine production rates increased, indicating an elevation in the activity of alanine transaminase.

Substitution of glutamine

The replacement of glutamine in cell culture media can eliminate most of the ammonia generation in culture. Although the idea of replacement is a good one, it is difficult to implement in most cases because the cells seem to be strongly dependent on glutamine for energy and biomass production. Cells need to adapted or genetically altered to grow in the absence of supplemented glutamine. Mammalian cells can use other amino acids as a substitute for glutamine. Studies were conducted to investigate the replacement of glutamine by glutamate, α-ketoglutarate, and asparagine. The efficiency of glutamate for supporting cell growth is very low, and high concentrations of glutamate (up to 20 mM) are required. The replacement of glutamine by glutamate was possible for mouse LS cells and for the McCoy cells. On the other hand, MDCK cells could not be adapted to glutamine-free media. Cells can convert glutamate to glutamine via the glutamine synthetase reaction, and the success of growing the cells on glutamate can depend on the concentration and activity of this enzyme.

However, McDermott and Butler observed that the key factor in cell adaptation to glutamine-free media is not the glutamine synthetase, but the uptake rate of glutamate by the cells. Asparagine is another amino acid used to replace glutamine. Although asparagine is also unstable in media, the degradation rate is much lower (half-life, 87 days). Kurano et al. observed the growth of CHO cells on asparagine after an initial lag phase. The asparagine-containing cultures generated 40% less ammonia than cultures in standard media. The use of *a*-ketoglutarate instead of glutamine was also effective in reducing ammonia generation. The success of growing cells in the absence of glutamine can be increased by genetic engineering.

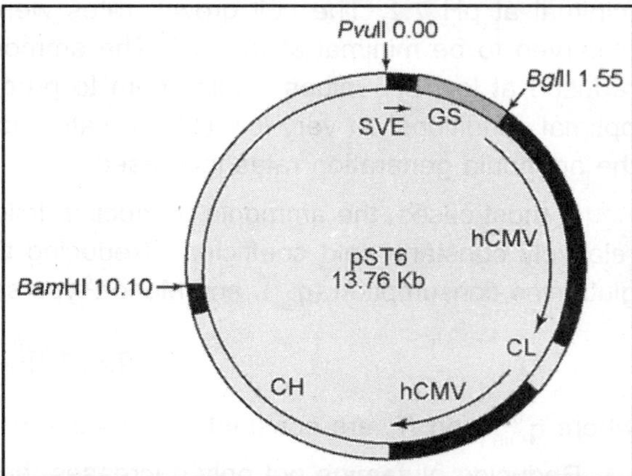

Figure 14.5 Expression plasmid for cB72.3 antibody containing glutamine synthetase. Expression vectors: GS, glutamine synthetase; CH, heavy chain; CL light chain.

Scientists at Cell-tech Ltd. developed a vector containing glutamine synthetase and infected NSo myeloma and CHO cells with plasmid containing this vector. The cells use glutamate as a substrate in glutamine-free media. The genes for protein expression were also integrated into the vector; thus, the glutamine synthetase gene was used as an amplifiable, selectable marker; in the glutamine-free media only the cells containing plasmid could grow. This resulted in the selection of high-producing clones. The system works best for NSo myeloma cells because these cells lack any endogenous glutamine synthetase activity. CHO cells, on the other hand, contain endogenous glutamine synthetase genes, and the selective pressure induced by the absence of glutamine does not work effectively. The use of a specific inhibitor, methionine sulphoximine (MSX), increases the efficiency of selection and amplification.

Adaptive role

The adaptation of cells to grow at high ammonia concentrations can result in more ammonia-tolerant cultures. Adaptation is a complex process, and it is not clear whether the cells alter themselves or a particular clone is selected as a result of this process. Regardless of the mechanism, the cells can tolerate higher ammonia levels after the adaptation, and this method of ammonia adaptation can be used to minimize the ammonia inhibition. The adaptation of hybridoma cells to ammonia has been demonstrated by several investigators.

Technique of Ammonia Removal

Several techniques were investigated to remove ammonia during cell culture and to improve culture performance. These techniques involve the use of adsorbents, gas exchange and ion exchange membranes, and electrodialysis. Although the systems can be placed in the bioreactor, they are normally used as a loop system. Cell culture fluid is recycled through a column, where ammonia is removed. In the case of adsorbents, the recycle is halted routinely to allow the regeneration of the column. The column is regenerated by stripping the ammonia from the adsorbents, using a stripping solution. The methods using gas and ion exchange membranes and electrodialysis, on the other hand, can be run continuously. In these cases, the ammonia fixing–stripping solution is continuously recycled on the other side of the membrane.

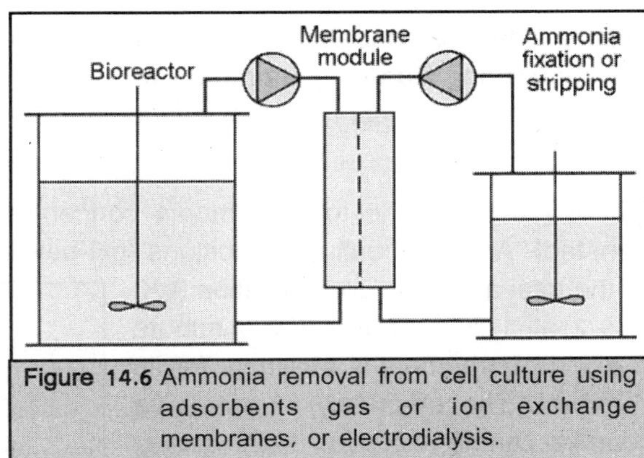

Figure 14.6 Ammonia removal from cell culture using adsorbents gas or ion exchange membranes, or electrodialysis.

Natural adsorbents

Several resins and natural adsorbents were investigated to selectively remove ammonia from cell culture media. Carbonell et al. and Capiaumont et al. used natural adsorbents such as Clinoptilolite to adsorb ammonia in a column where cells culture media is cycled through. The operation of the adsorbent had to be interrupted periodically to regenerate the column. Although these columns removed ammonia from the culture, the cell culture performance did not change significantly. Similar results were obtained when Zeolite (Phillipsite-Gismondine) and ion exchange resins were used as adsorbents.

On the other hand, Nayve et al. and Matsumura and Nayve obtained good results in hybridoma cultures by combining the adsorption method and other removal systems. In fact, Matsumura and Nayve obtained a viable cell concentration of 25 million cells/mL in a perfusion system with ammonia removal. The use of adsorbents to remove ammonia requires complicated systems. The regeneration of the column prohibits continuous operation and requires complicated process control. These systems have to be proven to be reliable and effective for commercial scale operation.

Role of polypropylene

Another method of ammonia removal from cell culture uses gas exchange membranes (such as polypropylene) where ammonia is stripped from the solution because of its gaseous properties. These systems were first developed for microbial systems and for chemical processes. The efficiency of ammonia removal in these systems depends on the concentration gradient for gaseous ammonia across the membrane. The ammonia removal rate (R) is given by:

$$R = k \cdot A \cdot ([NH_3]_c - [NH_3]_s) \quad \ldots(10)$$

where k is the mass transfer coefficient, A is the surface area, and $[NH_3]_c$ and $[NH_3]_s$ are the concentrations of gaseous ammonia, in the cell culture and receiving side, respectively. The concentration gradient can be maximized by maintaining a very low gaseous ammonia concentration on the receiving side of the membrane ($[NH_3]_s \rightarrow 0$). In most cases, the transported ammonia is irreversibly converted by either enzymatic or chemical reactions (e.g., protonization). On the cell culture side, the concentration of gaseous ammonia ($[NH_3]_c$) is determined by the total ammonia concentration. Gaseous ammonia is at equilibrium with ammonium ion, and the culture pH determines the concentration of ammonia in gaseous form:

$$[NH_3]_c = [NH_4^+]_T/(1 + 10^{(pKa - pH)}) \qquad ...(11)$$

where $[NH_4^+]_T$ is the total ammonia concentration (ionized and gas) and pK_a is the ionization constant. At typical culture conditions (pH below 7.5), gaseous ammonia constitutes less than 1% of the total ammonia concentration (pK_a = 9.2 at 37°C). Thus, only a minute fraction of ammonia is available for its removal. The efficiency of ammonia removal can be increased by increasing the pH on the cell culture side. Higher pH shifts the equilibrium to ammonia gas and increases the concentration gradient. However, this method cannot be utilized fully because the cells require a tight pH range for growth and maintenance. Increasing pH also affects the media components, and high pH can precipitate proteins, including the product. The removal of ammonia using gas exchange membranes was tested in a variety of hybridoma lines in an effort to increase antibody yield. Although the final cell density was reported to have increased, the effect on production was not significant. Ammonia removal, however, did result in an alteration in cell metabolism.

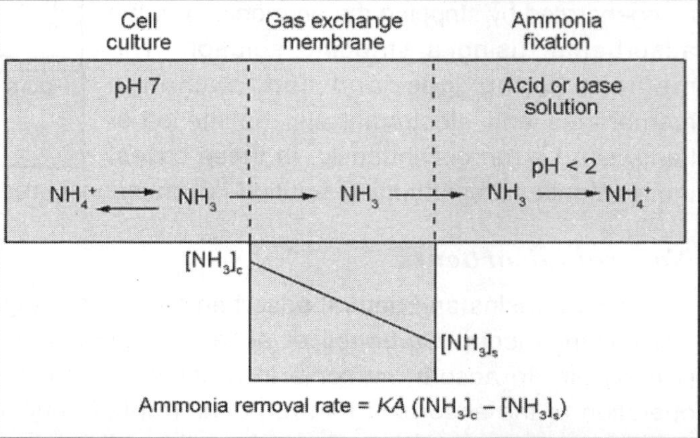

Figure 14.7 Ammonia removal from cell culture using gas exchange membranes. The rate of removal depends on the ammonia gas concentration difference.

Selectively removal of ammonia

Ammonia can be selectively removed by ion (cation) exchange membranes. Contrary to gas exchange membranes, ion exchange membranes remove ammonium ion instead of gaseous ammonia. As mentioned before, ammonia in the cell culture is almost 99% in ammonium ion form, so the removal via ion exchange membranes should be more effective.

The use of autoclavable cation exchange membranes was described by Thommes et al. Ammonium ions bind to the fixed anions in the membrane and pass into the strip–fixation solution, where they are deprotonized or stripped from the solution by pervaporation. The elimination of ammonia on the strip–fixation side and a high concentration of ammonium ions on the cell culture side establishes an efficient concentration gradient so the removal of ammonia is very effective.

Thommes et al. reported an increase in cell density and antibody production in a murine hybridoma culture. The problem with the ion exchange membrane is the simultaneous removal of some other cations from the culture. Some cations such as calcium and magnesium can also precipitate as bicarbonates as a result of ion exchange.

Electrokinetic mechanism

Another technique for ammonia removal involves the use of an electrokinetic mechanism utilizing electrophoresis. Chang and Wang applied a continuous DC electrical field to remove charged ammonia and lactate from the culture of ATCC CRL 1606 hybridoma cells. At a current density of 50 A/m^2, almost all ammonia in the culture could be removed. This ammonia removal system allowed the use of 4 x concentrated media. The system allowed enhancement of cell density and antibody production significantly. The applied current did not cause any detrimental effect on the cells. Removal of ammonia increased the glutamine consumption rate. The electrodialysis system also removed lactate and minimized lactate related inhibition.

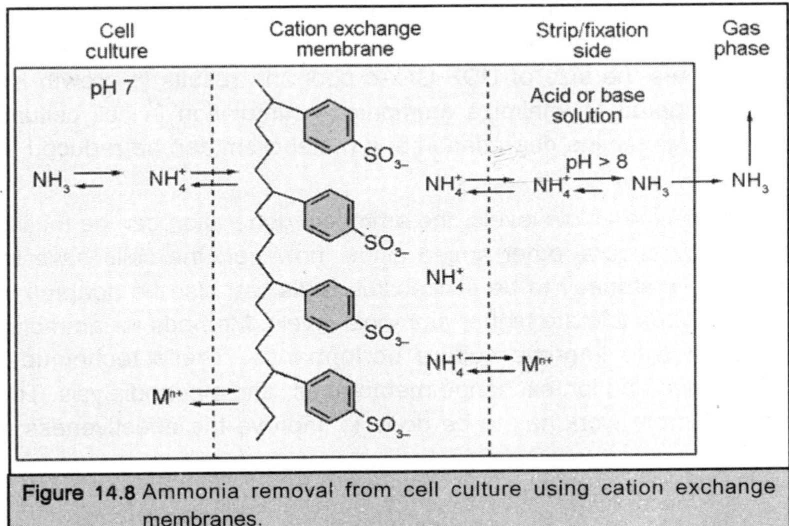

Figure 14.8 Ammonia removal from cell culture using cation exchange membranes.

CONCLUDING REMARKS

In this chapter, we reviewed several aspects of ammonia inhibition. Ammonia accumulation in cell cultures can be a serious problem for cell growth, productivity, and product quality. Ammonia is generated mainly by glutamine degradation and glutamine metabolism. Cell physiology can be affected by ammonia concentrations between 2 and 10 mM. Different cell lines display varying tolerance to ammonia inhibition. Cell growth rate and final cell density in batch and fed-batch cultures decrease at high ammonia levels. Under normal conditions, ammonia does not *directly* influence the death rate or the decrease in viability of the cultures. Ammonia is involved in several metabolic pathways and can influence the metabolic rates.

Specific metabolic rates for glucose, glutamine, ammonia, lactate, amino acids are accelerated at higher ammonia concentrations. The ammonia yield from glutamine decreases, and more alanine is produced at higher ammonia levels. Although ammonia does not seem to alter the specific oxygen consumption rate, the ATP production rate increases at elevated ammonia concentrations. Cells become more active metabolically and generate more ATP for maintenance when their growth is suppressed by ammonia. Specific rates for product secretion are not influenced by ammonia concentration. Cultures at higher ammonia levels result in lower product concentrations because of the low cell concentrations achieved. Ammonia can influence product quality, glycosylation, and sialyation; this was demonstrated for a number of cell lines and products. Several mechanisms have been hypothesized and examined experimentally in an attempt to understand ammonia inhibition.

Alteration of intracompartmental pH and membrane potential, futile cycles, metabolic inhibition, and alteration of critical ribonucleotides are identified as potential mechanisms for ammonia inhibition. Although the dynamics of ammonia transport to and from the cells were studied and intracellular pHs were measured, the data on metabolic rates could not be explained by internal pH hypothesis. Energy dissipation caused by futile cycles and disturbance of membrane potentials are viable explanations for ammonia inhibition. Ammonia was shown to elevate levels in several ribonucleotide pools. Ammonia increases the size of UDP-GNAc pool and results in growth inhibition. Several techniques were investigated to minimize ammonia accumulation in cell culture. Generation of ammonia resulting from glutamine degradation and metabolism can be reduced by controlling the environment of the cells.

By maintaining glutamine at low levels, the ammonia generation can be minimized. Glutamine can also be replaced by several other amino acids; however, the cells have to be adapted or genetically altered for this strategy to be successful. Cells can also be adapted to high ammonia concentrations and can thus tolerate higher ammonia levels. Methods for ammonia removal in cell culture were investigated to improve culture performance. These techniques involve use of adsorbents, gas exchange and ion exchange membranes, and electrodialysis. These systems are fairly complicated, and more work has to be done to improve the effectiveness and reliability for commercial production.

15

RECOMBINANT PROTEIN DRUGS

For the last twenty years, many biotechnology-derived pharmaceuticals have been manufactured for the treatment of diseases such as diabetes, cardiovascular diseases, cancers, dwarfism, anemia, AIDS, cystic fibrosis, chronic granulomatous disease, and kidney diseases. Some of these recombinant protein drugs have already been approved by the U.S. Food and Drug Administration (FDA) for marketing, others are still being tested in clinics or are under review for approval. The success of these protein pharmaceutical drugs greatly depends on the delivery of their biologically active forms to the target site. Therefore, the development of a stable formulation, in which the protein can maintain its native conformation and bioactivity, is one of the important steps in the production of protein pharmaceuticals. In fact, the FDA usually requires the pharmaceutical manufacturers to provide real-time stability data to demonstrate that a pharmaceutical drug contains at least 90% potency throughout the entire shelf life period (usually $\geq$ 2 years) before it can be approved to be a marketed product.

In addition, the manufacturers are required to demonstrate that the degradation products that may cause a loss in the potency of the drug do not have any adverse effects on the safety and efficacy of the drug. In developing a protein drug formulation and designing a delivery system, formulation scientists must consider the physicochemical properties, clinical indication, site of action, pharmacokinetics, and toxicity of the drug. The physicochemical properties of proteins such as the amino acid composition, molecular weight, isoelectric point (pI), glycosylation, and conformation can affect the pharmacokinetics and toxicity as well as the clinical indication. The potential clinical application of a protein drug depends upon the biological function and potency of the product, and the physical and chemical properties of a protein determine its biological function and potency. The unique physical and chemical properties of each protein also determine its in vitro and in vivo stability.

To obtain in vivo information on the pharmacokinetics and toxicity of a drug, the drug must be administered with a stable formulation. In considering the best formulation for the intended applications of a protein drug, the formulation scientist must also consider the route of administration. For initial animal testing, protein drugs are usually administered systemically via an intravenous (i.v.) injection. However, some indications may require a high local drug dose that cannot be achieved by i.v. administration due to toxicity issues. In this case, an alternative route of delivery is necessary. For example, Pulmozyme (rhDNase) is delivered to patients with cystic fibrosis via the pulmonary route of administration.

The drug is best administered directly into the lungs of the patients as an aerosol using nebulizers in order to achieve an efficacious dose at the target site (lungs). Thus, the development of both a stable formulation and suitable delivery route or system is critical in determining the success of a final product. During the past 10 years, several review articles and texts related to protein formulation and stability have been published. The purpose of this article is to provide a comprehensive overview of formulation and delivery aspects of protein drugs. Protein degradation pathways, analytical characterization of proteins, the process of formulation development, and drug delivery systems are reviewed.

CHEMICAL AND PHYSICAL DEGRADATION OF PROTEIN

Chemical and physical degradation is the major cause of instability for biopharmaceutical products. There are many degradation routes for proteins. Aggregation, deamination, and oxidation are the three most common ones, others include isomerization, cleavage, thiol disulfide exchange, and β-elimination.

To develop a stable protein formulation, protein degradation in the formulation must be inhibited or minimized throughout the shelf life of the product. Fortunately, with the advances in analytical techniques for protein analysis, most of these expected degradation routes can be identified, and the degradation products can also be characterized by formulation scientists. In addition, with the continued success of biotechnology, more recombinant proteins have become available for studying degradation mechanisms, thus increasing the general knowledge in formulating new proteins.

Aggregation Pathways

Protein degradation via aggregation pathways is considered a physical reaction that does not involve the breakage or formation of covalent bonds. For a protein to retain its biological function and stability, it must maintain its conformational or tertiary structure. Protein conformation is stabilized mainly by hydrophobic interactions. Thus, monomeric globular proteins often exist in native or folded conformations such that the hydrophobic groups are not exposed on the surface. Loss in the tertiary globular structure or unfolding of proteins due to denaturation results in exposure of hydrophobic residues to the environment.

Protein aggregation or denaturation in a formulation (liquid or lyophilized) can be caused by a change in temperature, extreme pH, ionic strength, pressure, or presence of denaturants. Usually, the native, N, and unfolded, U, states of a protein are in equilibrium during denaturation and folding in a two-state process, as shown in equation 1. The reversibility of the unfolding process depends upon the formulation conditions.

$$N \rightleftharpoons U \qquad \ldots(1)$$

$$\begin{array}{c} N \rightleftharpoons I \rightleftharpoons U \\ \downarrow \\ A \end{array} \qquad \ldots(2)$$

Sometimes, a protein denatures via a different pathway that involves the formation of at least one stable, partially unfolded intermediates. As shown in equation 2, the native protein unfolds to form an intermediate, I, that has the internal hydrophobic residues exposed to the environment. This hydrophobic intermediate on the refolding pathway may then form aggregates, A, due to intermolecular hydrophobic interactions.

According to this scheme, the protein must be denatured or unfolded before nonnative aggregates can be formed. The formation of a stable intermediate during denaturation and folding has been observed for many proteins. For example, the dimeric native form of interferon-*g* (IFN-*g*), unfolds to form a partially denatured monomeric intermediate, below pH 4.5 and in the absence of NaCl. Upon dialysis, both native IFN-*g* and aggregates were obtained. The aggregates lost most of the native tertiary structure as well as bioactivity.

Catalytic Hydrolysis

Deamination of protein is the acid- and base-catalyzed hydrolysis of the side-chain amide on glutamine (Gln) and asparagine (Asn) residues to form a carboxylic acid. Studies of deamination in peptides by Robinson and his coworkers suggested that Asn is more susceptible to deamination than Gln and degrades more readily in the presence of an adjacent glycine (Gly) residue on the C-terminal side of the sequence. It was also discovered that the deamination of Asn-Gly was accelerated at pH $\geq$ 7.0. In addition, the studies found that the amino acid sequence plays a major role in determining the rate of deamination.

$$\text{-NH-CH(CH}_2\text{-C(=O)-NH}_2\text{)-C(=O)-NH-CH(R)-C(=O)-} + H_2O \longrightarrow \text{-NH-CH(CH}_2\text{-C(=O)-O}^-\text{)-C(=O)-NH-CH(R)-C(=O)-} + NH_3 + H^+$$

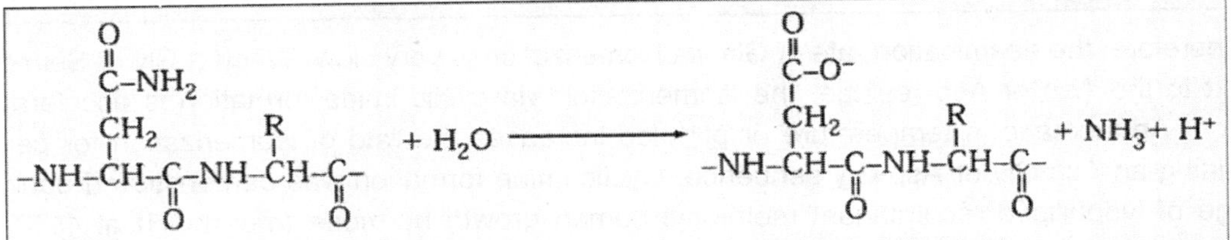

L-Asparaginyl peptide → L-Aspartyl peptide

Figure 15.1 Deamination of asparagine residue in a peptide by direct hydrolysis.

For example, the presence of polar amino acid residues adjacent to an Asn or Gln enhances the rate of deamination, whereas adjacent bulky hydrophobic residues decrease the deamination rate. Serine (Ser) or threonine (Thr) next to an Asn or Gln can act as a general acid by providing a proton in the acid-catalyzed reaction, thereby enhancing the rate of deamination. The major factors influencing the rate of protein deamination include pH, temperature, ionic strength, and buffer species. Degradation of proteins via deamination has been shown to exist in recombinant protein pharmaceuticals such as insulin and human growth hormone in liquid formulations.

Isomerization

Deamination of an asparagine (Asn) residue to yield aspartate (Asp) can also lead to the formation of an isoaspartate (iso-Asp) via intermediate succinimide formation. Nucleophilic attack of the main-chain amide nitrogen on the carbonyl carbon of the Asn side-chain amide group results in a five-membered succinimide ring. This cyclic imide can be hydrolyzed to form either aspartate or isoaspartate. An aspartate residue can also undergo deamination via intermediate succinimide formation to yield isoaspartate. In this case, nucleophilic attack of the main-chain amide nitrogen on the carbonyl carbon of the Asp side-chain carboxylic group results in the formation of a cyclic imide. The additional CH_2 group in the glutamine (Gln) side chain makes the formation of a cyclic imide intermediate by nucleophilic attack less favorable than for an Asn or Asp residue.

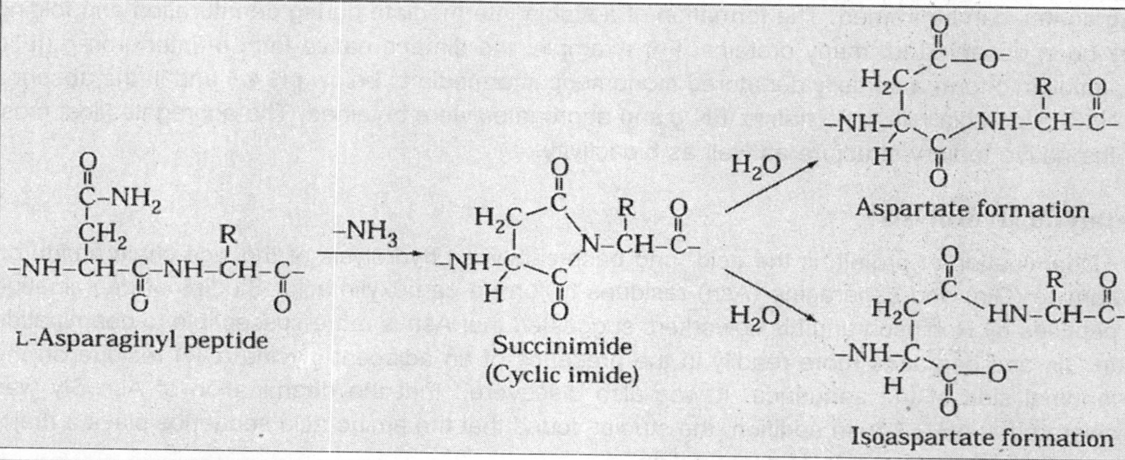

Figure 15.2 Pathway for asparagine deamination in peptides or proteins via intermediate succinimide formation.

Therefore, the deamination rate of Gln via isomerization is very slow. When a Gly or Ser residue is next to the Asn or Asp residue, the isomerization via cyclic imide formation is accelerated in solution. An increase in temperature or pH also increases the rate of isomerization for peptides containing an Asn-Gly or Asp-Gly sequence. Cyclic imide formation was demonstrated during the storage of lyophilized recombinant methionyl human growth hormone (met-rhGH) at 45°C for 4 months, and the Asp-130 residue in the protein underwent isomerization.

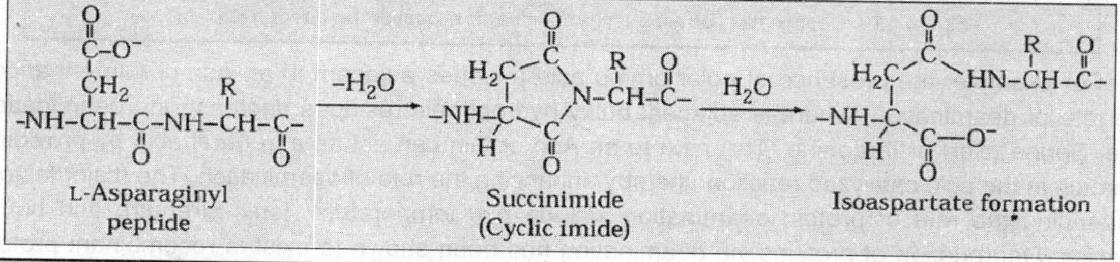

Figure 15.3 Deamination of aspartate to yield isoaspartate via succinimide formation.

Peptides Breakage

Cleavage of peptides at an asparagine residue can result from the nucleophilic attack of the side-chain amide nitrogen on the main-chain peptide carbonyl to form a C-terminal succinimide. The reaction is spontaneous and dependent on protein sequence. Asparagine having an adjacent Gly or Ser residue on the C-terminal side is more labile to spontaneous cleavage. Cleavage at an Asn-Ser linkage has been demonstrated in proteins such as bovine and porcine somatotropins.

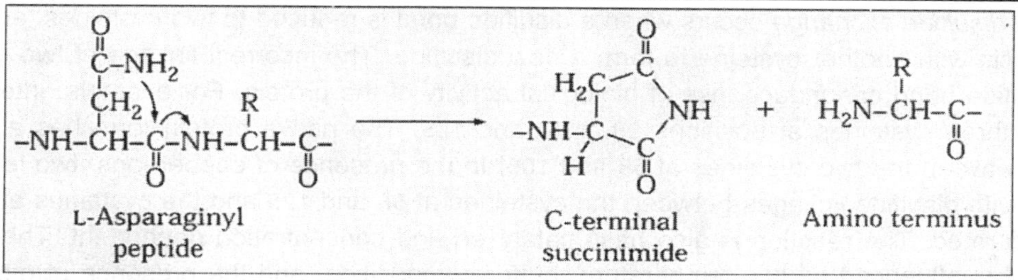

L-Asparaginyl peptide → C-terminal succinimide + Amino terminus

Figure 15.4 Spontaneous cleavage of peptide as asparagine residue.

Oxidation

Oxidation is one of the major degradation pathways for protein pharmaceuticals. Amino acids that can undergo oxidation include methionine, cysteine, histidine, tryptophan, and tyrosine. Oxidation of methionine has been demonstrated in many recombinant proteins such as interleukin 2, relaxin, parathyroid hormone, and human growth hormone. At low pH, the thioether group of methionine is not protonated and is susceptible to oxidation, resulting in methionine sulfoxides. Under extremely oxidative conditions, the sulfoxides can be further oxidized to sulfones.

Methionine can react with a variety of oxygen-reactive species such as hydrogen peroxide, alkyl-hydroperoxides, molecular oxygen, and singlet oxygen generated by heat and light. Methionine oxidation can be catalyzed by the presence of transition metal ions such as Cu^{2+}, Fe^{2+} or Fe^{3+}. Although there are many oxidative pathways for methionine reported in literature, oxidation via a free radical, singlet oxygen, or nucleophilic substitution are the three common mechanisms.

In contrast to methionine, oxidation of cysteine usually occurs at higher pH where the thiol is deprotonated. Cysteine can react with molecular oxygen to form cystine disulfide. The reaction can also be catalyzed by transition metal ions, especially Cu^{2+}. In some proteins cysteine can be oxidized to give sulfenic, sulfinic, or sulfonic acids. Histidine, tryptophan, and tyrosine are susceptible to photooxidation via the singlet oxygen pathway. The rate of photooxidation is pH dependent. At neutral pH, histidine and tryptophan photooxidize faster than tyrosine. At low pHs, tryptophan and methionine are the most photo-reactive amino acids.

Oxidation of methionine:

$$RSCH_3 \rightarrow RS(O)CH_3 \rightarrow RS(OO)CH_3$$

methioine sulfoxide   methionine sulfone   methionine

Oxidation of cysteine by oxygen:

$$RS^- + O_2 \rightarrow RS\cdot + O_2 -$$
$$RSH + O_2\cdot \rightarrow RS\cdot + HOO^-$$
$$RS\cdot + RS\cdot \rightarrow RS\text{-}SR$$

Thiol Disulfide Exchange

Thiol disulfide exchange occurs when a disulfide bond is reduced to two cysteines, and one of them reacts with another cysteine to form a new disulfide. The incorrect linkage of two cysteines in a disulfide bond can induce loss of biological activity of the protein. For example, interleukin 2 contains three cysteines at positions 58, 105, and 125. The native protein form has a disulfide linkage between the two cysteines at 58 and 105. In the presence of copper ions, two less active isomers with disulfide linkages between the cysteines at 58 and 125 and the cysteines at 105 and 125 are formed. The reaction is also base catalyzed and concentration dependent. The reaction rate can be influenced by pH, temperature, buffer composition, and the presence of other metal ions.

β-Elimination

Inactivation of proteins in an alkaline solution can result from *β*-elimination of the cystine residue, with the heterolytic cleavage of the disulfide and the formation of dehydroalanine and persulfide (equation 3). The persulfide can further react with hydroxide to form hydrosulfide (HS), as shown in equation 4. The rate of *β*-elimination increases with increase in temperature and pH. The presence of metal ions also enhances the reaction rate.

$$R'CH_2SSCH_2R'' + OH^- \rightarrow R'CH_2SS^- + CH_2 = CR'' + H_2O \qquad ...(3)$$
$$R'CH_2SS^- + OH^- \rightarrow R'CH_2SO^- + HS^- \qquad ...(4)$$

ANALYTICAL FORMULATION

After gathering all the background information (physicochemical properties, application, site of action, etc.) on the protein to be formulated, the first step that a formulation scientist usually takes is to characterize the protein in the initial formulation by analytical methods. During this step, analytical methods are also developed or optimized for use in formulation and stability studies. Because protein degradation is often dependent on pH and temperature, the second step of formulation development is to determine the relationship between the major degradation pathways of the protein and these parameters.

Formulation scientists usually perform short-term studies of pH and elevated temperature (e.g., 25-40°C) on initial liquid formulations. After determining the major degradation pathways in these formulations, the formulation scientist will select the most stable formulation that can achieve an optimum balance between the different degradation pathways. Finally, long-term stability testing is performed on the final chosen formulation to obtain real-time stability data. This section reviews the process of developing a stable formulation for protein pharmaceuticals.

Methods for Assessment

The selection of formulations for protein pharmaceuticals greatly depends on the stability results obtained in the formulation screening. Therefore, it is important to have suitable and sensitive analytical methods to study the degradation of proteins in formulations. Review articles by Jones provide detailed principles and applications of many analytical methods that can be used for protein characterization. In the following section, the commonly used analytical methods for determining protein formulation stability during formulation development are discussed.

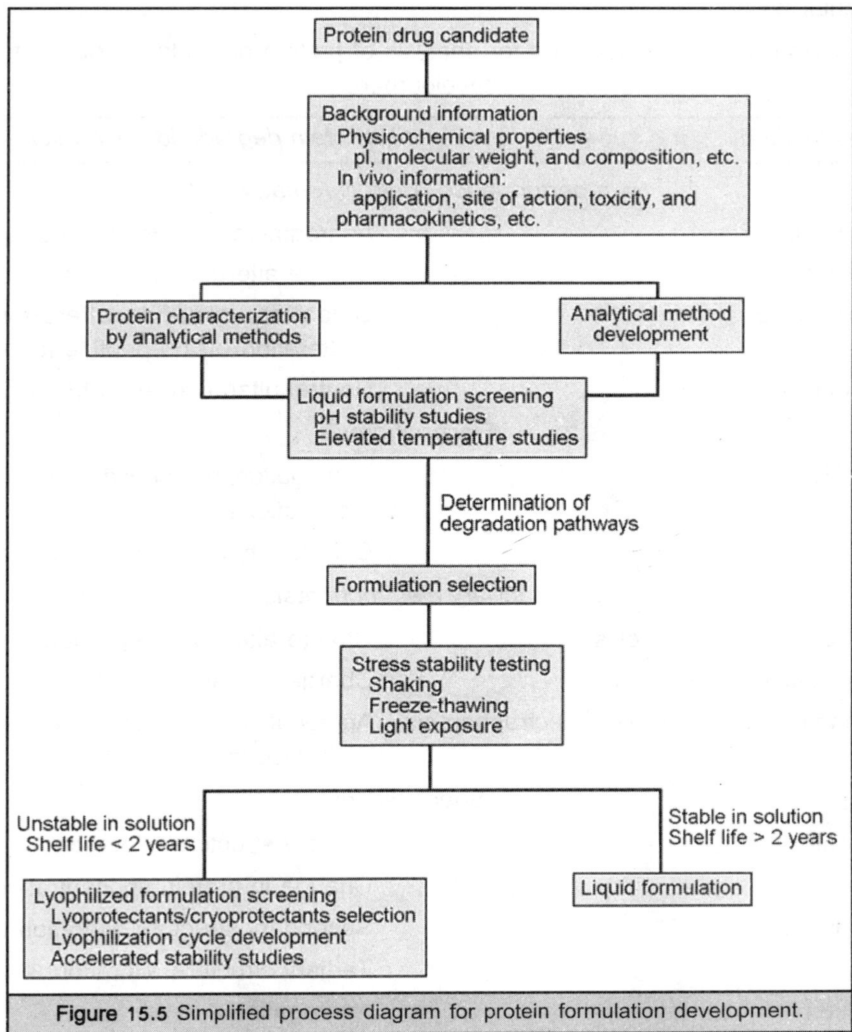

Figure 15.5 Simplified process diagram for protein formulation development.

High performance liquid chromatography (HPLC)

High performance liquid chromatography (HPLC) methods are widely used in the pharmaceutical industry for the analysis of peptides and proteins. The common HPLC methods for use in formulation development to assess protein stability are size exclusion chromatography (SEC), ion exchange chromatography (IEC), reversed-phase chromatography (RPC), and hydrophobic interaction chromatography (HIC). Degradation of proteins via aggregation and cleavage results in the formation of soluble and insoluble aggregates (oligomers) and fragments, respectively. These soluble aggregates and fragments have different apparent molecular weights and can be separated from the intact protein on a size exclusion column that contains porous particles with different pore diameters and has been calibrated with molecular weight standards, usually globular proteins of known molecular weight.

Table 15.1 Analytical methods used for analysis of protein degradation during formulation development

Analytical method	*Protein degradation/alteration*
	High performance liquid chromatography
Size exclusion	Aggregation, polypeptide cleavage
Ion exchange	Charge alteration (e.g., deamination)
Reversed phase	Oxidation, disulfide alterations, charge alteration (using peptide map)
Hydrophobic interaction	Neutral alteration (e.g., Met oxidation)
	Electrophoresis
SDS-PAGE	Aggregation, polypeptide cleavage, disulfide alterations
Isoelectric focusing	Charge alteration (e.g., deamination)
	Capillary electrophoresis
Capillary zone electrophoresis	Charge alteration (e.g., deamination)
Capillary isoelectric focusing	Charge alteration (e.g., deamination)
SDS-dynamic sieving capillary electrophoresis	Aggregation, polypeptide cleavage, disulfide alterations
	Spectroscopy
UV absorption	Tertiary structure alteration
	Change in protein concentration
Circular dichroism	Secondary structure alteration (far-UV CD)
	Tertiary structure alteration (near-UV CD)
Light scattering	Aggregation

IEC is often employed during formulation and stability studies for the separation of degraded proteins with charge heterogeneity resulting from deamination or isomerization. Charged protein tends to bind to ion-exchanged resin (e.g., sulfopropyl, DEAE) of opposite charge by ionic interaction. Bound proteins can be eluted at their own critical salt concentration, allowing separation of the various charge forms. The amount of salt required to elute a charged species depends on its affinity for the resin (e.g., number of binding sites), the net charge, and the ability of the salt to displace the bound protein. By this method, deamidated forms of a protein that exhibit different net charge from each other and the native protein can be separated and quantified. RPC methods are commonly used for the separation of small molecules and peptides, especially for peptides that result from proteolytic digestion of proteins in peptide mapping.

A hydrophobic surface on the protein binds to the hydrophobic site of the resin (e.g., alkyl groups such as butyl [C4], hexyl [C6], or octadecyl [C18] or aromatics such as phenyl) in a reversed-phase column by hydrophobic interaction. Such hydrophobic interactions can be weakened by increasing the content of organic modifier in the mobile phase. Thus, proteins can be eluted in the order of their hydrophobic interaction strengths by an organic modifier such as acetonitrile. Although the application of RPC for analysis of recombinant proteins is less common than for peptides due to difficulties in resolution of large molecules, it has been used for characterization or separation, as in the case of met-rhGH and rhGH, which differ by the only one amino acid. This method is also useful in detecting the formation of more polar degraded proteins resulting from oxidation. Similar to RPC, HIC is another useful technique for the separation of proteins based on their differences in surface hydrophobicity.

Most proteins have some hydrophobic groups exposed to the surface. When the hydrophobic surface of a protein is in contact with the hydrophobic ligands on the beads of a column, water is released from the hydrophobic regions, and the entropy of the system is increased. The release of water increases the entropy of the system, causing the free energy to decrease on adsorption. Protein adsorption on an HIC column can be adjusted by changing the salt concentration. At high salt concentrations, protein adsorption is mainly due to hydrophobic interactions. By running the column with decreasing salt concentration, protein will be eluted from the column when the salt concentration is not high enough to maintain or stabilize the interaction.

Separation of proteins

Electrophoresis is a common method for the separation of proteins based on their net charge and size. Upon applying an electric field to a protein, it migrates toward the anode or cathode, depending on its net charge. There are two basic forms of electrophoresis: sodium dodecyl sulfate polyacrylamide gel electrophoresis (SDS-PAGE) and isoelectric focusing (IEF). In SDS-PAGE, proteins are complexed with the denaturing agent SDS to give similar mass-to-charge ratios so that they can have free electrophoretic mobility. When SDS-complexed proteins are electrophoresed in a polyacrylamide gel with the appropriate pore size, their migration rate will be based on the size of the proteins. Thus, a mixture of proteins with different molecular weights can be separated using this method.

Separation of proteins by IEF is based on their isoelectric points (pIs), where the mobility of each protein in an electric field is zero and the net charge of the each protein is also zero. Because the mobility of nondenatured protein is dependent on its net charge, and this charge is determined by the pH of the solution, an electric field in the presence of a pH gradient causes the protein to migrate until the pH is the same as the pI. This technique is usually employed by preparing a polyacrylamide gel containing ampholytes of appropriate pI and buffer capacity to produce a pH gradient that is stable in the applied electric field. This method has been used to assess protein deamination during stability studies of many pharmaceutical products such as recombinant human growth hormone.

Capillary electrophoresis (CE)

Due to recent advances in technology, *capillary electrophoresis* (CE) is being more widely used in the biotechnology industry for protein analysis. The sensitivity and speed of CE make it a useful technique in early protein formulation development when only a small amount of material is available. CE provides a wide range of separation modes that can be performed with a single instrument. The basic separation principles are the same as with conventional electrophoresis techniques.

Capillary zone electrophoresis (CZE) is the most common CE separation method, and it employs a single buffer system in free solution. Separation of proteins using CE is based on their differences in mass-to-charge ratio. Separation takes place in free solution inside a capillary (usually made of fused silica) filled with electrolyte. Proteins with different mass-to-charge ratios are separated into individual zones as electrical force drives them at different rates through the capillary. As the mass-to-charge ratio increases, the electrophoretic mobility decreases, resulting in a longer migration time. Coated capillaries and buffer additives are often used to prevent protein adsorption and electroendosmotic flow. CZE has been used for the detection of the heterogeneity of recombinant proteins as well as the separation of the deamination products of growth-hormone-releasing factor.

Capillary isoelectric focusing (cIEF) is another separation mode in CE for the separation of proteins based on their isoelectric points. There are three basic steps in cIEF: sample injection, focusing, and mobilization. Proteins are premixed with ampholytes and injected into the capillary by pressure. High voltage is applied to the capillary during focusing so that sample ions can migrate to their neutral charge point in the pH gradient formed by the ampholytes along the capillary. After focusing, the focused protein bands are moved to the detection point to generate a signal. The on-line UV detection of cIEF provides a more accurate quantitation than the slab gel technique. A coated capillary is recommended for cIEF because the proteins stay in the capillary for focusing and mobilization.

cIEF has been applied to the characterization of charge heterogeneity of recombinant monoclonal antibodies as a result of deamination. The high resolving power of the method also makes it useful for the characterization of the heterogeneity of the glycoforms of glycoproteins such as recombinant tissue plasminogen activator.

SDS-dynamic sieving capillary electrophoresis (SDS-DSCE) is a useful separation mode in CE for the molecular weight determination of proteins. Because all proteins have the same mass-to-charge ratio after complexation with SDS, the migration rate of proteins depends only on their molecular size. Although SDS-DSCE is equivalent to the conventional SDS-PAGE, it provides a rapid and automated system for quantitative analysis of proteins based on their molecular weight. Monoclonal antibodies degraded by the formation of aggregates and fragments can easily be separated by this method.

UV spectroscopy

Ultraviolet (UV) absorption spectroscopic measurement is the most convenient and accurate method for the determination of protein concentration. Proteins with aromatic amino acids have a maximum in UV absorption between 275 and 282 nm at pH values below 8. Thus, the protein concentration in a sample can be determined by measuring its absorbance at the wavelength maximum near 280 nm, and this absorbance is then compared with the absorbance of a solution with a known protein concentration or, if known, the molar extinction coefficient or absorptivity of the protein may be used. The absorptivity (often incorrectly described as the extinction coefficient) of a protein is defined as the absorbance of a 1 mg/mL protein solution through a 1-cm path at the wavelength maximum (e.g., near 280 nm). The molar extinction coefficient is the absorptivity of a 1.0 M solution of the protein through a 1.0-cm path.

It is important to measure the protein concentration accurately for extinction coefficient determination. Quantitative amino acid analysis, nitrogen determination, or dry weight measurement are often used for this purpose. Circular dichroism (CD) spectroscopy is a useful technique to evaluate the secondary and tertiary structures of protein. The secondary structure of a protein such as α-helices, β-sheets, and random coils can give rise to CD signals in the far-UV region (170–250 nm). CD spectroscopy is commonly employed for the study of denaturation or unfolding of proteins. The tertiary structure of a protein affects the local environment of the aromatic amino acids such as trytophan, tyrosine, and phenylalanine, as well as the disulfide bonds, and these effects contribute to CD signals in the near-UV region (240–320 nm). Because CD signals can be either positive or negative, it is not easy to obtain detail from CD spectra. However, a change in the far- or near-UV CD spectrum can represent a change in the secondary or tertiary structure of the protein, respectively.

Therefore, CD is a useful tool to measure protein denaturation as well as aggregation, both of which result in structural or conformational changes. Light scattering spectroscopy can be used to detect protein aggregation due to the fact that larger particles scatter more light per weight unit than small ones. Aggregated proteins can be either soluble or insoluble. Insoluble aggregates usually cause opalescence (turbidity) in a protein liquid formulation, yielding light scattering at wavelengths even in the visible region of the spectrum. Turbidity can also be conveniently measured in the near UV(340–360 nm) if no other chromophores are present in the same region. Thus, an increase in absorbance in this region often indicates the formation of aggregates. Correction for light scattering errors is necessary for the measurement of protein concentration made by the absorbance at the peak wavelength near 280 nm.

Pharmaceutical final product solutions are required by the FDA to be inspected for clarity and degree of opalescence. Therefore, light-scattering determination may be substituted for the visual inspection method. Because native protein in solution also scatters light, light-scattering spectroscopy can be used to estimate the molecular weight of a protein. Dynamic laser light scattering, also known as photon correlation spectroscopy (PCS) or quasi-elastic light scattering (QELS), can be used to calculate the average hydrodynamic diameter and determine the diffusion coefficient of a protein by autocorrelation methods.

Protein Formulation Screening

The methods commonly used by formulation scientists for protein formulation screening to select the best candidate formulation include the short-term accelerated pH and temperature stability studies. Accelerated conditions are defined as the extremes of the parameter range (e.g., 40°C and high (≥ 9.0) or low (≥ 3.0) pH).

pH dependent pathways

Solution pH often plays an important role in the stability of the protein product because most of the major protein degradation pathways are pH dependent. Thus, the stability of a protein in different formulations can be compared by performing stability studies on the protein formulated at a range of pH using a variety of buffer solutions. Acetate, succinate, citrate, and phosphate are the most common buffers used in recombinant protein formulations. For rapid formulation screening, a pH stability study is usually performed in a short period of time and at an elevated temperature of 40°C because the rate of degradation is slow at low temperature. Often, a pH stability profile for each formulation can be developed by plotting the rate constants of all degradation reactions versus pH. The most stable protein formulation selected should be the one with a pH at which the overall degradation reactions are minimal.

Table 15.2 Examples of buffers commonly used in recombinant protein formulations for pH maintenance

Buffers	*pH*	*Commercial products*
Sodium acetate	4.0	Neupogen (Amgen)
Sodium succinate	5.0	Actimmune (Genentech)
Sodium citrate	6.0	Nutropin AQ (Genentech)
	6.5	Rituxan (IDEC, Genentech)
	6.9	Epogen (Amgen)
L-Histindine HCl	6.0	HERCEPTIN (Genentech)
Sodium phosphate	7.3	Intron-A (Schering-Plough)
	7.4	Protropin (Genentech)
	7.5	Humatrope (Lilly)
	7.0	Orthoclone OKT3 (Ortho)
	7.5	Proleukin (Chiron)
	7.4	Nutropin (Genentech)
Phosphoric acid	7.3	Activase (Genentech)

Effect of temperature

Because temperature also has effects on protein degradation reactions, stability testing at accelerated temperatures is often carried out by formulation scientists for formulation screening, and the degradation rate profiles in various formulations are compared using Arrhenius plots. An Arrhenius plot allows one to predict the degradation rate constants at storage temperatures other than those actually studied, assuming that the protein does not undergo physical changes at a given temperature. Formulation scientists often use the Arrhenius plot to extrapolate results obtained at higher temperatures to 5°C for product shelf life prediction. However, at least three temperatures are required to obtain an Arrhenius plot, and extrapolation of these data should be done with understanding of the assumptions made regarding the linearity of the response and the accuracy of the rate constants.

Small-molecule drugs that often degrade via a single degradation pathway usually follow Arrhenius behavior. However, due to complexity in protein structure, more than one degradation reaction can occur in a protein product. This may result in nonlinear Arrhenius behavior. Sometimes, the degradation pathways for proteins at higher temperatures are not the same as those at lower temperatures. For example, interleukin 1β undergoes deamination at or below 30°C and aggregation at or above 39°C. In this case, the formulation selected based on accelerated stability results may not be stable enough to protect the protein from degradation at lower temperatures. However, accelerated stability testing can still be a useful tool for protein formulation screening if the protein primarily degrades via only one degradation pathway within a selected temperature range, or if the degradation pathways are not influenced by one another (e.g., deamination rate not affected by oxidation).

Stabilizing Tests

After a stable protein formulation is selected from the formulation screening studies, stress tests are usually performed to determine whether additives are required to stabilize the protein against stresses such as agitation, freeze-thawing, and light exposure.

Method of agitation

Proteins are susceptible to denaturation by mechanical force such shaking or agitation. Shaking of liquid protein products can occur during filling, shipping, and handling. The mechanism of denaturation may involve an increase in area of the air-liquid interface during shaking. Proteins have a tendency to unfold and expose their hydrophobic groups at an air-liquid interface, resulting in denaturation and the formation of aggregates. Proteins also tend to concentrate at the air–liquid interface due to the presence of both polar and nonpolar side chains. Irreversible unfolding or aggregation of the protein occurs when the interface surface area increases. Henson explained that the turbidity observed in some protein solutions during vigorous shaking could be due to the formation of aggregates when the air-liquid interface increases. Formulation scientists often perform agitation studies on liquid protein formulations to determine if a surfactant is required to protect the protein from denaturation when agitation is unavoidable during manufacturing, shipping, and handling. The addition of surfactants could either reduce the interfacial tension or solubilize the concentrated protein at the interface. Surfactants commonly used as stabilizers in the pharmaceutical industry include polysorbate 20, polysorbate 80, pluronic 68, and brij, which are polymers composed of various combinations of ethylene oxide, propyl oxide, sorbitan, and fatty alcohols.

Low temperature pathways

Under many circumstances, such as processing, storage, shipping, and stability testing, freezing and freeze-thawing stresses may be applied to proteins. For instance, it is very common for a liquid protein bulk to be manufactured and stored as frozen bulk for long-term storage and then thawed before filling into containers for packing as a liquid protein product or for freeze-drying as a lyophilized product. Sometimes, a protein formulation is frozen as a concentrated bulk for storage and then thawed and diluted with placebo before use. The risks of accidental freezing and thawing of proteins in liquid formulations may also exist during shipping and handling. Formulation scientists often perform short-term stability testing such that samples at each time-point are kept frozen and then thawed at the end of the study for analysis.

In lyophilized formulations, proteins are subjected to freezing during the lyophilization process. Therefore, freeze- thawing stability studies (at least three cycles of freezing and thawing) are usually performed on potential formulation candidates during formulation development to ensure the final product is stable upon freeze-thawing. Protein denaturation during freezing is mainly caused by the phase separation due to the formation of ice. When ice is formed in a liquid protein formulation, the local concentration of protein as well as all excipients increases dramatically. The concentrated excipients in the non-ice phase may become destabilizing agents and cause the protein to denature. Some buffer species such as sodium phosphate and sodium succinate exhibit a decrease in pH during freezing. Freezing-induced pH changes may cause instability of recombinant proteins, as observed in IFN-γ formulated in succinate buffer. If a protein formulation is evaluated to be denatured during freeze-thawing studies, formulation scientists may consider adding cryoprotectants such as sugars, polyols, amino acids, and inorganic and organic salts to prevent denaturation.

The preferential exclusion mechanism of protection of proteins by these cryoprotectants during freeze-thawing was proposed by Timashelf and his coworkers. They suggested that the cryoprotectants, acting as cosolutes in the solution, are preferentially excluded from the protein. Consequently, the protein itself is preferentially hydrated by water. Hydration of a protein plays an important role in maintaining the conformational structure. Water contributes to the hydrogen bonds and hydrophobic interactions that are responsible for the formation of the secondary structure as well as maintaining the protein in a folded or native form (tertiary structure).

Photo-oxidation technique

Some amino acid residues in proteins can absorb energy from UV light and become modified to form photooxidized degradation products, especially in the presence of photosensitisers such as molecular oxygen, riboflavin, dyes, and polymers containing ether linkages (e.g., polysorbates). For example, the methionine residue in recombinant human relaxin can undergo photooxidation to form methionine sulfoxide, and the tryptophan residue in monoclonal IgG is converted to kynurenine and *N*-formylkynurenine upon exposure to UV light. The degree of photolytic degradation in the protein formulation can be influenced by many factors, including the buffer species and its concentration, excipients, and the formulation pH. Light intensity, duration of exposure, and storage temperature can also affect the rate of photolytic degradation. Light stability testing is often performed during protein formulation development.

The aim of a photostability study is to accelerate the formation of photodegradation products, if any, in a particular formulation upon exposure to light. FDA guidelines call for light stability testing of finished drug formulations but do not provide specifics of how to perform the test. A recent review article by Nema et al. provides an extensive overview of the protocols currently being used to perform photostability studies in pharmaceutical industry. Fluorescent lamps are usually used by pharmaceutical formulation scientists to simulate sunlight in photostability studies. Finished drug products are stored in a light cabinet with light intensity of 2-180 klx. The total exposure is between 8 and 4,500 klx days. Temperature inside the light cabinet is maintained at 25 ± 2°C. Control samples, wrapped in aluminum foil, are exposed concurrently with the test samples. Photolytic degradation of protein can be prevented by adding antioxidants to the formulation. Antioxidants can be classified into four main categories: chelating agents, reducing agents, oxygen scavengers, and chain terminators. If the addition of antioxidants in a protein formulation does not inhibit oxidation, the use of light- resistant containers is an alternative way of protecting light-sensitive protein drugs.

Freeze-drying Method

Protein products that are unstable in solution or whose most-stable liquid formulations have insufficient long-term storage stability for marketing require lyophilization. However, the lyophilized products are usually limited to reconstitution and use in a short period of time. Lyophilized products also require longer time and more labor for formulation development than the ready-to-use liquid products. Lyophilization is also expensive due to the added costs of drying. Therefore, lyophilization is usually considered as an alternative method in protein formulation development when success with a liquid formulation cannot be achieved. Lyophilization, also known as freeze-drying, is a process usually employed in the protein pharmaceutical industry to convert protein liquid formulations into solids of sufficient stability for storage.

The process involves three major steps: freezing, primary drying, and secondary drying. During the initial freezing step, solvent (usually water) from an aqueous solution of protein drug that has been filled into glass vials is frozen and crystallized into solid ice at a low temperature such as –40°C in a freeze dryer. The solutes including the protein and excipients are usually converted into an amorphous solid phase, which consists of uncrystallized solutes and uncrystallized water below their glass transition temperature, although some excipients such as buffer salts and mannitol may crystallize. The second step of the freeze-drying process, in which ice is removed from the frozen product by sublimation and condensation, is called primary drying. After the entire protein solution has been solidified, vacuum is applied to the drying chamber, and the shelf temperature is increased to supply heat to sublime ice from the frozen solid. Water vapor formed as a result of sublimation is condensed on cold surfaces in the condenser chamber. Primary drying is the longest part of the freeze-drying process and may require several days.

After the bulk water (ice) has been removed from the partially dried product, secondary drying begins. During the stage of secondary drying, the final step of the freeze-drying process, the shelf temperature is increased to provide an elevated product temperature to remove the unfrozen water remaining in the amorphous solid phase. Secondary drying is usually performed over several hours at a shelf temperature between 25 and 50°C, and the product temperature is maintained at about

25–35°C. The successful development of a lyophilized protein product requires the development of a liquid formulation that is stable to the freeze-drying process. Proteins can undergo denaturation or aggregation during both freezing and drying. Formulation scientists often consider adding cryoprotectants and lyoprotectants to formulations to protect the protein from damage by freezing and drying, respectively, in the lyophilization process. Often, lyoprotectants improve the stability of the lyophilized products during subsequent storage.

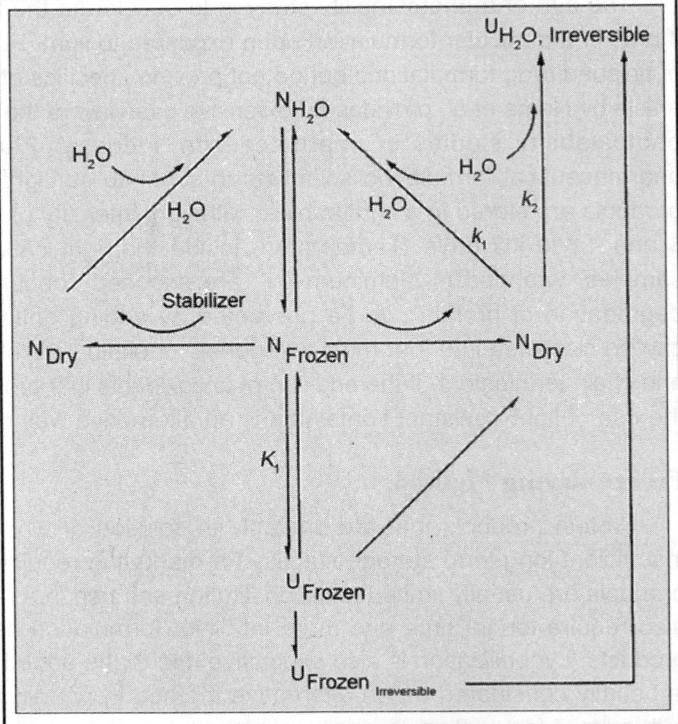

Figure 15.6 Schematic representation of the conformation of a protein during freezing, drying, and rehydration.

During freezing, all solutes in a formulation are concentrated when ice is formed, and buffer salts may crystallize, resulting in a pH shift. These freezing stresses can induce irreversible conformational changes that lead to protein instability. Thus, the use of buffers at low concentration and with minimal pH shift during freezing is recommended for formulations to be prepared by the freeze-drying process. The addition of cryoprotectants such as polymers, carbohydrates, and amino acids can provide cosolutes to preserve the native structure of protein in the frozen state.

The mechanism for solute-induced protein stabilization during freezing proposed by Arakawa et al. is applicable to these conditions as just described. Amorphous sugars such as trehalose, sucrose, and lactose are the most effective lyoprotectants for stabilizing proteins during drying and storage. Carpenter and his coworkers have found that the formation of an amorphous solid is required for preservation of proteins by sugars, and some sugars such as mannitol fail to protect labile proteins during freeze-drying because the sugar forms a separate crystalline phase during freezing. It has been hypothesized that lyoprotectants stabilize proteins by acting as a water replacement through hydrogen bonding to the dried protein, preventing conformational changes during drying. In developing a freeze-drying cycle for lyophilized protein product, temperature control is important in each step of the process.

During freezing, the protein solution is frozen so that most of the water is converted to ice and all solutes are converted to either crystalline solids or amphorous solids below their glass transition temperature. The temperature at which the concentrated solute crystallizes during freezing is called the eutectic temperature. If the solute crystallizes, the shelf temperature must be reduced to well below the lowest eutectic temperature to ensure complete crystallization. For an amorphous solute, the shelf temperature must be lowered below the glass transition temperature. Typically, the final product temperature during freezing is about –40°C. Usually, rapid ice growth as a result of low freezing temperature produces a high degree of supercooling. The degree of supercooling determines the number and size of ice crystals and therefore determines the drying characteristics and appearance of the dried cake. A very high degree of supercooling will result in very small pores, which means slower primary drying. Therefore, a very high degree of supercooling is undesirable.

The ideal freezing method is to provide a moderate supercooling and rapid ice growth. The time required for primary drying decreases by a factor of 2 for a 5°C increase in product temperature. For minimizing process time, it is important to maintain product temperature as high as possible during primary drying without affecting the product quality. Usually, the product temperature in primary drying is controlled at a safe margin (2-5°C) below the collapse temperature. After all ice has been removed, the product temperature is usually increased to about 25-35°C in the secondary drying step to allow the removal of unfrozen water remained in the product. To avoid collapse, the product temperature should not be raised until all the ice has been removed during primary drying.

Pre-marker Stability

Long-term pre-market stability studies of protein formulations are designed to establish the specific storage conditions and expiration dating for the drug product. It is usually required by the FDA that a pharmaceutical product contain not less than 90% of the label claim of active ingredient at the expiration date. Although real-time stability data are required by the FDA to demonstrate that the pharmaceutical product can meet this requirement prior to approval, an extrapolated expiration date based on the stability results obtained from clinical batches is usually allowed. Thus, stability-indicating assays are important for determining the stability of drug products. Analytical methods such HPLC, electrophoresis, and spectroscopy have been widely used for this purpose. In designing a long-term stability protocol for drug potency testing, one should consider the storage conditions and storage intervals. For refrigeration (5°C) and room temperature (15-30°C, by U.S.P. definition) storage, samples should be stored at 5, 15, and 25-30°C and monitored at 0, 1, 2, 3, 6, 9, 12, 18, and 24 months, and once yearly after. For lyophilized products, samples can be assayed at longer storage intervals and at higher storage temperatures such as 40°C.

Pharmaceutical drugs are often monitored at high temperatures for the estimation of the degradation constants at low temperatures by Arrhenius plots. For high-temperature storage, samples are usually stored at 37 or 40°C and assayed at 1, 3, and 6 months. Samples may also be stored at 45 or 50°C and assayed at 2 weeks and 1 and 2 months. However, these elevated temperatures may approach the denaturation temperature of the protein, resulting in physical denaturation of the protein that would not be encountered in real situations, or extrapolate to lower-temperature behavior. In addition to determining the amount of active ingredient in a drug product, evaluation of the product should also include the following: pH, ionic strength, appearance, color, clarity, particulate matter, sterility, and pyrogenicity.

CLINICAL STUDIES

Typically, proteins and many peptides must be administered by injection because of their poor bioavailability (fraction of administered drug in the circulation) by other routes. Initial clinical studies of these molecules is often conducted by dosing the drug i.v., intramuscularly (i.m.), or subcutaneously (s.c.). These clinical studies assess the safety and tolerability of systemic exposure to the new drug. The results from early clinical trials combined with preclinical studies in animal models may indicate the need for a drug delivery system. Drug delivery systems are often required when the efficacious dose is not well tolerated, indicating a requirement to localize the drug at the site of action or maintain a low systemic steady-state level (e.g., provide same total drug dose at a lower maximum serum concentration).

Localization of the drug to the site of action may also be necessary due to biological barriers preventing sufficient delivery of the drug via the systemic circulation (e.g. blood–brain barrier, blood–ocular barrier). In some situations, the disease state may be more readily accessible to an alternative delivery method than traditional par- enteral administration. In the case of cystic fibrosis, a disease of the lungs, aerosol delivery of rhDNase has proven to be an effective therapy. Another rationale for using a drug delivery system is patient compliance or ease of use. For example, if a patient is required to have multiple injections of a drug per day or per week, it will be difficult to maintain proper compliance unless the drug effects are immediately felt by the patient and/or the disease is debilitating. These issues along with the in vitro (physicochemical properties) and in vivo (pharmacokinetics, pharmacodynamics, and toxicology) characteristics of the drug must be well understood prior to embarking on the development of a drug delivery system.

Drug Delivery

Several potential options are available for drug delivery. These options include both invasive methods involving injections or noninvasive approaches that may or may not provide sufficient bioavailability. In each case, a stable formulation must first be developed, as described in the previous sections. For invasive methods, drug delivery may be applied with either devices or biodegradable depots. Delivery devices such as syringe pens and needle-less injectors (injection by high-pressure gas) are gaining acceptance for many frequently administered drugs. In addition, implantable pumps are currently used with protein drugs such as insulin. Several injectable depot formulations made from the biodegradable polymer, poly(lactic-coglycolic acid) (PLGA), are either approved or in clinical trials. These depot systems offer the advantage of eliminating multiple injections and localizing drug to the site of action, if necessary. However, a major challenge in the development of these systems is the production of sterile injectable depots, usually microspheres on the order of 20 μm or greater in diameter.

Approaches to developing microsphere depot systems for clinical and commercial use have recently been reviewed. For noninvasive administration, aerosol delivery of proteins and peptides is the most developed and accepted delivery option. Aerosol delivery offers the unique option of delivering the drug through a single inhalation instead of a single injection. The lung provides a large surface area for absorption of the drug, but the aerodynamic properties of the drug (liquid or solid) play a key role in the deposition and, ultimately, the amount of drug delivered systemically.

This work demonstrates that if the key parameters are analyzed and the appropriate drug formulation and device are selected, then aerosol delivery of proteins may be an effective method of administration. In contrast, the bioavailability of the protein or peptide when delivered by the other noninvasive routes is usually quite low (<1%), indicating that they are not an effective method for achieving significant systemic levels of the drug. However, they may be very effective for local delivery of protein drugs if a low dose is efficacious.

System of Stable Formulation

After selecting a delivery system, the formulation scientist is required to develop a stable formulation that is compatible with the system and its intended use. These studies usually involve the formulation development worked described in the previous sections, as well as protein–delivery system interaction studies. The type of studies undertaken to assure delivery of biologically active and nonimmunogenic protein depend upon the system characteristics. For injection devices, it is necessary to assess the protein stability in the container used to store the protein (e.g., cartridges) and the effect of the injection itself on the protein stability. These studies often involve storage stability studies of the protein in the device container as already described for protein filled into vials.

The protein delivered from the device is also assessed by injecting into an empty sample tube or a buffer solution. Both recombinant human growth hormone (rhGH) and insulin are currently available in injection devices. Typically, high-pressure devices such as the needle-less injectors may cause denaturation of the protein due to the air-water interface and the high shear force applied to the liquid. Surface-induced denaturation may also occur at the device surface or at an air-water interface in the device. This degradation can be reduced and potentially eliminated by the addition of surfactants (e.g., polysorbates). In addition, implantable pump devices require studies to demonstrate maintenance of protein stability during the protein's residence time in the pump at physiological temperatures (–37°C).

Several studies have been performed with insulin in these pumps, indicating the difficulty in designing stable formulations for these pumps. These systems pose a formidable challenge due to the potential for accumulation of degraded protein over time with repeated use of a refillable pump. Similar issues are also encountered in the development of the injectable depot formulations. In this case, the protein remains in a biodegradable depot at physiological conditions for prolonged periods. The protein must maintain its potency during its in vivo release from the depot. Recent studies with rhGH revealed that while it degrades during incubation at physiological pH, temperatures, and ionic strength, it maintains its biological activity and is not immunogenic. This result, however, cannot be assumed for each protein. Therefore, stability studies at physiological conditions should be performed prior to the development of a depot formulation to define the feasibility of the system and the acceptable duration of release (i.e., what level of degradation can be tolerated in terms of both potency and immunogenicity).

These systems also pose a challenge in protein stability during the manufacture of the depot. For example, PLGA formulations require the dissolution of the polymer in organic solvents or use of high temperatures to melt the polymer, followed by dispersion of the protein in the PLGA and extraction of the organic solvent or reduction of the temperature, respectively. This exposure to harsh conditions often causes denaturation of the protein. Screening studies are performed to select excipients that will stabilize the protein against denaturation under these conditions. An example of this type of study was recently described using the model proteins rhGH and recombinant human IFN-γ. These studies revealed that excipients (e.g., sugars), which cause preferential hydration of the protein (cause an increased water layer at the protein surface), also increase protein stability during exposure to organic solvents.

If these stability issues are overcome, an injectable depot system such as PLGA microspheres may be developed and tested in clinical trials. Unlike depot systems, aerosol formulations for proteins do not require long-term stability at physiological conditions or stability in organic solvents. However, the aerosolization of proteins in their liquid state often leads to some protein degradation due to the air-water interface as well as the shear stress generated during the process. Several studies have been performed with traditional nebulizer devices that generate aqueous droplets of the appropriate diameter for efficient delivery to the lung. For rhGH, a surfactant was required to stabilize the protein during the process of aerosol formation, whereas rhDNAse was very stable in an unbuffered isotonic solution. In many cases, it may not be possible to develop an aqueous protein formulation that both prevents degradation and is not an irritant to the lung. In addition, devices required to generate liquid droplets are usually not easily portable, making them less attractive to patients that are not hospitalized or bed-ridden. However, dry powder protein formulations may be delivered in small hand-held devices.

To produce a protein powder with the proper aerodynamic properties, spray-drying of the protein is usually performed. This process usually involves high temperatures and aerosolization of the protein, both of which may cause denaturation, as observed for rhGH. Once again, a formulation must be developed to prevent this denaturation. In addition, the dry powder protein formulation may be required to have stability at controlled room temperature (15-30°C) and variable levels of residual humidity. The formulation scientist then carries out formulation screening experiments at different temperatures and residual humidities. If a protein powder is feasible, it offers advantages in ease of use and, perhaps, greater long-term stability.

Commercial Complexities

The development of drug delivery systems may also require elaborate manufacturing procedures due to the complexity of both the system and the protein. When considering the manufacture of injection devices such as pens and pumps, some of the key issues are the ability to perform a sterile filling operation, the stability in the device, and the interaction between the drug, excipients, and the device or manufacturing components. The level of complexity is further increased in the case of pulmonary delivery. A nebulized liquid formulation is usually filled into ampules or other containers that allow easy addition of the liquid to the nebulizer by the patient. The filling of plastic ampules also has unique challenges in the ability to maintain sterility of the final product, the permeability of the material to gases and volatile solvents, and the physical properties of the container. In contrast, if a dry powder is used as the aerosol, the manufacturing process requires large-scale spray-drying equipment and machines for filling the dry powder into blister packs that can be used for metered-dose inhalers. Manufacturing complexity is further increased when considering depot formulations such as PLGA micro- spheres. These microspheres are typically 20-90 μm in diameter and are therefore not readily sterilized by the traditional filtration methods (e.g., 0.22-μm filters).

Terminal sterilization of these microspheres causes degradation of the polymer backbone, altering the release properties and often causing significant degradation of the protein within the microspheres. Microspheres must then be produced aseptically to avoid terminal sterilization while still maintaining the delivery of a sterile product to the patient. An aseptic process may be achieved through the use of barrier technology (isolators) for raw material input into the process and final product removal from the process. The remainder of the process operations is then designed as steam-in-place systems to assure a sterile environment throughout the production process. If the unique challenges of a drug delivery system are overcome for a given protein, protein delivery may be successfully achieved. To reach commercial use, the drug delivery system must be well matched with the clinical and marketing requirements as well as the physiological requirements of the protein.

16

ANAEROBIC BACTERIA

Anaerobes, in the simplest form, are microorganisms that are unable to grow at the surface of a medium exposed to air, where the oxygen may either be bacteriocidal or bacteriostatic. These organisms do not use oxygen for catabolic reactions, and oxygen is not required for their survival. Anaerobes, with the exception of autotrophic methanogenic and acetogenic bacteria, sulfur-metabolizing archaea, and sulfate-reducing bacteria, obtain their energy by fermentation, a redox process in which organic substances serve as electron donors and terminal electron acceptors. *Obligate anaerobes* share two common characters: an extreme sensitivity to molecular oxygen, and a capacity to produce energy and perform essential biosyntheses without molecular oxygen.

The sensitivity to O_2 is extremely variable from one group to another, and evaluation of O_2 sensitivity requires that all organisms be cultivated under identical conditions, both for medium composition (including reducing substances) and gaseous environment. The physiological, ecological, and genetic diversity of anaerobic bacteria makes them very interesting organisms. Unlike aerobic microorganisms whose anabolic features are largely exploited for production of single-cell proteins, antibiotics, and amino acids, anaerobes offer the potential for utilizing catabolic features in the production of specialty and bulk chemicals, fuels, and unique enzymes that are stable and active under extreme environmental conditions. These properties of anaerobes make them highly desirable for certain industrial uses. Today, a large body of information is available on the metabolic diversity of anaerobes.

Anaerobes seem to have adapted from the primordial atmospheric conditions to inhabit both moderate and extreme environments with metabolic machinery that is active from about 0°C to over 100°C, in very little salt to saturated salt conditions, in highly acidic (pH < 2) to alkaline (pH > 9) conditions, and even in the presence of highly toxic chemicals. In this chapter, we will only cover procaryotes and archaea, not anaerobic eucaryotes (i.e., yeasts and fungi). We will include examples of pure cultures used to produce specific products and mixed cultures or consortia used in environmental processes.

PHYSIOLOGICAL APPROACH

Microbial Diversity

Knowledge about microbial diversity is essential in understanding the relationship between environmental factors and metabolic functions. Such knowledge can be used to assess the effects of environmental conditions on production of microbial products. However, relatively little is known about the entire spectrum of anaerobic bacteria that may be of importance to various industries. What is known about anaerobic species diversity is presented in *Bergey's Manual*, where over 170 genera and over 800 species of anaerobic bacteria and archaea are described. Cultivation of microorganisms greatly helps in determining the full complement of genetic and physiological diversity of newly isolated species. Strains available in pure and mixed culture can be examined to determine what novel features they have, including those that might be of special interest to industry. Thus, this aspect of microbial diversity is particularly important.

Pure cultures are suitable for determination of morphological, physiological, and metabolic characteristics and commercial potential. The 16S rRNA sequence can provide a powerful means to organize the phylogenetic relationships among different microbial species in a meaningful way, but this analysis is often considered too conservative to provide species-level distinctions because any two different species can show 97–100% 16S rRNA sequence homology. Initially, only mesophilic anaerobic species were isolated and characterized from moderate environments. A good example is the rumen, which provides a great variety of anaerobic species with different metabolic properties. However, in recent years, studies on microbial diversity in extreme ecosystems, such as thermal springs and deep-sea vents, have offered an unparalleled diversity in extremophile communities.

The Yellow stone National Park ecosystem holds the world's greatest diversity of accessible, extreme microbial habitats. Not only are there temperature and pH extremes, but also areas with high levels of sulfur, heavy metals, radionuclides, and UV radiation, as well as a range of organic nutrients. Since oxygen solubility in water decreases as the temperature increases, anaerobes are common at the highest temperatures. Interest in novel microorganisms thriving under environmental extremes reflects a growing awareness of their value in biotechnological and industrial processes. Not only are the enzymes and metabolites often more stable, but in some cases, novel enzymes, not known in organisms from moderate habitats, have been found in extremophiles. The range of anaerobe diversity has proven to be extraordinary. For example, following the discovery of the deep-sea hydrothermal vents, hyperthermophilic anaerobes from hot springs and volcanic fields, as well as from shallow and abyssal marine hydrothermal vents were identified. Numerous *Archaea* have been identified and classified, and some species produce enzymes that have, or are expected to have, commercial utility.

Cytogenetic Approach

Microbial diversity at the biochemical level is more important from the industrial perspective. During the past decade, molecular biological and genetic approaches have come to dominate the study of microorganisms. Although fruitful, exclusive reliance on such approaches poses both practical and philosophical problems for microbiology. After an early phase of discounting the importance of physiology, people in industry are learning from the "biotechnology revolution" that gene cloning and biochemical manipulations do not solve all the product-development problems that may arise. Some of the more valued intellectual benefits anticipated in the field will emerge from understanding the physiology of peculiar and diverse microorganisms from unusual environments.

As long as oxygen is present as the electron acceptor, denitrification, sulfate reduction, and methanogenesis are inhibited. Denitrification begins after consumption of oxygen, and sulfate reduction begins after consumption of nitrate. Finally, methanogenesis occurs, which is partly a dissimilatory carbon dioxide reduction. The sequence of these reactions agrees with the sequence of their free-energy changes, which decrease from respiration to methanogenesis. Table 16.1 lists some representative anaerobic bacteria that display different metabolic types based on their catabolic pathways. These different biochemistries explain, in part, the great diversity of anaerobic microorganisms.

Table 16.1 Potential industrial fermentation products from anaerobic bacteria

Fermentation product	*Substrate*	*Typical species*	*Concentration (g/L)*	*Status*
Ethanol	Glucose	*Zymomonas mobilis*	84.5	Bench-scale
	Paddystraw	*Clostridium thermocellum*	23.6	Bench-scale
	CO	*Clostridium ljungdahlii*	48	Bench-scale
n-Butanol	Starch/glucose	*Clostridium acetobutylicum*	20	Bench-scale
	CO	*Butyribacterium methylotrophicum*	2.7	Bench-scale
1,3-Propanediol	Glycerol	*Clostridium butyricum*	58	Pilot-scale
Lactic acid	Glucose	*Lactobacillus acidophilus*	82	Industrial
Acetic acid	Glucose	*Clostridium thermoaceticum*	83	Bench-scale
Butyric acid	Glucose	*Clostridium tyrobutyricum*	62.8	Bench-scale
Propionic acid	Glucose	*Propionibacterium acidipropionici*	57	Bench-scale
	Glycerol	*Propionibacterium acidipropionici*	42	Bench-scale
Succinic acid	Glucose	*Actinobacillus succinogenes*	105	Pilot-scale
Caproic acid	Cellulose & ethanol	*Clostridium kluyveri* plus *Fibrobacter succinogenes*	4.6	Bench-scale

Methanogenic bacteria

Methanogenic bacteria are strictly anaerobic archaea with a unique form of energy metabolism involving the generation of methane. Biological methanogenesis is an important component of the carbon cycle in a variety of anaerobic habitats. It represents the terminal step in the anaerobic breakdown of organic matter under sulfate-limiting conditions. The substrate range for methanogenesis is limited to carbon dioxide, formate, carbon monoxide, methanol, methylamines, and acetate, but no single isolate is able to utilize all these carbon sources. The metabolic pathways of methane formation are unique and involve a number of enzymes and coenzymes that occur only in methanogens. A primary Na^+ gradient and H^+ gradients are formed by interesting and unique membrane-bound enzymes, and the reactions by which these ion-motive forces are generated are also novel.

The organisms and the methanogenic reactions, and the pathways of energy conservation, have been extensively reviewed. In recent years, bioconversion of agricultural and industrial wastes to methane has been considered important from two perspectives, as a waste-disposal method for residual organic matter and as a source of energy. Anaerobic digestors for treatment of sewage, agricultural and animal waste, and industrial effluent are being used worldwide. More recently, methanogenic microbial communities have been found to be very efficient in treating toxic organic wastes. For example, transformation of C_1 and C_2 halocarbons was observed in the early 1980s and, as one of the mechanisms responsible for these transformations, was proposed as a biologically mediated reductive dechlorination. Anaerobic dechlorination and degradation of chlorinated aromatic compounds such as chlorophenols, chlorobenzenes, and polychlorinated biphenyls (PCBs) occurs in a wide variety of environments.

Methanogenic microbial consortia that dechlorinate tetrachloroethylenes (PCE) and trichloroethylenes (TCE), and dechlorinate, degrade, and mineralize chlorophenols and PCBs have been developed. The potential of these consortia in bioremediating the contaminated soil and sediment systems is being examined. DDE [1,1-dichloro-2,2-bis(*p*-chlorophenyl)ethane], a metabolite of the pesticide DDT, has been observed to reductively dechlorinate to DDMU at a relatively high rate under methanogenic conditions as compared to sulfidogenic conditions. The catabolic reactions carried out by methanogens yield very little energy compared to those of aerobes. For example, use of H_2 to reduce CO_2 to CH_4 has a $\Delta G^{0'}$ of –34 kJ /mol H_2, whereas the corresponding energy value for an aerobe oxidizing H_2 using O_2 as an electron acceptors is –237 kJ/mol H_2. Most of the known species of methanogens can use H_2 to reduce CO_2 to CH_4.

Many of the hydrogenotrophic methanogens also utilize formate. In contrast, only species belonging to the genera *Methanosarcina* and *Methanosaeta* use acetate (acetotrophic) and convert it to methane. Acetate is an important end product of many fermentative anaerobes, and is a primary methanogenic substrate in anaerobic digestors. The methylotrophic methanogens can utilize several simple methylated compounds. These methanogens include *Methanosarcina*, which can also use acetate and usually H_2–CO_2, and *Methanolobus* and *Methanococcoides*, which are only known to grow on methylated compounds. Carbon monoxide is also converted to methane by methanogenic species. Methanogenesis has been observed at low, moderate, and high temperatures, and in neutral to mildly acidic or alkaline conditions. In summary, methane is produced under a variety of environmental conditions and is considered a by-product of the anaerobic treatment of industrial wastes.

Clostridium species

Ethanol production is a very well- known fermentation process since it has been an outgrowth of the alcoholic beverages industry. During World War I, shortages of acetone in the manufacture of munitions led to the development of an acetone–butanol– ethanol process involving the fermentation of starch by *Clostridium acetobutylicum*. This fermentation gave approximately 2 parts butanol to 1 part acetone and 1 part ethanol. During the 1940s and 1950s, lower costs of petrochemical processes stopped butanol production by fermentation routes in the U.S. and Europe. The process was also stopped in South Africa, but continued in China into the early 1990s. The present market for ethanol is met by yeast fermentation, but the industrial market for isopropanol, acetone, and butanol could be met by other anaerobic fermentations. Biomass-derived solvents produced by fermentation can enter into the current petrochemical synthetic pathways through a number of reactions.

The most important of these is the dehydration of alkanols to alkene to form ethylene, propylene, butylene, and butadiene. Therefore, production of chemical feedstocks from biomass via fermentation is becoming increasingly attractive because biomass production costs are not as tightly bound to energy costs. A great diversity exists among the solvent-producing microorganisms. These organisms span several groups of yeasts and a broad range of both mesophilic and thermophilic, and aerobic and anaerobic, bacteria. By far, the ethanol-producing organisms are the most abundant solventogens. However, industrial use is mainly limited to strains of yeasts, but great potential exists for recombinant *Zymomonas mobilis*. Several clostridia are capable of producing butanol and either acetone or isopropanol with ethanol as a by-product. The *C. acetobutylicum* and *Clostridium beijerinckii* strains are of prime importance for these industrial fermentations. *Clostridium thermocellum*, *Thermoanaerobacter ethanolicus*, *Clostridium thermosaccharolyticum*, *Thermoanaerobacterium thermosulfurigenes*, *Thermoanaerobacter brockii*, and *Thermobacteroides acetoethylicus*, use a wide range of substrates, from polymeric carbohydrates such as cellulose, pectin, xylan, and starch, to mono- and disaccharides such as glucose, cellobiose, xylose, and xylobiose.

The primary fermentation product is ethanol, with the production of acetate, lactate, carbon dioxide, and hydrogen in various ratios. *T. acetoethylicus* has not been shown to produce lactic acid. At present, one of the major limitations in the use of thermoanaerobes is the variability of end- product ratios, yield, and low ethanol concentration. These are affected by species, enzyme complement, and environmental conditions. Certain strains of *T. ethanolicus* have the best conversion of carbohydrates to ethanol, forming 1.6-1.9 mol/mol of glucose fermented. High ethanol and hydrogen concentrations also reduce the yield of ethanol owing to the flexibility of the carbon and electron pathways, which may possess many reversible enzyme systems. In addition to ethanol, several saccharolytic clostridia produce butanol and acetone besides volatile fatty acids and gaseous products from carbohydrate fermentation. In some cases acetone is further reduced to isopropanol. In most species, the production of solvents only occurs late in the fermentation cycle, following a shift from the pathways leading to acetate and butyrate production. The production of butanol and ethanol is usually associated with the uptake and reutilization of acids, and the production of acetone or butanol. The ability to produce solvents is influenced by the type and concentration of substrate, the pH and the buffering capacity of the culture medium, and the environmental conditions. Several strains that have the ability to produce solvents undergo degenerative changes, resulting in the loss of their ability to produce solvents.

The solvent-producing clostridia are found in a wide variety of natural habitats; however, the strains of *C. acetobutylicum* and *Clostridium beijerinckii* are the most frequently studied species.

C. acetobutylicum, which characteristically produces butanol, acetone, and ethanol in the ratio of 6:3:1, has been used extensively for the industrial production of solvents. Most butanol producers are mesophiles with a temperature optima for fermentation between 30 and 37°C. Recently, solvent yield was shown to increase at lower temperatures and acetone but not butanol yield to decrease at elevated temperatures. At present, many ethanol- and butanol-producing strains of *Clostridium* remain poorly classified, and there is still no accepted standard classification for the clostridia group as a whole.

Strains of *C. beijerinckii* (formerly classified as *C. butylicum*) constitute a second group of solvent producers that do not show DNA homology with the *C. acetobutylicum* group. This group contains both high- and low-solvent-producing strains that produce either acetone or isopropanol in addition to butanol. The strains of *Clostridium aurantibutyricum* group also include butanol-producing strains that produce both acetone and isopropanol. *Clostridium tetanomorphum* produces butanol and ethanol, but not acetone or isopropanol. As obligate anaerobes, butanol producers require anaerobic conditions. However, vegetative cells of *C. acetobutylicum*, for example, survived several hours exposure to oxygen and formed fruiting-body-like structures on agar plates when exposed to oxygen.

Acetate forming bacteria

The term *acetogen* is commonly applied to a bacterium that forms acetate, whether the acetate is produced by a catabolic process such as fermentation or by an autotrophic-type synthesis. Acetogenic bacteria are ubiquitous in anaerobic ecological systems. Homoacetogens have the ability to grow well on H_2 plus CO_2 and poorly on CO, and to form acetyl CoA by a CO-dependent pathway involving CO dehydrogenase. *Clostridium thermoaceticum* and *Clostridium thermoautotrophicum* are homoacetogens and synthesize acetate from C_1 compounds by the recently established Wood pathway of acetyl CoA synthesis. This pathway was established by studies with *C. thermoaceticum* that can grow on CO or CO_2–H_2, utilizing them as carbon and energy sources. In industrial fermentations, the formation of a single product is advantageous because its recovery would be simplified. In this respect, homoacetogenic bacteria essentially form only acetate in the fermentations of hexoses and pentoses. They appear to be ideal for the microbial production of acetate. Energy for cell growth is obtained by the reduction of CO_2 to acetate via the Wood pathway.

The homoacetogens also grow on other one-carbon compounds, including CO, formate, and methanol with acetate as the product. In the acetyl CoA pathway for synthesis of acetate, acetyl CoA is the first two-carbon intermediate of the autotrophic fixation of CO_2, and is either used for the synthesis of cell carbon or converted to acetate. *Butyribacterium methylotrophicum* is an acetogenic anaerobe that can grow on multicarbon compounds as well as on one-carbon compounds (e.g., CO or methanol). It produces acetate or butyrate using the Wood pathway, but acetyl CoA can be condensed and reduced to butyrate. Higher levels of NADH were found in butyrate-producing than in acetate-producing cells. In *B. methylotrophicum*, butyrate production is regulated by the carbon source and is dependent on cellular NADH/NAD ratios, and the levels and direction of ferredoxin- and NAD-linked oxidoreductases. Also, the growth pH regulates both hydrogenase and FD-NAD oxidoreductase activities such that, at acid pH, more intermediary electron flow was directed towards butyrate synthesis than H_2 production.

Acidogenic bacteria

The *acidogenic bacteria* include those bacteria that only form acetate by fermentation. These bacteria invariably produce other acids in addition to acetate. The acidogenic bacteria grow heterotrophically on a variety of carbohydrates, producing a mixture of acids such as lactic, propionic, butyric, succinic, and caproic. The acidogens differ from the acetogens in that they do not synthesize acetate from CO_2 or other C_1 compounds by the Wood autotrophic pathway. Many of the species can utilize hexoses such as glucose, fructose, galactose, and mannose; disaccharides including cellobiose, lactose, and maltose; pentoses represented by xylose; and other substrates such as glycerol, mannitol, sorbitol, inositol, and polymeric carbohydrates. For example, *Clostridium butyricum*, the type strain of the genus *Clostridium*, is saccharolytic in nature and metabolizes glucose, producing butyrate, acetate, CO_2, and H_2 as fermentation products. Glucose is converted to pyruvate by the Embden-Meyerhof-Parnas glycolytic pathway. Pyruvate is then simultaneously decarboxylated and oxidized by the enzyme complex pyruvate-ferredoxin oxidoreductase to yield acetyl CoA, CO_2, and reduced ferredoxin.

The reduced ferredoxin is reoxidized in several reactions, the most important of which involves H_2 evolution catalyzed by hydrogenase. Acetyl CoA is condensed and reduced to butyrate. In some other acidogens, phosphoenolpyruvate (PEP) serves as a branching point in the pathway producing pyruvate and oxaloacetate. The PEP-oxaloacetate step involves CO_2 fixation mediated by PEP carboxykinase. Depending upon the species, products such as succinate and propionate are produced as fermentation products with acetate, formate, or H_2. For example, oxaloacetate is converted to propionate in species like *Propionispira arboris*, whereas, oxaloacetate is further converted to succinate in species like *Anaerobiospirillum succiniciproducens*. Lactic acid bacteria such as *Lactobacillus acidophilus* convert pyruvate to lactic acid, whereas *Clostridium kluvyri* produces caproic acid from ethanol and acetate.

Sulfate reducing bacteria

Sulfate, a chemically inert, nonvolatile, and nontoxic compound, is widespread in rocks, soil, and water. In contrast, hydrogen sulfide, because of its chemical properties and physiological effects, is a far more conspicuous substance and is chemically reactive. The presence of black-colored sediments due to the formation of ferrous sulfide from iron-containing materials, accompanied by a distinctive smell, is indicative of H_2S production. Hydrogen sulfide acts as a reductant, and is toxic to plants, animals, and humans. It is produced by reduction of sulfate by sulfate-reducing bacteria, which are present in ecological niches where oxygen has no access. In the absence of oxygen, the oxidized form of sulfur (SO_4^{2-}) is used as an electron acceptor by sulfate-reducing bacteria (sulfate reducers). Since reduction of the inorganic compound serves for energy conservation, the process is distinguished as dissimilatory reduction from the assimilatory reduction of sulfate in plants and bacteria. Under anaerobic, reduced conditions, hydrogen sulfide is the energetically stable form of sulfur, as sulfate is the stable form under aerobic conditions.

The most important reducers of elemental sulfur in nature are probably special anaerobes that may be designated as true sulfur-reducing bacteria; these carry out the dissimilatory reduction of sulfur as their primary or even obligate metabolic reaction during oxidation of organic substrates.

The true sulfur- reducing bacteria do not reduce sulfate, and only some reduce other oxygen-containing sulfur compounds. The majority of the eubacterial sulfur reducers are mesophilic, as are most sulfate reducers. In contrast, all described archae-bacterial sulfur reducers are hyperthermophiles with temperature optima (up to 110°C) the highest known in the living world.

Sulfate-reducing bacteria include a rather heterogenous assemblage of microorganisms having in common merely dissimilatory sulfate metabolism and obligate anaerobiosis. The genus *Desulfovibrio* includes species that are still the best-studied sulfate reducers. These are nonsporing, having curved motile cells and growing on a relatively limited range of organic substrates, preferably lactate or pyruvate, which are incompletely oxidized to acetate and H_2S. Spore-forming sulfate-reducing bacteria with a similar metabolism were classified within the genus *Desulfotomaculum*. Other types of sulfate-reducing bacteria differ markedly physiologically and morphologically from the known *Desulfovibrio* and *Desulfotomaculum* species.

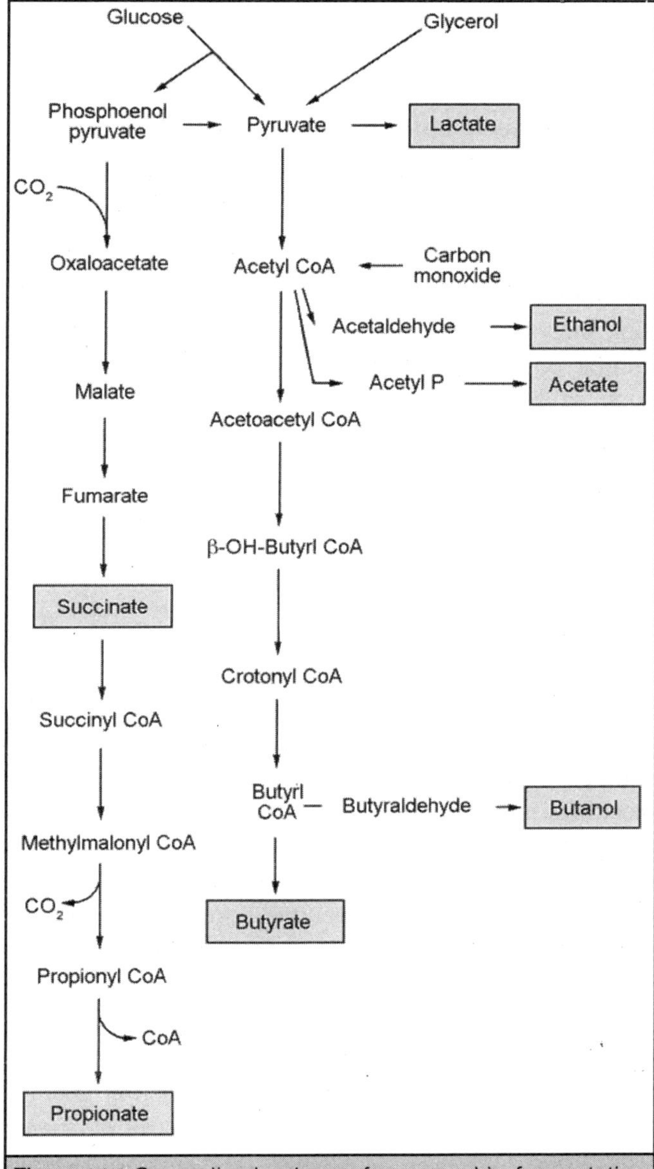

Figure 16.1 Generalized scheme for anaerobic fermentative production of common organic acids and alcohols.

The reactions by which sulfate-reducing bacteria are involved in anaerobic degradation are indicated by their metabolic capacities. As far as tested, these bacteria use low molecular weight compounds as electron donors and therefore depend on fermentative bacteria that degrade the original polymers from biomass. Thus sulfate reducers are terminal degraders and their role is analogous to that of methanogenic bacteria that form methane and carbon dioxide as final anaerobic products. The recognition of dissimilatory sulfate reduction and methanogenesis as two alternative terminal carbon degradation processes has contributed a great deal to our understanding of anaerobic mineralization.

Gene Transfer Technology

The field of genetics has been restricted for many years to *Escherichia coli* and few other genera of aerobic or facultative anaerobic bacteria such as *Pseudomonas, Bacillus,* and *Salmonella*. Anaerobic bacteria were known and studied since before 1900, but work on the genetics and molecular biology of anaerobic bacteria began to emerge only in the 1970s. The volume by Sebald is the most recent comprehensive review on genetics of anaerobes. Improvements in basic techniques for culturing anaerobes and recent advances in molecular biology techniques have helped in making rapid progress in understanding genes and genetic systems in anaerobic bacteria. Thermoanaerobic bacteria produce many thermostable enzymes but the yields are low. However, their enzymes can be overproduced by cloning the genes into mesophilic, industrially important aerobic bacteria. The genes can be over-expressed in both gram-positive and gram-negative hosts such as *Bacillus subtilis* and *E. coli*. One major advantage of cloning genes for thermostable enzymes in mesophiles is in the recovery of enzymes in greater than 95% purity by a single-step high-temperature treatment that inactivates host proteins, including protease enzyme.

Aerobic bacteria produce about 10-fold higher biomass yields than the anaerobic bacteria. Therefore, there is no advantage, except for biotransformation, in producing the enzymes in an anaerobic system. Several species of the genus *Clostridium* are of biotechnological interest because of the end products of their fermentative metabolism, the stereospecific reductions that they undertake, and the potentially important enzymes that they produce. Other species, such as *C. botulinum*, produce some of the most powerful toxins known today, which have been commercialized for medical treatments such as the correction of crossed eyes. Various elements of gene-transfer technology have been developed in several species. Two gene-transfer procedures should prove widely applicable throughout the genus—electroporation and conjugal plasmid mobilization. In clostridia, mutants defective in a variety of functions, including purine, pyrimidine, vitamin, and amino acid biosynthesis, pathways of fermentative metabolism, and sporulation, have been isolated with ethyl methane sulfonate, *N*-methyl-*N*′-nitro-*N*-nitrosoguanidine, and ultraviolet light, and have been characterized. These mutant strains provide valuable information about gene regulation and function. Conjugal transfer of plasmids and transposons as well as phase transfection/plasmid transformation has now been documented in several species.

In addition, recombinant DNA technology has also been employed to gain insights into the gene-transfer systems in *Clostridium*. Protoplasts of *C. acetobutylicum* can be transfected with phage DNA. The whole cells of *C. acetobutylicum* can be transformed with plasmid DNA using an electroporation procedure. Permealized cells of *Thermoanaerobacter thermohydrosulfuricus* have been transformed with plasmid pUB110 (encoding resistance to kanamycin, Km^R) and a derivative, pGS13, carrying a chloramphenicol resistance (Cm^R) marker. A variety of cloning vectors have been developed for use in *C. acetobutylicum*. To date only pMTL 500E has been used to introduce cloned heterologous genes into *C. acetobutylicum*. These genes included the *Clostridium pasteurianum leu*B gene and the *C. thermocellum cel*A gene. Acquisition of the *cel*A gene was demonstrated by an in situ plate assay using Congo red. In most cases, screening of clostridial gene banks has relied on expression of the heterologous genes in *E. coli*. Appropriate *E. coli* mutants have been employed for the isolation of *C. acetobutylicum* genes involved in the acetone–butanol fermentation. The butyraldeyde dehydrogenase gene of *C. acetobutylicum* has been isolated by complementation of an *E. coli* aldehyde dehydrogenase–deficient mutant. Endo-β-glucanases and xylanse of *C. thermocellum* and other cellulolytic clostridia are easily detectable on plates containing the appropriate substrate by the Congo red assay. Upon staining with Congo red, positive clones are surrounded by a yellow hydrolysis zone on a red background.

INDUSTRIAL PRODUCTS

Industrial biotechnology has developed into a more than $25 billion business of enzymes and of chemically defined compounds produced by fermentation, microbial transformation, and enzymatic conversion. These include organic acids, amino acids, alcohols, vitamins, microbial polymers, antibiotics, industrial enzymes, biopesticides, and pharmaceutical specialities. Of these, ethanol, citric acid, L- lactic acid, gluconic acid, monosodium L-glutamate (MSG), and L-lysine are examples of bulk fermentation products.

Organic Solvents

Solvents are compounds that are essentially neutral in character and comprise organic alcohols and ketones. The major solvent products of microbial carbohydrate catabolism include ethanol, acetone, isopropanol, butanol, acetoin, 1,3-propanediol, and butanediol. These are the most useful oxychemicals produced by fermentation. Solvents have various uses in the chemical industry, including the direct use as organic chemicals and indirect use as fuel additives, extenders, or feedstocks for further chemical synthesis. Butanol is primarily used as a feedstock chemical in the manufacture of lacquers, rayon, plasticizers, coatings, detergents, and brake fluids. It can also be used as a solvent for fats, waxes, resins, shellac, and varnish. In addition, butanol may also be used as an extractant and solvent in the food industry.

Industrial production of butanol

The acetone-butanol fermentation currently has potential because butanol has many characteristics that makes it better than the currently used liquid-fuel extender, ethanol. Higher alcohols have several characteristics that are favorable for motor fuel use, either in gasoline blends or, in some cases, when used directly as fuel. They are miscible in gasoline; their heats of combustion per gallon are greater than those of methanol, and they have good octane-enhancing properties. Butanol, for example, has a heat of combustion 54% higher than methanol and 83% of that for gasoline. Although the octane numbers are less than that for methanol (RON 100 vs. 110), the research octane number (RON) value of 100 is still above that of gasoline (RON 92), and therefore useful for fuel blends. In addition, butanol has low miscibility with water and exhibits high miscibility with both diesel and gasoline.

Owing to its high heat of combustion butanol solutions containing as much as 20% (v/v) water have the same combustion value as anhydrous ethanol. In the early 1900s, butanol was used to produce butadiene, the most desirable raw material for synthetic rubber. The annual production of fermentation-derived butanol was over 45 million pounds during 1945. But the fermentation-derived butanol process declined after World War II in the U.S. due to both changes in availability of renewable feedstocks (molasses, sugarcane) and the increase in availability of inexpensive petrochemical feedstocks. Butanol fermentation of beet molasses continued through the 1970s in the Soviet Union, and fermentation of sugarcane molasses continued through the 1980s in South Africa. In China, butanol was still produced fermentatively until recently, but the last viable industrial acetone-butanol-ethanol (ABE) fermentation in the Western world was carried out by National Chemical Products in Germiston, South Africa, using *C. acetobutylicum*. Butanol is now synthesized chemically from petroleum- derived ethylene, propylene, and triethylaluminium, or carbon monoxide and hydrogen. The major domestic producers of butanol and its derivatives are BASF, Chem Service Inc., Dow Chemical, Eastman Chemical, Hoechst Celanese, Shell, Union Carbide, and Vista. The current U.S. production of butanol is more than 1.2 billion pounds per year and is experiencing a growth of 3-4% annually.

Butanol pricing has been rising steadily in recent years due to an increase in demand. The bulk butanol pricing in the United States is $0.40–$0.50 per pound. The price of butanol in Europe and the Far East is higher than $0.50 per pound. If the production cost of fermentation-derived butanol could be in the range of $0.25–$0.30 per pound, the market for butanol will likely expand: The market penetration of fermentation-based butanol will drive the utilization of corn from 7.2 million bushels in 1995 to 153 million bushels in 2010. In addition, the supply of other alternative feedstock, such as biomass, is more available than corn. However, biomass requires effective pretreatment technology that has not yet been developed. Biomass-derived solvents produced by fermentation can enter into the current petrochemical synthetic pathways through a number of reactions, the most important of which is the dehydration of alkanols to alkenes to form ethylene, propylene, butylene, and butadiene. The ABE fermentation in batch culture is a sequential process characterized by a primary acid-producing or acidogenic phase that is coupled to growth, and a secondary solvent-producing or solventogenic phase. Thus, solventogenesis appears as a secondary metabolic process caused by uncoupling of growth.

Acetate and butyrate are produced as soluble fermentation products during the acidogenic phase. As the acids accumulate, the growth rate and the culture pH decreases, and solventogenesis begins. The acids and H_2 are consumed and are reduced to solvents in the solventogenic phase. The "switch" from acidogenesis to solventogenesis is accompanied by a large number of physiological and biochemical changes, and is clearly a multi-factorial process. The exact mechanism that regulates this switch has not been clearly elucidated, but the concentration of undissociated butyric acid within the cytoplasm functions as a "bioregulator." The end products of the ABE fermentation alter and inhibit growth and metabolism. Acids are more toxic than solvents, and the more hydrophobic a product, the higher is its toxicity. It should be noted that toxicity of acids depends upon the pH. At neutral pH, high concentrations of organic anions (i.e., butyrate, acetate) are not toxic because they are dissociated.

In normal batch fermentation conditions, organic acids are produced, and then consumed during solventogenesis to prevent dissipation of the cellular proton-motive force. Butanol inhibits the fermentation at a relatively low concentration. Butanol producers use a wide variety of carbohydrates, including hexoses, pentoses, oligosaccharides, and polysaccharides as fermentation substrates, and including starch, molasses, whey, wood hydrolysates, pentosans, inulin, and sulfite liquor.

Cellulosic materials are one of the most attractive substrates for chemicals, as they are the most abundant and least expensive raw material. Few attempts have been made to use cellulosic substrates for the production of butanol. A simultaneous saccharification fermentation process was studied to produce butanol from alkali-pretreated wheat straw using *Trichoderma reesi* cellulase and *C. acetobutylicum*.

The final butanol concentration was 10.7 g/L from about 140 g of straw. Pentoses did not accumulate during the fermentation, suggesting simultaneous use of both hemicellulose and cellulose components. Different substrates alter fermentation parameters of *C. acetobutylicum*, such as growth rate and the solvent production ratio. Higher growth and substrate consumption rates were obtained when glucose, cellobiose, or mannose were used than when xylose, arabinose, or galactose were used. The solvent production ratio with the first sugar group was 1:4:10 (ethanol:acetone:butanol), whereas ratios of 1:2:5 were obtained with the latter. Starch is at present the most useful raw material for fermentation because it is abundant, has high density, and requires limited pretreatments to be made from biomass.

Starch is produced from corn, cereals (wheat, oats, rice), and starchy roots (sweet potatoes, cassava). Typically, *C. acetobutylicum* produces 12-13 g/L butanol with about 20-22 g/L total solvents. In recent years, the majority of investigators have focused their interests on the type culture strain *C. acetobutylicum* ATCC 824. Other cultures included DSM 792 strain and NCIMB 8052. The *C. acetobutylicum* ATCC 4259 strain and the equivalent DSM 1731, NCIMB 619, and NRRL B-530 strains are all derived from the original Weizmann industrial strain. An asporogenous mutant ATCC 39236 derived from this strain has been patented by scientists at the Moffett Technical Center CPC International. Recently, a limited sporulating *C. acetobutylicum* E604 mutant strain has been developed by treating the parent *C. acetobutylicum* strain (ATCC 4259) with ethane methane sulfonate. This mutant strain has been used to ferment a high-carbohydrate substrate concentration in a multistage, continuous-temperature programmed-fermentation process followed by batch fermentation, to yield an extraordinarily high concentration of butanol (>20 g/ L) and total solvent (>30 g/L) production. This mutant strain has higher butyrate uptake rate (0.33 g/L) in comparison to the parent strain (0.26 g/L) at a lower temperature of 30°C. Butyrate up to an externally added concentration of 11.4 g/L did not inhibit butyrate uptake.

Optimization studies for butyrate uptake by *C. acetobutylicum* suggested a direct correlation between minimum pH and butyrate concentration or temperature. These developments make it possible to commercially produce butanol fermentatively with no residual butyrate. The saccharolytic strain *C. acetobutylicum* P262, one of the NCP production strains, has been studied by groups working in South Africa and New Zealand. The various solvent-producing strains now tend to be regarded as varieties of *C. acetobutylicum* or *C. beijerinckii*. For batch ABE fermentation to be industrially viable, and to compete with the chemical process, concentrations of between 22 and 28 g/L must be obtained in 40–60 h. A large number of publications on butanol-producing microbial cultures and the process demonstrate the extensive research work that has been done in past years.

Simultaneous recovery of butanol from the broth during fermentation can eliminate product inhibition. This will allow complete utilization of sugar and continual fermentation of the substrate at relatively higher rates. In contrast to the traditional distillation processes, removal of butanol can be accomplished by integrating a pervaporation process with the multistage fermentation process. In this process, the solvents will pass through the membranes preferentially and the organic acids and cells will remain on the liquid side and return to the fermentor. The permeate stream that is enriched in solvent relative to the feed will separate into a low-butanol aqueous phase and a butanol-rich solvent phase. This aqueous phase could be recycled back to the feed tank. Pervaporation also provides for the retention of cells in the fermentor, which improves volumetric productivity.

Industrial production of ethanol

In the United States, use of ethanol as a fuel is almost wholly confined to its use as an octane enhancer in the higher grade of gasoline. Currently, this fuel ethanol is obtained from both fermentation and chemical synthesis. All of the fermentation ethanol produced in the United States is made by yeast fermentation. Interestingly, the substrate range for both yeasts and *Zymomonas* for ethanol production is similar. Traditionally, yeasts have been used to produce alcoholic beverages and, more recently, to produce a considerable portion of industrial ethanol all over the world. Among yeasts, *Saccharomyces* is the most important and most commonly used organism. *Saccharomyces cerevisiae* converts glucose to ethanol and carbon dioxide. Other yeasts include *Pachysolen tannophilus* and *Kluyveromyces marxianus*. Many facultative and anaerobic bacteria produce varying amounts of ethanol as one of the metabolic end products.

Bacterial species such as *Zymomonas mobilis*, *Sarcina ventriculi*, and *Erwinia amylovorans* utilize pyruvate decarboxylase to convert pyruvate to CO_2 and acetaldehyde, which is then reduced to ethanol. *Zymomonas* grows on glucose, fructose, and sucrose, but is unable to use starch, maltose, or pentose sugars. *Z. mobilis* is a promising bacterium with some potential to replace yeast as the major organism for production of industrial ethanol. *Z. mobilis* was metabolically engineered to broaden its range of fermentable substrates to include the pentose sugar xylose. Two operons encoding xylose assimilation and pentose-phosphate-pathway enzymes were constructed and transformed into *Z. mobilis* in order to generate a strain that grew on xylose and efficiently fermented both glucose and xylose to ethanol. Subsequent metabolic-pathway engineering further expanded the fermentation substrate range of the ethanologenic bacterium *Z. mobilis* to include the pentose sugar L-arabinose.

Direct fermentation of cassava starch to ethanol by *Z. mobilis* was obtained in the medium containing amylase-rich culture filtrate of Endomycopsis fibuligera. The ethanol concentration of 105 g/L could be increased to 132 g/L by further addition of glucoamylase enzyme at 0.01%. A number of different groups of bacteria, including clostridia, produce ethanol via the reduction of acetyl CoA, which is generated by the cleavage of pyruvate produced during glycolysis. At least 30 species of *Clostridium* have been reported to produce ethanol in amounts varying from trace to close to the theoretical maximum of 2 mol ethanol/mol glucose fermented. In most cases, the extent of ethanol production is dependent upon the nature and concentration of fermentation substrate and the fermentation conditions, such as pH and temperature. For a number of mesophilic clostridial species, ethanol yields ranging from 1.7-1.9 mol/mol hexose fermented have been observed. In a single-step biomass to ethanol conversion process, *C. thermocellum* has been shown to produce

23.6 g/L ethanol from an alkali-treated paddy straw. The bacterium *C. ljungdahlii*, a fast-growing bacteria, has been shown to produce ethanol from carbon monoxide in concentrations up to 48 g/L from synthesis gas in a CSTR with cell recycling. In recent years, thermophilic bacteria have been shown to produce solvents and are considered potential candidates for use in process development because of certain advantages over mesophiles.

Thermophiles have the ability to use complex plant polymers such as cellulose or starch, and offer high growth rates, fast fermentation, reduced contamination, increased process stability, and eliminate high energy demand in cooling the media and product recovery. The major disadvantage is production of end products in low concentrations and yield due to lower end- product tolerance. However, this can be partially overcome if the solvents are removed simultaneously. There are increasing number of patented processes concerning the use of thermophilic microorganisms, including ethanol fermentation. The thermophilic anaerobic bacteria have provided a great deal of information on ethanol-fermentation pathways via acetyl-CoA. There are two different ethanol pathways in thermophiles. The type I pathway uses NADH-linked alcohol dehydrogenase and produces ethanol from acetaldehyde. The Type II pathway uses NADPH-linked alcohol dehydrogenase to produce ethanol from both acetyl CoA directly or acetaldehyde. In Type II ethanologens, the NAD-linked alcohol dehydrogenase functions to consume ethanol. The Type I pathway is applicable in anaerobic bacteria such as *C. thermocellum*. The Type II pathway is present in *Thermoanaerobacter thermohydrosulfuricus, Thermoanaerobacter brockii*, and *Thermoanaerobacter ethanolicus*. Ethanol yields of heterofermentations vary considerably with the specific growth conditions used.

The biochemical basis for different reduced end-product ratios of thermophilic ethanol producers that contain the same glycolytic pathways is related to subtle differences in the specific activities and regulatory properties of the enzymes that control electron flow during fermentation. Similarly, specific changes in culture conditions such as temperature and pH influence the rate and direction of the enzymatic machinery responsible for end-product formation. *T. ethanolicus* 39E has a low tolerance for ethanol, with growth inhibition occurring at 2% (wt/vol) ethanol. An ethanol-tolerant strain (39EA) that was tolerant to 4% (wt/vol) ethanol at 60°C, produced ethanol under these conditions, and lacked the NAD-linked alcohol dehydrogenase was selected. Another strain (H8) can grow at 8% ethanol but produces lactic acid at high solvent concentrations. The mechanism of high (i.e., 8%) ethanol tolerance in *T. ethanolicus* is related to its ability to produce unique transmembrane lipids (C_{30} to C_{34} fatty acids) that provide solvent tolerance. Alcohol increases membrane fluidity, and these transmembrane lipids may serve to reduce the fluidity and maintain membrane integrity.

Industrial production of 1,3-propanediol

1,3-Propanediol (PD) is a versatile intermediate compound for the synthesis of heterocycles and as a monomer for the production of polymers such as polyethers, polyesters, and polyurethanes. PD can also be used as a solvent and an additive for lubricants. The microbial conversion of glycerol to PD is simple in comparison to the chemical conversion of acrolein. It has been shown that some *Clostridium* species are able to convert glycerol to PD with an appreciable yield. 1,3-Propanediol from glycerol was produced by *C. butyricum* DSM 5431 at 50-58 g/L with productivities of 2.3-2.9 g/L per hour. The fermentations were conducted in 300- and 3000-L fermentors with almost similar results, indicating that the scale-up of this microbial fermentation process should not cause a major problem. The substrate glycerol is relatively cheap and its conversion to PD could help reduce glycerol surpluses in the market.

Industrial production of organic acids

Currently, organic acids such as acetic, propionic, butyric, fumaric, succinic, malic, and lactic acids can be produced by anaerobic fermentation technology. The key technical problems blocking the rapid advances in developing the bioprocess technology for organic-acid fermentations have been low product concentration, low specificity to desired product(s), low productivity or rate of fermentation, and inefficient or energy-intensive recovery processes. However, recently, significant technological advances have occurred in anaerobic fermentations for production of organic acids by fermentation of carbohydrates. Acidogenic fermentation of carbohydrates to volatile organic acids is well known.

In general, anaerobic bacterial fermentations are of interest because of the wide range of products formed and substrate fermented, the high substrate-to-product conversion yields and rates, and the potential for enhanced process stability and product recovery due to physiological diversity of species (e.g., extreme thermophily, acidophily, halophily). For acetate fermentation, homacetogenic fermentation producing 3 mol of acetate from 1 mol of dextrose (e.g., by using *C. thermoaceticum*) has been described by many authors. Currently, the available organisms produce salts of organic acids in low concentrations (400-800 mM for acetate, 200-400 mM for propionate and 200-300 mM for butyrate) with low productivity of 0.2-2 g/L per hour. New fermentation processes with better yield and productivities have been developed, making it feasible to produce volatile as well as nonvolatile organic acids economically.

Industrial production of lactic acid

Lactic acid is a natural organic acid with a long history of use in the food industry. Lactic acid, because of its flavoring and preservation properties, is used as an acidulant, particularly in dairy products, confectionery, beverages, pickles, bread, and meat products. It is also used in the pharmaceutical and cosmetic industries. The cyclic dimers of lactic acid are used as raw materials in the synthesis of biodegradable polymers for a variety of uses, including absorbable surgical sutures, slow-release drugs, and prostheses. Ethyl lactate is the active ingredient in many anti-acne preparations. Crude grades of lactic acid are used for the deliming of hides in the leather industry, and it is used for fabric treatment in the textile and laundry industries. Its ability to form polymeric polylactic acids finds application in production of various resins. Because of its structure and its two functionally reactive groups, hydroxyl and carboxyl, lactic acid can be used to make numerous value-added functional chemicals. These include biodegradable polymers and copolymers, propylene glycol, propylene oxide, and acrylates.

Lactic acid-based polymers and copolymers have potentially large-volume uses as biodegradable thermoplastics, pesticide formulation, and environmentally benign polymers, with a potential market exceeding several hundred million dollars per year. The market for fermentation lactic acid is growing every year because it is now feasible to produce large-volume chemicals from lactic acid. Homofermentative lactic-acid bacteria produce none or only trace amounts of end products other than lactic acid, and are used for industrial processes. Numerous organisms are known to produce lactic acid with high (>90%) yield and productivity (>2 g/L per hour) from carbohydrate fermentation. Some of the common homolactic acid bacteria are: *Lactobacillus casei, L. pentosus, L. leichmannii, L. acidophilus, L. delbrueckii, L. bulgaricus, Streptococcus cremoris, S. lactis, S. diacetylactis, Sporolactobacillus* sp. and *Pediococcus* sp.

Homolactic acid-producing bacteria ferment starch, hexoses, pentoses, and cellobiose. Depending upon the cultural conditions, the yield of lactic acid by the homolactic acid bacterium *L. casei* and heterolactic acid bacterium *T. brockii* are nearly equivalent. In non-energy-limited batch culture, *L. casei* produces lactic acid as the sole end product, whereas in energy-limited, continuous culture, nearly equivalent amounts of lactic, acetic, and formic acids, and as well as ethanol, are produced. Lactate is the major fermentation product of *T. brockii* grown in high-yeast extract, batch fermentation. It should be noted that lactate yields in homolactic-acid bacteria depend upon specific growth conditions and that additional products (e.g., formic acid, acetic acid, and ethanol) can also be formed. Lactic acid yields are generally highest during glycolysis via the homolactic-acid-fermentation pathway. Theoretically, 2 mol of lactate and 2 mol of ATP are formed per 1 mol of glucose fermented. Anaerobic fermentations for production of organic acids operate optimally at pHs where salts of organic acids, rather than free acids, are produced.

The free acids and their derivatives are required for the manufacture of functional chemicals. The development of new technologies to recover and purify the fermentation acids from broth have reduced the costs associated with product recovery and purification, thus making the commodity production of organic acids such as lactic acid practical and economically attractive. One of the approaches that has been traditionally applied is simultaneous fermentation and neutralization with calcium salt to obtain calcium lactate precipitate. The precipitate can be separated easily, and free lactic acid can be obtained after treating the salt with acids such as sulfuric acid. The disadvantage of such a system is in handling the solids and slurry. Another emerging technology for separation and recovery of lactic acid is electrodialysis (ED). ED is a rapid process for altering the composition and/or concentration of electrolytes in aqueous solutions by transferring ions across ion-exchange membranes under the influence of a direct electric current. The recovery and purification of lactic acid using ED has been successfully demonstrated. In this process, sodium lactate is first selectively separated from proteins, amino acids, and cells, then, using water-splitting membrane, the lactic acid salt is converted to free lactic acid and base. The traditional process using lime for neutralization followed by cell filtration, carbon adsorption, evaporation, acidulation, gypsum filtration, carbon adsorption, evaporation, filtration, esterification, and final purification is uneconomical. However, separation and recovery of lactic acid by ED seems a cost-effective and attractive approach.

Industrial production of succinic acid

Succinic acid is a relatively new non-hygroscopic acidulant. Succinic acid is listed by the FDA as a GRAS additive for miscellaneous and/or general-purpose uses. It readily combines with proteins in modifying the plasticity of bread doughs. It is also an dibasic acid for producing edible synthetic fats with desirable thermal properties. Succinic acid is manufactured by the catalytic hydrogenation of maleic or fumaric acid. It has also been produced commercially by aqueous alkali or acid hydrolysis of succinonitrile, which is derived from ethylene bromide and potassium cyanide. Succinic acid is a common intermediate in the metabolic pathway of several anaerobic microorganisms. For example, succinate is a key intermediate for anaerobic fermentations by propionate-producing bacteria, but it is only produced in low yields and in low concentrations. Succinate is also produced by anaerobic rumen bacteria. These bacteria include *Ruminobacter amylophilus* and *Prevotella ruminicola*. As such, accumulation of succinic acid in fermentation broth has been observed in a number of rumen microorganisms.

Although the rumen bacteria give higher yields of succinate than do the propionate-producing bacteria, the reported fermentations were generally run in very dilute solutions and gave a variety of products in generally low yields. Moreover, the rumen organisms tend to lyse after a comparatively short fermentation time, thereby leading to unstable fermentations. *Anaerobiospirillum succiniciproducens*, an anaerobic non-rumen bacterium, has been demonstrated to produce succinic acid in higher concentrations and yield. The pathway for succinic acid production in *A. succiniciproducens* has been established. Two processes to produce succinic acid as sodium succinate and calcium succinate were developed using *A. suciniciproducens*, as this organism produced succinate in a relatively higher concentration and yield with fewer co-fermentation products, had a faster growth rate, and could grow in an industrial grade medium. In order to develop a commercially attractive process to produce succinic acid by fermentation, the product should be produced in a high concentration and in a high yield (wt%) using inexpensive raw materials and nutrients.

A. succiniciproducens produces succinate and acetate in a ratio of about 4:1, thus a good amount of carbon is lost as acetate and the separation of succinate and acetate becomes very critical. Therefore, a fluoroacetate-resistant variant FA-10 of *A. succiniciproducens* was developed that produced very minute amounts, if any, of acetic acid. The organism produces enough succinate to allow simultaneous fermentation and calcium succinate precipitation under strictly controlled conditions and when the correct calcium salt is used for neutralization. However, the process seems to have a major disadvantage in handling gypsum at a commercial scale. The succinic acid fermentation in *A. succiniciproducens* is regulated by fermentation pH and the amount of carbon dioxide. For example, lactate is the major fermentation product at pH 6.5 and higher with limiting levels of CO_2, whereas succinate is the major product at a pH of 6.0-6.2 with an abundance of CO_2. Phosphoenolpyruvate carboxykinase has been purified and described for *A. succiniciproducens*, a non-ruminal anaerobic bacterium, as well as from a ruminal anaerobic bacterium, *Ruminococcus flavefaciens*.

In *E. coli*, overexpression of PEP carboxikinase had no effect on succinic acid production, but succinic acid was produced as the major fermentation product by weight (3.5-fold increase in the concentration) by overexpression of PEP carboxylase. Recently, a rumen-facultative anaerobic bacteria, *Actinobacillus* sp. strain 130Z, has been shown to produce succinic acid in much higher concentration (60-80 g/L) than any of the previously known succinic acid-producing microorganisms.

Acetate and small amounts of pyruvate and formate are produced as co-fermentation products. The variants of *Actinobacillus* sp. strain 130Z have been developed to produce succinic acid in a concentration of >105 g/L and yield of 85–95 wt % under neutralization with magnesium. The organism is reported to be robust with a wide pH range and tolerance to high substrate and product concentrations. The levels of key succinic acid-producing pathway enzymes, such as phosphoenolpyruvate (PEP) carboxykinase, a key CO_2-fixing enzyme, malate dehydrogenase, and fumarase were significantly higher in *Actinobacillus* sp. strain 130Z than in *E. coli* K-12. The key enzymes in end-product formation in *Actinobacillus* sp. 130Z were regulated by the energy substrates.

Industrial production of butyric acid

Butyric acid, butyrate esters, and other butyrate derivatives are important flavor ingredients in many natural and processed foods. For example, butyric acid is present in butter as an ester to the extent of 4-5%. Some of its esters serve as bases of artificial flavoring ingredients of certain liqueurs, soda water syrups, candies, and so on. Development of a bioprocess for production of natural butyric acid by anaerobic fermentation thus is likely to satisfy the potential market of natural fermentation products. In addition, ethyl and methyl esters of butyric acid can also be used as octane enhancers when added to gasoline. Some common anaerobic bacterial genera producing butyrate are *Butyribacterium, Butyrivibrio, Clostridium, Eubacterium, Fusobacterium, Haloanaerobium,* and *Sarcina*. Most butyric-acid producing bacteria ferment starch, hexose, pentose, and cellobiose, and form acetic acid, in addition to butyric acid, as the major (i.e., percentage of substrate weight converted to product) fermentation product. Butyric acid production is also related to cultural conditions of the specific species. For example, *C. thermosaccharolyticum* produces butyric acid as the fermentation product during exponential growth. *B. methylotrophicum* can produce either acetate or butyrate as the sole end product, or it can form a mixture of butyrate and acetate. *B. methylotrophicum* ferments glucose to acetate with small amounts of butyrate at near neutral pH; it produces high levels of butyrate at acidic pH. In addition, it will produce butyrate from methanol in the presence of carbon dioxide and acetate under the same pH conditions.

Products of CO fermentation by this organism are acetate and butyrate in a ratio of approximately 30:1. However, the fermentation can be shifted towards butyrate by decreasing the pH at the onset of the stationary phase. Theoretically, yields of 1 mol of butyrate and 2 mol of hydrogen and CO_2 are obtainable from 1 mol of hexose via the butyrate fermentation path. *B. methylotrophicum* or *C. butyricum* under cell-recycle can achieve this scenario. In general, the most studied butyrate-producing microorganisms belonged to the genus *Clostridium*. A large number of fermentation substrates, including hydrolysates of waste cellulosic material, lactose from whey, molasses, and cellulosic materials can be utilized to produce butyric acid. However, the low concentration in the fermentation broth did not make the process commercially attractive as a commodity product.

In addition, a major characteristic of this fermentation is the concomitant production of acetate, which is observed in butyrate-producing *Clostridium* and other bacterial species. Recently, fed-batch fermentation of glucose with *Clostridium tyrobutyricum* have yielded butyrate in a concentration of 42.5 g/L with a selectivity of 0.90, a productivity of 0.82 g/L per hour and a weight yield of 36%. In glucose-limited, fed-batch cultures, initially produced acetate was reutilized, resulting in exclusive production of butyrate. Because the ratio of butyrate to total acids was strongly influenced by the growth rate of bacteria, acetate being produced along with butyrate at higher growth rates achieved increases in butyrate concentrations, productivity, selectivity, and yield by controlling the substrate feeding by the rate of gas production. In a fermentation of wheat flour hydrolysate (380 g/L of glucose) with *C. tyrobutyricum*, a butyrate concentration of 62.8 g/L was obtained with a productivity of 1.25 g/L per hour, a selectivity of 91.5%, and a weight yield of 45%. Thus, production of butyrate for specialty uses by fermentation is feasible.

Industrial production of propionic acid

Propionic acid is a value-added specialty chemical used in various chemical and food-processing industries. It has wide-ranging applications, such as an antifungal agent in foods and feeds, and as an ingredient in thermoplastics, antiarthritic drugs, perfumes, flavors, and solvents. Currently, propionic acid is produced by chemical synthesis from petroleum feedstocks, but small amounts of natural propionic acid are produced by fermentation. Propionic acid is a major end product of fermentations carried out by a variety of anaerobic bacteria, many of which ferment glucose to propionate, acetate, and CO_2. Lactate can also be fermented to propionate either via the acrylate pathway in a stepwise reduction to propionate, or via the succinate pathway, where lactate is converted to propionate via pyruvate and succinate. Although several microorganisms can produce propionic acid, fermentations using propionibacteria have been studied most extensively. The slow growth rate of propionibacteria usually results in propionate production at a slow rate. *Propionispira arboris* ferments glucose to propionate, acetate, and CO_2 via the succinate pathway. The ratio of propionate to acetate is low, resulting in a carbon loss to co-fermentation products. However, the ratio of propionate to acetate can be changed from 2 to 16:1, approaching a homofermentation yield from glucose, by use of H_2 as a co-substrate because this species consumes H_2.

Various fermentation systems have been examined to increase the propionate fermentation rate, including batch, fed-batch, continuous, continuous with cell recycle, extractive, and immobilized cell fermentations. Also, various substrates such as glucose, xylose, lactate, glycerol, food-processing waste, and whey have been examined to produce propionate not only with increased productivity, but also economically. Using *Propionibacterium acidipropionici*, propionate has been produced at 42 g/L from glycerol and at 57 g/L from glucose in batch and fed-batch fermentation systems, respectively. It seems fermentative production of propionate is feasible if an appropriate fermentation system and recovery technology can be integrated to obtain the acid in high concentration, yield, and productivity.

Biological Catalysts

In recent years, *biological catalysts* or *enzymes* have gained wider applications in the biotechnology industry. The global industrial market has recently been valued at $1.4 billion, with an expected market increase of 4-5%. Enzymes are grouped into six major classes: oxidoreductases, transferases, hydrolases, lyases, ligases, and isomerases. The majority of commercialized enzymes are hydrolases such as amylase, cellulase, xylanases, pectinase, proteases, lipase, and collagenase. Other important enzymes are glucose isomerase and glucose oxidase. Anaerobic bacteria are very diverse and thus produce a wide variety of enzymes, excluding oxygenases. Enzymes are widely used in pulp and paper, textile, detergent, and food- and feed-additive industries. In addition, enzymes also find applications in pharmaceutical synthesis, therapeutic contexts, and clinical and chemical analysis. The rumen anaerobic microbial population presents a rich and, until recently, underutilized source of novel enzymes with tremendous potential for industrial applications.

The enzymes from these microorganisms include cellulases, xylanases, β-glucanases, pectinases, amylases, proteases, phytases, and tannases. High-molecular-mass complexes containing numerous cellulases have been identified in a number of rumen bacteria, including *Butyrivibrio fibrisolvens*, *Ruminococcus albus*, and *Fibrobacter succinogenes*. The most common enzyme-producing anaerobic bacteria are mesophilic and thermophilic clostridia, and moderate- and hyperthermophilic non-clostridial species (100,101). Among the amylolytic enzymes, α-amylase hydrolyzes internal α-1,4 linkages of starch at random in an endo fashion, producing oligosaccharides of varying chain lengths. Generally, it cannot act on α-1,6 linkages of starch. *C. butyricum, C. acetobutylicum, T. ethanolicus, C. thermoamylolyticum* have been reported to produce this amylase enzyme. β-Amylase hydrolyzes alternate α-1,4-glycosidic linkages of starch in an exo fashion from the nonreducing end, producing β-maltose. The β-amylase has been produced in high yield as a primary product during growth of *Thermoanaerobacterium thermosulfurigenes*. A hyperproductive mutant was isolated that produced eightfold more β-amylase than the wild type.

Synthesis of the enzyme was both constitutive and resistant to catabolite repression. The β-amylase has also displayed industrial potential for the production of high-maltose syrups from raw or soluble starch at 75°C. Various maltose-containing syrups are used in the brewing, baking, canning, and confectionery industries. Glucoamylase is an exoacting carbohydrase that cleaves glucose units consecutively from the nonreducing end of starch molecules. High levels of a thermostable glucoamylase activity has been reported in crude extracts of *T. ethanolicus* 39E, although it was purified and later described as an α-glucosidase activity. Glucoamylase is widely used in alcoholic fermentation of starchy materials and in the commercial production of glucose and high-glucose corn syrups. However, the current source of this enzyme is fungal. Pullulanase is a debranching enzyme that specifically cleaves α-1,6 linkages in starch, amylopectin, pullulan, and related oligosaccharides. It is generally used in combination with saccharifying amylases such as glucoamylase, fungal α-amylase, or fungal β-amylase for the production of various sugar syrups because it improves saccharification and yield.

Hyun and Zeikus found that *T. ethanolicus* produces highly thermoactive and thermostable cell-bound pullulanase. They also developed a hyperproductive mutant that displayed improved starch metabolism features. Pullulanase has also been reported to be produced by *C. thermosaccharolyticum*. Taking advantage of the high thermoactivity and acidoactivity of these pullulanases, it may be assumed that these enzymes might effectively replace α-amylase and pullulanase in both starch liquefaction and saccharification processes. α-Glucosidase hydrolyzes terminal nonreducing α-1,4-linked glucose residues of various substrates, releasing α-D-glucose. It is generally considered to be maltase, but has a wide specificity, being able to cleave glucosides of non-sugars in addition to maltose, maltotriose, and other malto-oligosaccharides, and to transfer α-D-glycosyl residues of maltose and α-D-glucosides to suitable acceptors.

α-Amylase, which is used in starch liquefaction, solubilizes α-1-4 linkages in starch-forming maltodextrin syrups. α-Amylase has been recently described in thermoanaerobes. For example, the *Pyrococcusfuriosus* α-amylase gene has been cloned and expressed in *E. coli*. The *P. furiosus* α-amylase is twice as active as *Bacillus litcheformis* commercial α-amylase, but does not require Ca^{2+} for stability at high temperature (i.e., >100°C). The cellulase system in both bacteria and fungi comprises three different classes of enzymes: (i) endo-1,4-β-glucanases; (ii) exo-1,4-β-D-glucanases, including both 1,4-β-D-glucan cellobiohydrolases and 1,4-β-D-glucan glucohydrolases; and (iii) 1,4-β-D-glucosidases, also referred as cellobiases. Several species of cellulolytic clostridia have been described in the literature. These include *Clostridium cellobioparum, C. acetobutylicum, Clostridium cellulovorans, Clostridium stercorarium*, and *Clostridium thermocellum*.

Most of the work on industrial cellulases has been accomplished using aerobic fungal systems. However, due to the high specific-enzyme activity on the one hand and the high thermostability on the other, the cellulases from *C. thermocellum* have been considered for potential industrial utilization in direct alcohol fermentations, but not for saccharification *per se* (i.e., glucose production). Recently, endoglucanases have been used in laundry detergents. Collagenases are endopeptidases that hydrolyze native, insoluble fibrous collagen. One of the anaerobic bacteria producing collagenase that has been extensively studied is *C. histolyticum*. The collagenase of *C. histolyticum* is available commercially. Other collagenase-producing anaerobic bacterial species include *C. collagenovorans* and *C. proteolyticum*. Pure collagenase can be applied as a sensitive probe for biosynthetic studies and sequence determinations. It is useful in prevention or cure of keloids.

Table 16.2 Enzymes of potential industrial importance from anaerobic extremophiles

Enzyme	*Optimal temperature (°C)*	*Localization*	*Organism*
Endoglucanase	65	extracellular	*Clostridium therm ocellum*
α-Amylase	100	extracellular	*Pyrococcusfuriosus*
β-Amylase	75	extracellular	*Therm oanaerobacterium thermosulfurigenes*
Xylose/glucose isomerase	95	intracellular	*Thermotoga neopolitana*
Amylopullulunase	85	extracellular	*Thermoanaerobacter ethanolicus* 39E
	105	extracellular	*Pyrococcusfuriosus*
Endo-xylanase	70	extracellular	*Thermoanaerobacterium saccharolyticum* B6A-RI
1° Alcohol dehydrogenase	80	intracellular	*Thermococcus litoralis*
2° Alcohol dehydrogenase	90	intracellular	*Thermoanaerobacter ehtanolicus* 39E
β-Xylosidase	70	intracellular	*Thermoanaerobacterium saccharolyticum* B6A-RI
Alkaline phosphatase	85	intracellular	*Thermotoga neopolitana*

Collagenase is useful for the dispersal and dissociation of animal tissues in the laboratory. It is routinely used to separate cells from their parent tissues. *C. histolyticum* also produces an extracellular sulfhydryl proteolytic enzyme called clostripain (clostridiopeptidase). This enzyme possesses amidase, esterase, and proteolytic activity, which is directed toward the carboxyl peptide linkage of arginine. Pectin-degrading enzymes are produced by a variety of microorganisms, including clostridia. These have been produced by *C. aurantibutyricum, C. felsineum, C multifermentas, C. roseum,* and *T. thermosulfurigenes.* An active thermostable polygalactunonate hydrolase and pectin methylesterase have been produced by *T. thermosulfurigenes.* These thermostable pectinolytic activities may have application in fruit juice clarification and for processing food or agricultural/forestry products. Several xylose (glucose) isomerases have been isolated from thermoanaerobes including from *Thermotoga neapolitana.* This enzyme is more stable and active than commercial enzymes. An alkaline phosphatase that is more stable and active than commercial calf enzyme has been purified and characterized from *T. neapolitana.* This serves to illustrate the diversity of enzymes from anaerobes.

Bacterial Biotransformations

Diverse species of acetogenic and methanogenic bacteria grow on CO, H_2-CO_2, or methanol. These bacteria have been studied for conversion of synthesis gas (i.e., CO + H_2) or CO to methane by pure cultures and consortia, or into organic acids and alcohols by pure culture. Some of these bacteria include *B. methylotrophicum, Eubacterium limosum, Peptostreptococcus productus, C. thermoaceticum, C. ljungdahlii,* and *Methanosarcina barkeri.* Synthesis gas represents a cheap feedstock for microbial conversion to higher-value commodity products. Recent work has focused at syngas fermentation to liquid-fuel additives. The feasibility of ethanol production from syngas fermentation by *C. ljungdahlii* and ethanol plus butanol production by *B. methylotrophicum* has been established. However, these processes need to be developed further for their commercialization. Recently, anaerobes have also been examined for a wide range of specialty fermentation products including antimicrobials, bioflavors, biopigments, biopesticides, and anticancer agents. Antimicrobial compounds such as antibiotics and bacteriocins are produced by some of the anaerobic species.

The most common examples are that of nisin and pediocin production by *Lactococcus lactis* and *Pediococcus acidilactici,* respectively. Some strains of lactobacilli catalyze the decarboxylation of glutamate to γ-aminobutyrate. Many chiral compounds can be synthesized by microbial hydrogenation using hydrogen (or formate) and hydrogenase-containing microorganisms. Oxidoreductase enzymes are involved in electron-transfer reactions and can be applied in a stereoselective catalysis. The best known alcohol dehydrogenase is that present in both *T. ethanolicus* and other thermoanaerobes, the most notable being *T. brockii.* This NADP-linked secondary alcohol (aldehyde/ketone) dehydrogenase, found in thermoanaerobes, exhibits a wide substrate specificity toward linear and cyclic secondary alcohols and thioesters. The biotransformation reactions can be carried out either by whole-cell fermentation (which eliminates coenzyme regeneration but may be subject to interference by competing enzymes), or by crude or purified enzyme in batch or continuous-flow systems using an immobilized enzyme column. *C. sporogenes* performs the stickland reaction, in which pairs of amino acids are fermented, one amino acid acting as an electron donor (e.g., valine, leucine, isoleucine), and the other acting as an electron acceptor (e.g., proline, glycine). Synthesis of pyruvate and other 2-oxacids from acyl phosphate derivatives have been shown with permeabilized cells of *C. sporogenes*. In amino acid–fermenting anaerobic bacteria, a set of unusual dehydratases is found that use 2-hydroxyacetyl-CoA, 4-hydroxybutyryl-CoA, or 5-hydroxyvaleryl-CoA as substrates. These anaerobic bacteria include *C. propionicum, C. aminobutyricum,* and *C. aminovalericum*.

Table 16.3 Novel oxidoreductase in *Clostridium* species

Reduction	*Electron donor*	*Organism*
Aldehyde/ketone dehydrogenase		
Steroids	Unknown	*C. paraputrificum*
		C. bifermentans
Methyl ketones	NADPH	*C. thermohydrosulfuricum*
Ketones	Unknown	*C. pasteurianum*
	Unknown	*C. tyrobutyricum*
2-Oxoacid synthase		
Fatty acids	Ferredoxin	*C. sporogenes*
Acetate	Ferredoxin	*C. kluyveri*
Linoleic reductase		
Linoleic acid	Unknown	*C. sporogenes*
Enoate reductase		
Cinnamic acid	NADH	*C. sporogenes*
Crotonic acid	NADH	*C. tyrobutyricum*
2-Oxoacid reductase		
Phenylpyruvic acid	NADH	*C. sporogenes*
Nitroaryl reductase		
Chloramphenicol	Ferredoxin/flavodoxin	*C. acetobutylicum*
Metronidazole	Ferredoxin/flavodoxin	
Paro-nitrobenzoate	Ferredoxin/flavodoxin	
2-Nitrobenzene	Ferredoxin/flavodoxin	
Lipoamide hydrogenase		
NAD/lipoamide	Lipoamide/NADH	*C. kluyveri*

Microorganism in Waste Treatment

Microorganisms excel in using organic substances as sources of nutrients and energy. The challenge of wastewater treatment is to remove (i) compounds with a high biochemical oxygen demand, (ii) pathogenic organisms and viruses, and (iii) a multitude of human-made chemicals. Anaerobic digestion (AD) is commonly used to treat materials with a high content of insoluble organic matter, such as cellulose, and to degrade concentrated industrial wastes, such as those from the food-processing industry. AD is a complex biological process that utilizes a consortium of anaerobic microorganisms to act in concert to hydrolyze complex organics to simple monomers and then to volatile fatty acids (VFAs). These VFAs are ultimately converted to methane and carbon dioxide in the final step in the anaerobic food chain. The degradative and fermentative reactions in the anaerobic treatment processes can be divided into two stages: acid-forming and methane-forming, which involve at least three groups of anaerobic bacteria—acidogenic, syntrophic, and methanogenic. In the acid-forming stage, complex organic polymers, including carbohydrates, fats, and proteins, are hydrolyzed and fermented to VFAs, alcohols, and ketones by selective anaerobic bacteria.

Organic acids and alcohols such as lactic acid, propionic acid, butyric acid, and ethanol are converted to acetic acid and carbon dioxide by acetogenic bacteria in the syntrophic association of, generally, methanogenic bacteria. The acetate and CO_2 are finally converted to CH_4 by methanogenic bacteria that may include acetate- and H_2-CO_2-consuming methanogens. Anaerobic micro-organisms in general show a high degree of metabolic specialization. The success of the anaerobic digestion process therefore depends upon cooperative interactions between micro-organisms with different metabolic capabilities. Since the rate of VFA production can be significantly higher than that of VFA conversion to methane, an imbalance between these two rates can occur in a single-stage digestor, resulting in VFA accumulation, a concomitant pH drop, cessation of methane fermentation, and ultimate process failure. A two-stage AD process that separates the acid-forming and methane-forming stages has been practiced with a variety of wastes. Alternatively, a high-rate syntrophic methanogenic microbial consortium can be developed. The leach-bed two-phase digestion process, consisting of an acid-phase, solid-bed reactor operated in tandem with a separate, packed-bed, methane-phase digestor (anaerobic filter), for high solids waste has been described.

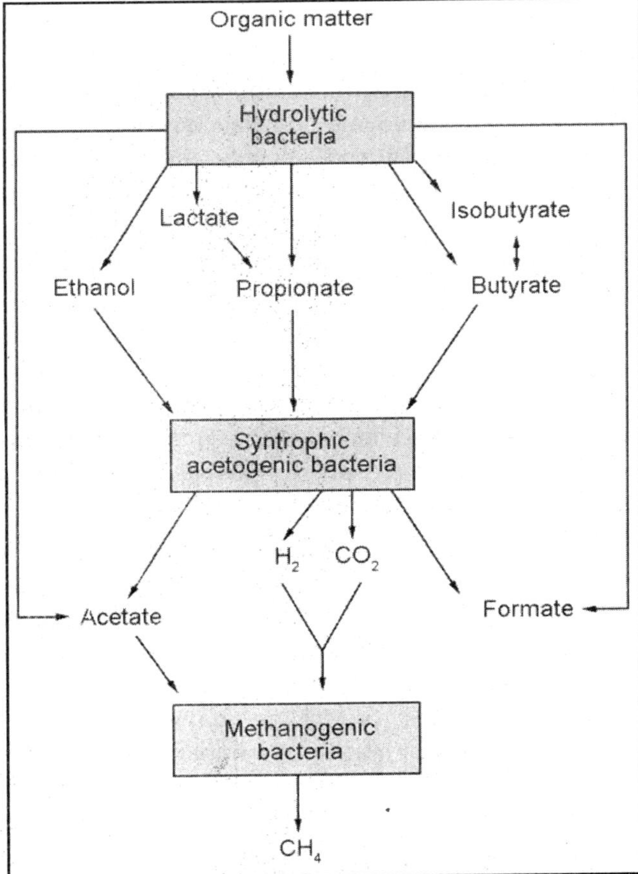

Figure 16.2 Anaerobic degradation of organic matter to membrane in relation to microbial trophic groups.

Low-strength wastewaters can now be treated, even under psychrophilic conditions, by using specific rheologic conditions in the expanded granular sludge bed (EGSB) reactor. Solid wastes can be treated anaerobically using the thermophilic "high-solids" fermentation technology. New reactor designs permit S° recovery from SO_2-rich waste gases. Many human-made organic compounds are degraded during sewage treatment. Since 1986, full-scale UASB reactors have been used to treat municipal wastewater throughout the world. Treatment costs are halved when anaerobic treatment (e.g., UASB reactor) is applied instead of just aerobic processes. In warm climates, a simple upflow sludge blanket (USB) septic tank with an aerobic posttreatment (e.g., a trickling filter) can be combined to remove the bulk of its suspended solids. AD is expected to reduce most of the pathogenic bacteria. Thus, the treated water may be considered reclaimed for use in crop production in the next decade.

Organic slurries containing particulate organic matter, such as animal manures and primary or secondary sewage sludges, are normally digested in completely mixed reactors. Because the hydrolysis rates of certain solids is slow, separation of hydraulic retention time (HRT) and solids retention time (SRT) may improve performance of digestors treating slurries. AD does not remove NH_4^+. In a new NH_4^+-removal pathway observed in the methanogenic reactors, NH_4^+ was oxidized anaerobically to N_2 in the presence of NO_2^-, with a lab-scale reactor achieving a removal rate of 0.7 kg NH_4^+-N/m^3 per day. The UASB technology is now widely used to anaerobically digest industrial wastewaters with concentrations in the range 2-20 g COD/L. UASB reactors are usually implemented when wastewater is rich in carbohydrates and relatively poor in other contaminants. Biomass retention through adequate granulation is of utmost importance in UASB technology, first to obtain a good effluent quality and, second, in order to ensure a minimal cell residence time of 7-12 day that is required to avoid the wash-out of the slowest-growing anaerobic bacteria.

The onset of sludge granulation after start-up of an anaerobic-sequencing batch reactor was enhanced by adding a cationic polymer, divalent cations, and granulation nuclei such as clay minerals. Consideration of surface thermodynamics is important in granulation. Some key methanogenic species have been identified and implicated in the granulation process, and exopolysaccharides produced by these methanogens, specifically *Methanobacterium formicicum, M. mazei,* and *Methanosaeta* sp. seem to be responsible for the stability of granular structure. Industrial wastes are often unreliable in terms of their composition. Inhibition of methanogens caused by toxic compounds present in industrial wastewaters can be alleviated by adding activated carbon directly in the anaerobic reactors.

Addition of GAC to UASB reactors treating textile wastewater prevented the chronic intoxication of anaerobic sludge that takes place in the absence of the GAC. Anaerobic treatment systems are gaining in popularity and finding new applications. As revealed in papers presented at the 8th International Conference on Anaerobic Digestion, Sendai, Japan, in May 1997, the databank of anaerobic reactors presently includes 1066 operating anaerobic reactors worldwide. Of these reactors, 956 (89.7%) are for the treatment of industrial effluents, while 78 (7.3%) treat domestic sewage, and only 32 (3.0%) of the total are anaerobic-treatment systems for organic solid waste (excluding the vast number of biogas plants for manure treatment that have been installed worldwide). It is estimated that there are about two million biogas plants for manure treatment just in India alone.

The largest number of municipal anaerobic treatment plants presently exist on a moderate, but remarkably growing, scale for sewage treatment in countries such as Brazil, Colombia, Mexico,

India, and China. In industrial countries, only sewage sludge is treated anaerobically on a large scale due to climatic factors. For industrial wastewater treatment, countries such as India, China, Thailand, Brazil, and Mexico have a growing interest in anaerobic treatment for the local industry; in industrialized regions the leading countries, especially The Netherlands and Belgium, have almost already saturated their home markets. For anaerobic solid-waste treatment (and co-fermentation), Germany and Denmark have significant numbers of plants so far. In the near future, anaerobic treatment of solid wastes will become a viable option provided that appropriate collection systems are introduced.

Xenobiotic compounds

In recent years, AD has been used to degrade xenobiotics present in wastewater, soils, and river sediments. Complex xenobiotics often require more than one species to be completely mineralized. Microorganisms need to adapt, maybe for several months, before the maximal conversion rate of a new xenobiotic compound can be achieved. Anaerobic tunnel reactors are used to treat soils and sediments polluted with chloroethene, BTEX, or TNT. Efficiency of removal of adsorbable organic halogens by a UASB treating kraft-mill bleach wastewater varied between 27-65%, depending on the residence time. *Desulfomonile tiedjei*, which can rapidly transform 3-chlorobenzoate (3-CB), has been shown to become established in an UASB reactor. This indicates that a specific strain or desired species can be incorporated in microbial granular culture of an UASB reactor. Incorporation of an adapted microbial sludge culture into syntrophic biomethanation granules has been demonstrated to treat PCP. These studies were further validated by developing PCE/TCE-dechlorinating methanogenic granules using a similar protocol. In this, the anaerobic granules, using methanol as an electron donor for reductive dehalogenation, completely converts highly oxidized PCE to ethylene.

In contrast, PCBs biodegrade very slowly and persist in the environment for decades. An effective microbial enrichment that was shown to remove *ortho-*, *meta-*, and *para*-chlorine was incorporated into the PCP-degrading methanogenic microbial granules to produce PCB-dechlorinating anaerobic microbial granules in a manner similar to achieving PCP-granules in the first place. The PCB-dechlorinating anaerobic methanogenic granules have shown PCB-dechlorination of defined PCB-congeners and Aroclor-contaminated river sediments. These granules are stable, culturable, and can be produced on a mass scale for bioaugmentation for application in both in situ and ex situ processes. These granules have also demonstrated degradation and mineralization of biphenyl, a non-chlorinated end-product of PCB-dechlorination process reductively (Natarajan, Wu, and Jain, unpublished data).

Since PCBs are a complex mixture of various PCB compounds, and each PCB compound varies in its degree of dechlorination, it is expected that a mixture of bacteria capable of removing chlorines from different positions on the PCB molecule must be involved in achieving a complete dechlorination of PCB. A similar assumption, however, is no longer true for dechlorination of perchloroethylene (PCE). Recently, Zinder and his group isolated a pure culture of an anaerobic bacteria, tentatively named *Dehalococcoides* strain 195, that dechlorinates PCE and TCE in a stepwise manner to ethene. This novel eubacterium is not methanogenic, acetogenic, or sulfidogenic.

2,4,6-Trinitrotoluene (TNT), one of the most widely used explosives, occurs as a pollutant of soil and ground water, especially at sites of ammunition factories. There have been several reports of degradation or transformation of nitro and aminoaromatic compounds under anaerobic conditions. A nonspecific reduction of the nitro group to the corresponding amine was assigned to a variety of methanogenic bacteria, sulfate-reducing bacteria, and clostridia.

In contrast, it was reported that 2,4-dinitrophenol, 2,4- and 2,6-dinitrotoluene, and TNT were used as sources of nitrogen by sulfate-reducing bacteria. Thus, it is clear that many nitroaromatic compounds can serve as growth substrates for anaerobic bacteria. Anaerobic treatment systems have been proposed as a means of avoiding the accumulation of partially reduced intermediates during degradation of TNT. Under strictly anaerobic conditions, TNT can be completely reduced to triaminotoluene. *C. acetobutylicum* transformed 2,4,6-TNT to undetermined end products via monohydroxylamino derivatives. In contrast to solventogenic cells, acidogenic cultures showed rapid transformation rates and the ability to transform TNT and its primary reduction products to below detection limits. Anaerobic treatment of explosive (primarily TNT)-contaminated soil was demonstrated in open bulk containers containing soil, phosphate buffer, and potato starch. The potato starch served as a rapidly degradable carbon source that allowed the rapid establishment of anaerobic conditions. Cresols and small organic acids were observed as end products. Similarly, Dinoseb, a very persistent herbicide that is highly toxic to virtually all living systems, has also been shown to be degraded by anaerobic microbial consortia using potato starch as a carbon source.

Mineral processing activities

Microorganisms degrade certain toxic constituents used in mineral processing, and concentrate and immobilize soluble heavy metals released as a result of mining and mineral-processing activities. The initial mechanism of metal binding by microorganisms is electrostatic attraction between charged metal ions in solution and charged functional groups on microbial cell walls. The cell walls are composed of macromolecules with functional groups (principally carboxylate, amine, imidazole, phosphate, sulfhydryl, and sulfate) that contribute a net negative charge to the surface of the microorganism. These functional groups remain active even when the microorganism is not viable. Sulfate-reducing bacteria (SRBs) are used in highly controlled reactor systems and in constructed anaerobic wetlands for removal of sulfate and heavy metals from acid-rock drainage and other aqueous, metal-contaminated, streams. SRBs such as *Desulfovibrio* and *Desulfatomaculum* oxidize organic matter or H_2 by using sulfate as an electron acceptor to produce hydrogen sulfide and bicarbonates. The sulfide immediately reacts with soluble heavy metal ions to form highly insoluble metal sulfides.

A system to evaluate anaerobic and aerobic treatment of mine effluents containing excessive sulfate and heavy metals has been described. In the continuous system, the H_2S is produced by the SRB biofilm on the dolomite pebbles in the anaerobic stage, which precipitates metal sulfides from a waste stream amended with an organic energy source. In the aerobic stage, a completely mixed reactor and a settling tank facilitate oxidation of residual H_2S and biodegradation of residual organics from the primary anaerobic column. Sulfate reduction has been considered as a method for permanent stabilization of sulfidic mine tailings. High sulfate concentrations are common in wastewaters from paperboard industries, molasses-based fermentation industries, edible oil refineries, and in acidic leachates of pyritic waste rock and tailings. Sulfidogenic UASB reactors have attracted some attention for treating such polluted streams, but the biological processes to remove high SO_4^{2-} contents have not yet been optimized. A new bioprocess in which sulfate and heavy metals are removed simultaneously from groundwater has been developed. In this two-step process, SO_4^{2-} bacteria produce sulfides that precipitate with the heavy metals after ethanol, an electron donor, is added to the first reactor (BIOPAQ UASB). In the second reactor (THIOPAQ-submerged fixed-film reactor), sulfur bacteria reoxidize the excess sulfide selectively to solid S^0 under controlled dosage of O_2.

With HRT of 4 h in the BIOPAQ reactor and 30 min in THIOPAQ reactor, removal efficiencies of 99% for heavy metals and of 85% for sulfate have been achieved. This technology has also been adapted for desulfurization of flue gas from power plants. In the modified process for flue-gas desulfurization, H_2 can be used as an electron source in lieu of ethanol. Minewaters and industrial effluents containing high sulfate concentrations create a disposal problem that requires an urgent solution in order to avoid excess mineralization of surface waters. Sulfate can be converted quantitatively to hydrogen sulfide by *D. desulfuricans*. Further conversion to elemental sulfur can be effected by the photosynthetic bacteria *Chlorobium limicola* and *Chromatium cinosum*. Two separate reactors were used for hydrogen sulfide and sulfur production. A possible way to increase the sulfate reduction rate is by making use of a packed-bed, instead of a completely mixed, reactor.

CONCLUDING REMARKS

The use of anaerobes in industrial processes has grown dramatically over the past 20 years. Traditionally, anaerobes were used routinely in the processing of fermented foods such as lactic acid and propionic acid bacteria in cheeses, yogurts, sausages, and pickles. In waste treatment, mixed methanogenic cultures were used in both industrial and municipal anaerobic digestors for removal of residual organic matter. Now, *C. botulinum* toxins are used as medicines. Numerous organic alcohols and acids are produced by industry as natural fermentation chemicals. These natural chemicals are used in flavors, fragrances, preservatives, and other specialty products. Fermentation-derived lactic acid is now produced by industry as a commodity chemical and is used as an acidulant, disinfectant, green solvent, and precursor for polylactide-based biodegradable plastics.

Anaerobic composting systems are used to treat food and municipal solid wastes because they produce less end product that must be used or land spread. Sulfite is also removed from coal-bearing flue gases in bioreactor systems comprising mixed sulfidogenic cultures. With the rapid growth in isolation and characterization of new anaerobic species from normal and extreme environments, the diversity of anaerobic species with unique biochemical attributes far surpasses that of aerobic microbes. This great diversity of anaerobic microbes should be expected since anaerobes were the first organisms to evolve on earth, yet they were the last large natural group to be studied in detail by biologists.

The vast array of biochemical diversity shown by anaerobes is related in part to the fact that they are not limited to use of O_2 as an electron acceptor, they can use fermentative metabolism or CO_2, SO_4, or other electron acceptors for anaerobic respiration. This biochemical diversity will undoubtably be exploited in the future since many of the fermentation products and enzymes of anaerobes are of interest to industry (i.e., food-feed, chemical, pharmaceutical, energy, and environmental companies). Several research areas on anaerobes showing special industrial promise include succinate fermentation, thermozymes, and dechlorinating methanogenic granules. Ethanol fermentations are limited in part because two moles of CO_2 are lost from the product per mole of glucose fermented.

On the other hand, one can derive more than one pound of succinate per pound of glucose fermented because the theoretical chemical yield is 1 glucose + $2CO_2$ + $2H_2$ $\rightarrow$ 2 succinate. Succinate has a wide variety of uses as both a specialty chemical and commodity intermediate chemical. Perhaps the largest markets for succinate includes their use as a feedstock to produce stronger-than-steel engineered plastics and polyesters, and as the chelator EDDS (ethylene diamine disuccinate) to replace non-biodegradable EDTA. The saccharolytic enzymes of thermoanaerobes are very active and stable. These thermozymes could be used to develop the next generation of enzymes used in the starch- processing industry (i.e., *a*-amylase, glucose isomerase, glucoamylase, and pullulanase) or to initiate a cellulose- processing industry for enhanced biomass utilization based on very active and stable cellulases and hemicellulases. Anaerobic sediments and soils are contaminated with a wide variety of chlorinated compounds. The use of dechlorinating methanogenic granules offers an alternative to expensive dredging and landfilling. Dechlorinating methanogenic granules can degrade a wide variety of toxicants (DDT, DIOXINS, PCE/TCE, PCBs, PCP, etc.) by in situ bioagumentation technology.

17

ARTIFICIAL SWEETENER

Aspartame, α-L-aspartyl-L-phenylalanine methyl ester (APM), is an artificial sweetener whose sweetness is about 200 times stronger than sucrose by weight. Its sweetness was found accidentally by a researcher of Searle Research Laboratories, where APM was produced as an intermediate of a C-terminal tetrapeptide of gastrin. Its strong sweetness allows reduction of the amount of calorie sweetener added to foods to achieve a sweetness equivalent to that from sucrose. Although other intense sweeteners often exhibit a slightly bitter aftertaste, APM has none. Its sweet taste is close to that of sucrose, a traditional sweetener and the standard of sweetness. Therefore, the demand for APM has grown rapidly since its approval by the Food and Drug Administration (FDA) in 1981. Worldwide APM consumption is currently over 10,000 tons per year, and it is the most common intense sweetener for beverages, tabletop sweetener, ice cream, chewing gum, and so forth. APM is produced by two methods: a chemical method, used by Nutrasweet in the United States and Ajinomoto in Japan, and an enzymatic method, used by Holland Sweetener Company, a joint venture of TOSOH (formerly Toyo Soda Manufacturing) in Japan and DSM in the Netherlands. In the enzymatic method, an APM precursor, *N*-benzyloxycarbonyl-APM (Z-APM) is produced by a protease-catalyzed condensation reaction of *N*-benzyloxycarbonyl-L-aspartic acid (Z-L-Asp) and L-phenylalanine methyl ester (L-PheOMe). The advantages of the enzymatic method are that a racemic substrate can be used and there is no production of β-aspartame (β-APM). The reduction of enzyme consumption was, however, desirable for further cost reduction. In the present article, the APM production process with the protease reaction and various efforts to reduce the enzyme consumption therein are described.

ENZYMATIC PRODUCTION OF ASPARTAME

In APM production by conventional organic synthesis, the side-chain carboxylic group (β-carboxylic group) of aspartic acid must be protected to avoid the formation of unwanted β-APM; additionally, optically pure substrates are required to avoid the production of stereoisomers of APM.

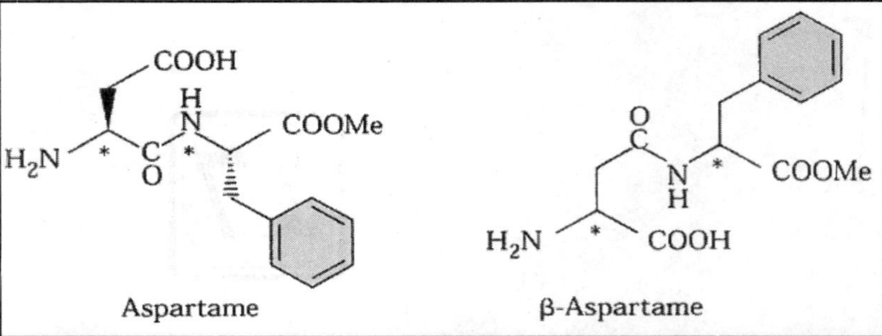

Figure 17.1 Structures of aspartame and its regioisomer. Asymmetric carbon atoms are indicated by an asterisk. The configuration of aspartame is L-α-L.

These APM-related compounds, such as β-APM and stereoisomers, exhibit a bitter taste. Use of a protease as catalyst does not give formation of these undesirable compounds, even if racemic substrates are used as starting materials and without protection of the β-carboxylic group of aspartic acid. The use of racemic phenylalanine in the enzyme process is very advantageous, because L-phenylalanine is more expensive than racemic phenylalanine. Proteases such as thermolysin, subtilisin, or papain were found to be useful tools for the peptide synthesis because of their high stereo and regio-selectivity, and they have been applied in the synthesis of various biologically active peptides, such as peptide hormones, neuropeptides, and insulin.

APM synthesis catalyzed by a protease was also investigated by Isowa et al. They reported that N-protected aspartic acid and Phe-alkyl ester are condensed to N-protected L-α-Asp-L-Phe-alkyl ester by thermolysin in high yield (83–96%) as a precipitate consisting of the addition compound of the condensation product and one molecule of Phe-alkyl ester, as shown in Fig. 17.2. In this scheme, R indicates N-protective group, such as Z or PMZ, and R1 indicates lower alkyl ester, such as methyl or ethyl ester. Proteases usually catalyze the hydrolysis of peptides in water as solvent; the condensation reaction is, in other words, the reverse reaction. Therefore, it is usually difficult to obtain products in high yield because of the occurrence of hydrolysis of the products by

Figure 17.2 Showing the production of Phe-alkyl ester.

the proteases. The synthesis of APM-related peptides can, however, be performed in high yield when the product is removed from the reaction system as a precipitate and is not hydrolyzed back by the protease, for example, by thermolysin. Although this method requires an expensive catalyst, such as thermolysin, the high stereo and regio selectivity is a great advantage in APM production.

The industrial APM production process based on the proposal of Isowa et al. was developed by Toyo Soda Manufacturing and was adopted for APM production by Holland Sweetener Company. In contrast to the enzymatic method, the chemical method produces an APM precursor where formyl-APM (f-APM) is derived from *N*-f-Asp anhydride and L-PheOMe, as is shown in Fig. 17.3. Although β-APM is produced as by-product in this method, α-selectivity of the reaction can be enhanced by selection of the optimal solvent for the reaction, such as alkaline aqueous medium or acetic acid partially replaced with an alkyl ester with a secondary or tertiary alcohol. It was reported that the production *of* β-APM was reduced to 20% of total peptide by selection of a proper solvent.

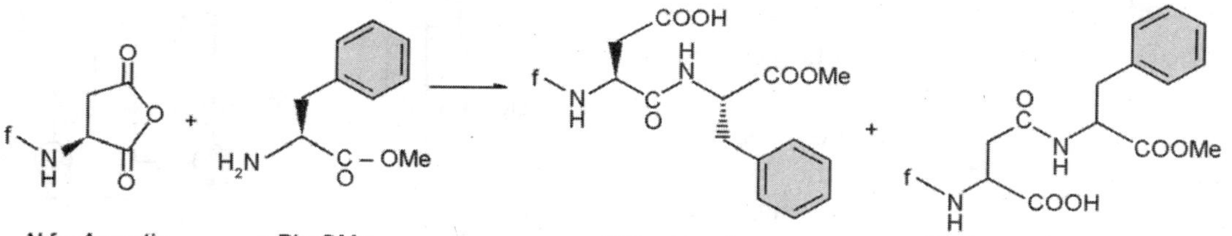

N-f-L-Aspartic anhydride + L-PheOMe → *N*-f-α-aspartame + *N*-f-β-aspartame

Figure 17.3 Enzymatic production of *N*-f-Asp and L-PheOMe.

The β-APM reportedly can be converted to APM by alkaline treatment in alcohol in the presence of metal ions such as zinc or copper as catalysts. Productivity of APM seems to be enhanced by these methods in the chemical procedure. As described earlier, the enzymatic and chemical methods both have individual advantages. An advantage of the enzymatic method is the high selectivity of the reaction, where APM is produced from cheap raw materials by a simple procedure in high yield. Not requiring expensive catalysts, such as enzymes, in the condensation step is an advantage of the chemical method; however, expensive starting materials must be used and by-product is produced. Attempting cost reduction in APM production and improving the condensation step are of great importance for APM production by Holland Sweetener Company, and many efforts have been made by both TOSOH and DSM. Hereafter, I describe the enzymatic APM production process and the efforts to improve the condensation reaction.

Role of Thermolysin

The principles of the thermolysin process were introduced by Oyama et al. in 1987. At first, L-PheOMe from racemic PheOMe is coupled with Z-Asp stereo and regio selectively by thermolysin. The reaction product, Z-APM, forms an insoluble addition compound with one molecule of remaining D-PheOMe immediately and precipitates. When the reaction is completed, the Z-APM·D-PheOMe addition compound is collected as a solid product, and thermolysin is recovered from the solution. D-PheOMe is separated from Z-APM as its hydrochloride by acid washing (with hydrochloric acid, for example). The Z-group is removed from the Z-APM by hydrogenolysis. Finally, APM is obtained by crystallization and drying. D-PheOMe is recycled after racemization by alkaline treatment and re-esterification in methanol. Although the β-carboxylic group of aspartic acid does not need to be protected, the *a*-amino group of aspartic acid and the carboxylic group of phenylalanine must be protected. This plays an important role not only for the selective reaction of the α-carboxylic group of L-aspartic acid and the amino group of L-phenylalanine, but also for being suitable to be used by thermolysin as substrates.

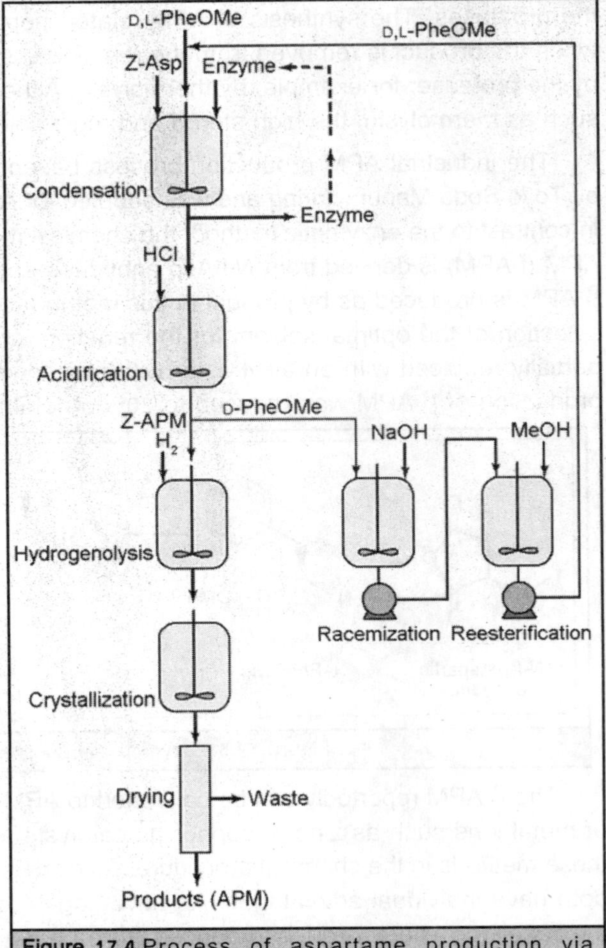

Figure 17.4 Process of aspartame production via thermolysin catalyzed reaction.

Thermolysin can not catalyze the condensation reaction without such protections. The methyl ester group is used preferably for the protection of the carboxyl group of phenylalanine because this is an essential group expressing the sweetness of APM. The amino group of aspartic acid is protected by the Z-group because it is easily removed by hydrogenolysis without removal of the methyl ester group at the C-terminal of APM. Additionally, a higher reaction rate is obtained when a hydrophobic and bulky group exists at this position. Although formyl or acetyl groups can also be used for the protection, the Z-group, which is hydrophobic and bulky, is more suitable for the reaction rate of the condensation. Thermolysin and similar microbial metalloproteases (thermolysin-like metalloproteases) were selected from among a number of industrially available proteases. Although other serine or cysteine proteases, such as subtilisin or papain, have been more widely used in various industrial fields, they are not very suitable for APM production because they tend to hydrolyze lower alkyl esters at the C-terminal of peptides and also the methyl ester group of L-PheOMe and Z-APM.

The cleaving sites of proteases in Z-APM are summarized, where residues of Z-APM are numbered as P1, P2, and P3, respectively, for those toward the N-terminal from the cleavage site, and as P'1 and P'2, respectively, for those toward the C-terminal from the cleavage site. Specificity of thermolysin and thermolysin-like metalloproteases is restricted to hydrophobic and bulky amino acids, such as phenylalanine or leucine at the P'1 position. The metalloproteases also have esterase activity, but this activity is restricted in peptide mimic esters like Bz-Gly-OPhe-Ala or furylacryloyl-

Gly-OLeu-NH_2. The methyl ester of L-PheOMe or Z-APM is not hydrolyzed by them at all, because the methyl ester group is too small at P'1 position. Z-APM is, therefore, produced by thermolysin or thermolysin-like metalloproteases without by-products.

Table 17.1 Cleavage sites on Z-APM

Cleavage site	*P3 - P2 - P1 - P'1 -P'2* ↑
Thermolysin	Z - Asp - Phe- OMe
Subtilisin	Z - Asp-Phe - OMe
Papain	Z - Asp - Phe- OMe
	Z - Asp-Phe - OMe
V8 protease	Z - Asp - Phe- OMe

Industrial Application

To improve the condensation reaction in the enzymatic process, reduction of enzyme consumption maybe of great importance. Although thermolysin is currently the most stable of the proteases industrially available, some portion of it is inactivated during the condensation reaction. This enzyme inactivation, of course, influences production costs. Stabilization of the enzyme during the condensation reaction thus was one of the most important aspects of cost reduction. It is well known that calcium ion stabilizes many microbial proteases, and thermolysin is also stabilized by it. If the concentration of calcium ions in the reaction mixture is increased, it will diminish the inactivation of thermolysin. However, this would not be suitable to improve this process, because high concentration of calcium will frequently cause troubles caused by scaling. If thermolysin can be replaced by a more stable protease, the enzyme loss can be reduced further. Extremely stable proteases such as archaelysin and a serine protease from *Desulfurococcus* strain SY were recently found in cultures of hyperthermophilic archaea (formerly called archaeobacteria).

The archaea, such as *Desulfurococcus, Pyrococcus,* or *Thermococcus,* showed an optimum temperature for growth around 90°C and even can grow above 100°C. Their enzymes exhibited extremely high stability against extreme conditions; archaelysin and the protease from *D.* strain SY retained 50% of their activity when they were heated at 95°C for 80 min and 450 min, respectively. These proteases are expected to have applicability in various industrial fields. Unfortunately, no metalloprotease has yet been found from hyperthermophilic archaea. The proteases separated from the archaea up to now belong to serine proteases or cysteine proteases. Because their specificity toward Z-APM is similar to that of subtilisin or papain, they are not appropriate for APM production. They will hydrolyze the ester bond of PheOMe or APM in the same way as subtilisn or papain. Although finding a metalloprotease from such archaea would be desirable, thermolysin remains the most stable protease currently available for APM production. Recently a metalloprotease was found from a culture broth of a hyperthermophilic aerobic archaeon "Aeropyrum pernix K1" that also exhibited extremely high stability against thermal denaturation with a half-life of 70 min at 125°C, although its specificity for APM relating peptides has not been solved.

Enzymatic Inactivation

Immobilization of thermolysin was also expected to reduce enzyme inactivation because it is protected from autolysis. However, an application of immobilized thermolysin in the current production process seems difficult, because the reaction for Z-APM production takes place in the presence of a dense solid precipitate, that is, an insoluble addition compound of Z-APM and PheOMe. If both the immobilized enzyme and the reaction product occur in the reaction mixture as solids, they cannot be separated by a simple procedure. Additionally, if the reaction would be performed in a continuous process equipped with an immobilized thermolysin-packed column, the precipitate would block the flow of the reaction solution in the column. Z-APM forms its insoluble addition compound with PheOMe quite easily in water. To overcome this problem, Oyama et al. attempted to perform the reaction in ethyl acetate. Because ethyl acetate dissolves Z-APM without formation of the insoluble addition compound, it was expected that Z-APM could be recovered in solution and that the immobilized thermolysin could be recovered as a solid, thus enabling complete recycling of enzyme.

Moreover, because the equilibrium state of the synthesis and hydrolysis of Z-APM shifts toward the synthesis side in organic solvents, it was expected that Z-APM could be obtained in high yield in ethyl acetate even though the product is not removed from reaction system in situ. Unfortunately, significant inactivation of immobilized thermolysin was observed in both batchwise operation and continuous operation. Additionally, Z-APM synthesis was disturbed by channeling of organic solvent and water in the immobilized thermolysin column when it was operated continuously; thus, high production yield could not be obtained by this system. Only if these problems can be solved, immobilized thermolysin can be applied in the APM production. Studies for APM production by immobilized thermolysin are being continued by various researchers.

Site-Directed Substitution

Improvement of enzymes by site-directed amino acid substitutions at selected sites has recently become possible by the progress in genetic engineering and of computer simulation of the three-dimensional structure of the protein molecules. These so-called protein engineering techniques are becoming an important tool for the improvement of enzyme character. A first example of improved thermolysin-like metalloproteases was shown by Imanaka et al. using a metalloprotease from *Bacillus stearothermophilus* CU21. This protease exhibited 85% homology in amino acid sequence with thermolysin. They substituted the 141st glycine of this protease to alanine and expected to find enhancement of the internal hydrophobicity and stabilization of the internal α-helix. The resulting mutant protease indeed exhibited higher thermostability than did wild-type protease. Another example of stabilization of this metalloprotease was reported by Hardy et al., who introduced proline at the 65th or 69th site.

Although the stability of this protease was lower than that of thermolysin even with introduction of these mutations, it may be expected that thermolysin can be improved by a similar procedure. Thus, studies for the improvement of thermolysin by protein engineering were started by a collaboration of TOSOH and Sagami Chemical Research Center. In this research, attention was focused to enhancing the activity of thermolysin rather than its stability. Even when the stability of thermolysin is enhanced until the enzyme inactivation during the condensation reaction is at a negligible level, the enzyme has to be recycled for reduction of enzyme costs. This requires additional equipment or procedures to recover the enzyme from the process solution, such as concentration of enzyme by ultrafiltration, salting out, or immobilization of enzyme. This will increase the fixed costs at an APM plant. On the other hand, if the activity is enhanced adequately and if the enzyme amount in the APM production can be reduced to a negligible level by application of highly active mutant thermolysins, it would be possible to use the enzyme without recycling. A highly active mutant thermolysin would be more effective than a highly stable one.

Additionally, useful knowledge would be gained about the mechanism determining the relationships among the proteins' structures, properties, and functions from the improvement of thermolysin activity. Thus, studies for improvements of thermolysin were started, and highly active mutant thermolysins were successfully constructed. One of these mutants exhibited a five times higher activity toward Z-APM synthesis when it is evaluated by the initial reaction rates. Details of the improvements of thermolysin are described in the article entitled Thermolysin. The stability of these mutant thermolysins was the same as that of wild-type thermolysins when it was determined by calorimetry. It is therefore expected that mutant thermolysin can be used for APM production in the same way as wild-type thermolysin, except that the enzyme amount to obtain the same reaction rate is lower with mutant thermolysin.

APM Production by Mutated Species

The highly active thermolysin exhibiting a five times higher activity than wild-type thermolysin was called TZ-1. TZ-1 has substituted amino acid residues at three sites, namely the 144th leucine to serine (L144S), the 150th aspartic acid to histidine (D150H) and the 227th asparagine to histidine (N227H). Although this mutant exhibits five times higher activity toward Z-APM synthesis, it should be noticed that its activity was evaluated at conditions different from real APM production conditions as described in the article on thermolysin referred to earlier. The evaluation conditions were chosen to reduce the viscosity of the reaction solution; the concentration of substrates was lower than that under conventional production conditions, and the ratio of Z-Asp and PheOMe was set at about 1:1. These conditions enabled to obtain exact data in a milliliter scale of reaction mixture. These conditions are referred to as "model conditions."

The model conditions enabled evaluating several candidates during molecular engineering of mutants. Because TZ-1 showed very high activity under the model conditions, its efficiency in small-scale APM production under the conventional production conditions was also examined. The reactions were performed in 200 mL of a mixture containing 0.2 M of Z-L-Asp, 0.5 M of D,L-PheOMe, 5.1 mM $CaCl_2$, 0.17 M NaCl and 50-150 mg of the thermolysin at 40°C and in a range of pH between 5 and 7. Mutant thermolysins used here were produced by fermentation of *Bacillus subtilis* MT-2, harboring recombinant plasmid pUBTZ2 in a jar fermentor equipped with a 16-L vessel. Other conditions for the fermentation were the same as described in the article on thermolysin. Enzyme amounts in the reaction mixture were determined by casein hydrolytic activity according to the method of Endo as are also described in the article on thermolysin.

The condensation reactions were monitored by HPLC equipped with TSK gel G2000SW with 5% acetonitrile in 0.05% trifluoroacetic acid as solvent for determining the amount of the product and the remaining substrates. Before HPLC, the samples were diluted by appropriate solution, such as 10 mM calcium acetate, and the insoluble addition compound of Z-APM and PheOMe was removed by centrifugation. The conversion rates were calculated from remaining Z-Asp using the following formula:

$$CV(\%) = ([\text{Z-Asp}]_0 - [\text{Z-Asp}]_t)/[\text{Z-Asp}]_0$$

Herein, $[\text{Z-Asp}]_0$ and $[\text{Z-Asp}]_t$ refer to the concentration of Z-Asp at the initial and at an appropriate later time, respectively.

The optimum pH of TZ-1 was 6.0, and that of wild type was 6.5. At pH 6.0, the activity of the TZ-1 was about fivefold that of wild type. The optimum pH shift of TZ-1 seems to result from the introduction of a plus charge of histidine instead of the 150th aspartic acid and the 227th asparagine. Because the 150th residue and the 227th residue exist at vicinity of the Zn ion and the 23 1st histidine residue in the active center of thermolysin, respectively, the plus charge of histidine may influence the electrostatic interactions in this center. Both Zn ion and the 23 1st histidine are regarded as important to express thermolysin activity. The mutation N227H especially seems to influence the pH profile of thermolysin activity, because the pK_a value of the 23 1st histidine is likely to be shifted toward the acidic side by introduction of a plus charge of histidine at the 227th site. This shift in optimum pH is not observed, however, when the activity is evaluated under the model conditions where the optimum pH of both wild type and TZ-1 are pH 7.0, as is mentioned in Thermolysin.

An explanation for the difference of the optimum pH between the model conditions described in there and the production conditions described herein have not yet been found, although important factors may be: (i) concentration of substrates, (ii) ratio of the two substrates, and (iii) optical purity of PheOMe. Reaction rate and production yield by 50 mg of TZ-1 are the same as those by 150 mg of wild type. When the reaction is performed with 150 mg of TZ-1, the initial reaction rate is three times higher than that by the same amount of wild type. It seems that activity of TZ-1 is three times higher than that of wild type under this condition. Remaining activities of both wild type and TZ-1 after the condensation reaction were 70%. This indicates that the stability of TZ-1 is the same as that of wild type. Moreover, there are no differences in thermostability between TZ-1 and wild type as has been demonstrated by heat capacity analysis with the calorimeter as described earlier. From these results, it follows that mutant thermolysins can be useful for cost reduction in APM production. Thus, studies for improvement of thermolysin were continued. Another important amino acid substitution, the 150th aspartic acid to tryptophan (D150W) was recently found. The activity of the single site mutant D150W was five times higher than that of wild type. This is the same as the effect of the combination of three mutations, L144S, D150H, and N227H in TZ-1. Combination of D150W with L144S and N227H resulted in a new mutant, L144S-D150W-N227H, which exhibits about 10 times higher activity than wild type under the model conditions. This mutant is called TZ-5 and will be useful for further reducing of costs in APM production.

Regulatory Process

Studies on stabilization of thermolysin by controlling the reaction conditions were also continued separately by a collaboration of TOSOH and DSM. Harada et al. reported such stabilization by an

N-protected amino acid, such as Z-Asp, Z-Glu or Z-Gly, during storage of thermolysin solution. Z-amino acids seem to act as inhibitors against the autolysis of thermolysin herein. The inhibitory effect of Z-Ala was reported earlier by Morihara and Tsuzuki after they had found that Z-Ala inhibited hydrolysis of Z-Ala-Leu-Ala catalyzed by thermolysin at 30 mM of Ki value. Although these results for Z-Ala are not conclusive for determining whether inhibition by protected amino acids is competitive, it now appears that Z-amino acids are likely to bind to the active site of thermolysin instead of substrates. Moreover, Harada et al. demonstrated that thermolysin in the reaction mixture was kept stable when Z-Asp concentration is kept at a high level during the condensation reaction. They studied effects of the ratio of Z-Asp and D,L-PheOMe at the start of the condensation reaction.

When the reaction was started at a 1:2 ratio of Z-Asp to PheOMe, the enzyme recovery was 92%, whereas it was 44% at a ratio of 1:2.5. Remaining Z-Asp after the reactions were 55.9 and 10.5 mmol/kg, respectively. At the ratio of 1:2, much more enzyme and Z-Asp were remaining than at the ratio of 1:2.5. The remaining Z-Asp apparently protects the enzyme from inactivation. Productivity of Z-APM at the ratio of 1:2 was 0.225 mol/kg; at 1:2.5 it was 0.238. Thus, significant decrease of productivity was not observed by higher concentration of Z-Asp. This indicates that the synthesis is not inhibited by Z-Asp, whereas Z-Asp is an inhibitor against autolysis or peptide hydrolysis. Nagayasu et al., however, reported that Z-Asp inhibited the condensation reaction by immobilized thermolysin. The inhibition effect toward the condensation reaction can be avoided by choosing the reaction conditions, as suggested by Harada et al. According to Harada et al., enzyme inactivation may be caused by adsorption of thermolysin on the surface of crystals of the addition compound of Z-APM and PheOMe. Apparently, Z-Asp is likely to prevent such adsorption.

Table 17.2 Condensation reaction at different Z-aspartic acid-to-PheOMe ratio

Z-Asp/ PheOMe ratio	*Initial [Z-Asp] (mol/kg)*	*Conversion rate (%)*	*Produced Z-APM (mol/kg)*	*Remained [Z-Asp] (mmol/kg)*	*Enzyme recovery (%)*
1:2	0.28	80.5	0.225	55.9	92
1:2.5	0.25	95.2	0.238	10.5	44

In contrast to the protection by Z-Asp, thermolysin is not protected from inactivation by L-PheOMe, although the latter is another substrate of the condensation reaction; this may be because of much higher K_m value of thermolysin toward L-PheOMe. According to results of earlier kinetic studies of the condensation reaction by Oyama et al., a double reciprocal plot toward L-PheOMe and D,L-PheOMe showed no apparent K_m value, whereas that toward Z-Asp is 10.3 mM. These authors presented a reaction model where thermolysin at first forms an enzyme-Z-Asp complex, and then Z-APM is produced by the attack of L-PheOMe on the complex. In this scheme, TLN indicates thermolysin. This model suggests that thermolysin cannot form an enzyme–L-PheOMe complex to prevent autolysis and adsorption on crystals of the addition compound. Therefore, thermolysin does not seem to be stabilized by L-PheOMe.

$$\text{Z-Asp + PheOMe + TLN} \rightleftharpoons \text{Z-Asp – TLN + PheOMe} \longrightarrow \text{Z-APM + TLN}$$

Figure 17.5 Showing effect of L-PheOMe.

The stabilization by Z-Asp is quite effective for thermolysin stabilization during APM production because the effect can be obtained quite simply by adjusting the concentration of the substrates. Moreover, Harada et al. mentioned that Z-Asp stabilized thermolysin without presence of calcium ions. This is another advantage of this stabilization method because various problems caused by calcium scaling can be avoided by this method. As described earlier, enzyme inactivation during the condensation reaction can be prevented by the adjusting the ratio of the two substrates. Recently another method to stabilize thermolysin during condensation was revealed by Harada et al.: an addition of toluene into the reaction mixture at about 30% (w/w) of low materials. It prevented adsorption of thermolysin onto the crystal surface of the addition compound of Z-APM and PheOMe and resulted in an increase of thermolysin recovery after the reaction.

IMPROVED PRODUCTIVITY

Various strategies have been used in the attempt to improve the APM production process. The possibility of reducing enzyme consumption was indicated by mutant thermolysins and by adjusting of the concentration of substrates. On the other hand, an alternative production system is desired to take long strides in reducing costs. Much of this research has been focused on avoiding N-protection of aspartic acid. If APM can be produced without N-protection, it will not only reduce raw materials costs but will also simplify the production process by eliminating the protection and de-protection steps. Although low yields have prevented industry adoption of a new method, the production costs of APM would be further reduced by N-protection-free methods if productivity could be improved. Hereafter, we summarize such attempts as appeared in patent applications. Direct production of APM by condensation of aspartic acid and PheOMe was proposed by TOSOH and Ajinomoto, respectively, where both starting materials are condensed by microbial cells belonging to *Pseudomonas, Alkaligenes,* and so forth. In this process, the reaction seems to be catalyzed by aminopeptidases in the bacterial cells. Although this is a very simple production method, its yield was very low.

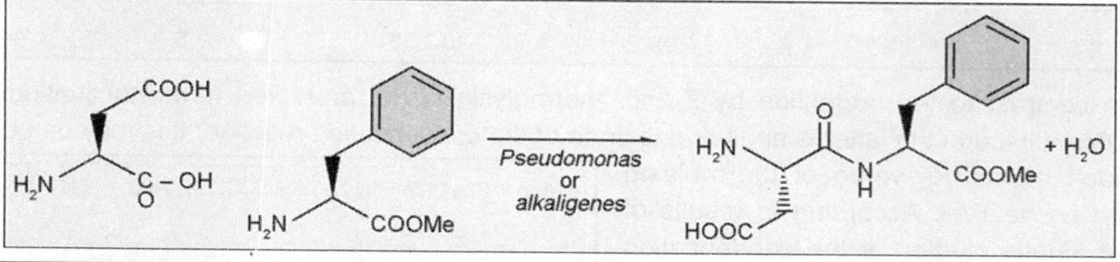

Figure 17.6 Showing the role of *Pseudomonas.*

According to the TOSOH patent application, only 0.06 g of APM was produced from 1.0 g of aspartic acid after 16 h of reaction. Because APM is not removed from the reaction mixture as a precipitate under this condition, it seems to be hydrolyzed back by the aminopeptidases. If the product could be separated from the reaction mixture in situ, the production yield will be increased. It is, however, difficult to remove APM from the reaction mixture as a precipitate, because APM is not only highly soluble (similar to the substrates) but also does not form an insoluble addition compound with remaining substrate. No other additive that would form an insoluble addition compound with APM has yet been found. Although separation of APM by electrodialysis was attempted by Snedecor and Hsu, sufficient separation yields do not seem to be obtained. Development of an effective method to separate APM from its starting materials and enzymes will be of importance in establishing a feasible production process based on those patent applications. Another APM production method without N-protection is the coupling of α-AspOR (R indicates a lower alkyl group) and PheOMe by an ester–amino exchange reaction catalyzed by a protease from *Staphylococcus aureus* strain V8 (V8 protease). Because this reaction is a dealcoholation reaction, the reaction rate is higher than that of a dehydration reaction.

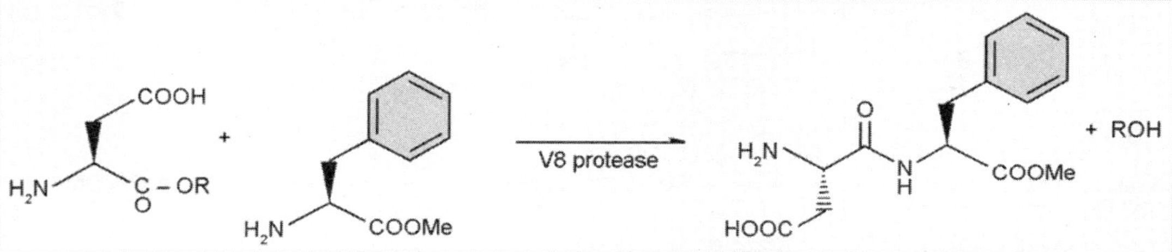

Figure 17.7 Role of V8 protease of *Staphylococcus aureus.*

Additionally, by-products due to misconfiguration, such as PheAspOR or PhePheOMe, are not produced because V8 protease is specific for a peptide bond or ester bond that involves the carbonyl group of amino acids with an acidic side chain, for example, aspartic acid at the P1 position. However, removal of APM from the reaction system encounters the same difficulties as in the first example of the APM direct production; therefore, a higher production yield cannot be expected. It will also be difficult to produce α-AspOR industrially without formation of β-AspOR and $Asp(OR)_2$, because the reactivities of the two carboxyl group of aspartic acid are not so different. Thus, various barriers exist to establishing a production process based on this method.

Other Methods of Production

The methods described in the previous section are aimed at coupling of aspartic acid and PheOMe without protection; the methods described in this section do not depend on aspartic acid as the starting material. Because aspartic acid can be produced by ammonia addition to fumaric acid by an ammonia lyase, APM synthesis from fumaric acid, PheOMe, and ammonia as starting materials was proposed by TOSOH. *Pseudomonasputida* TS-15001 was found to be catalyzing the coupling of the three substrates simultaneously. Although all details of this reaction system have not been studied, this system seems to consist of two reactions in the cells. At first, aspartic acid is produced from fumarate by the ammonia addition reaction by an ammonia lyase; second, aspartic acid is coupled with PheOMe by an aminopeptidase.

The low production yield seems to result from the hydrolysis of APM by the aminopeptidase in the same way as in the first example of the direct production. Production of APM from similar substrates was proposed by Showa Denko Corporation, where APM was derived from fumaric-PheOMe (FPM) by ammonia addition reaction by certain bacterial cells that produced an ammonia lyase. In this proposal, FPM was derived from its *cis-trans* isomer, maleic-PheOMe (MPM) which is more easily produced by the coupling reaction of maleic anhydride and PheOMe than by the direct production of FPM by the coupling reaction of fumaric acid and PheOMe. When fumaric acid and PheOMe are coupled directly, use of a coupling reagent is required, such as oxalyl chloride, thionyl chloride, or carbodiimide. However, the coupling of maleic anhydride and PheOMe proceeds spontaneously.

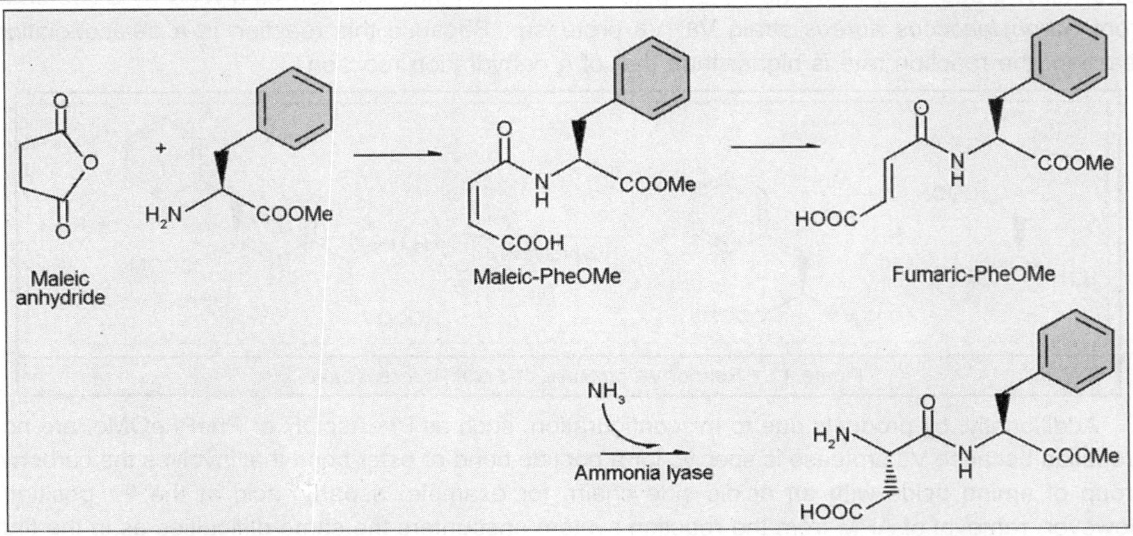

Maleic anhydride + H_2N–PheOMe (COOMe) → Maleic-PheOMe → Fumaric-PheOMe; NH_3, Ammonia lyase →

Figure 17.8 Showing the role of ammonia lyase.

Fumaric-PheOMe was reportedly produced in good yield by isomerization of MPM. Unfortunately, the yield at the step of ammonia addition is too low to be applicable in an industrial APM production process. A new microbial strain exhibiting higher lyase activity has recently been reported by Xu et al., but even with this strain, productivity seems to be insufficient. Further screening of bacteria and improvement of ammonia lyase thus appears to be important. Because APM is an excellent high-intensity sweetener, especially in its taste, demand for it will grow in the future, and APM will gradually become a commodity chemical. The drive for cost reductions will therefore increase. Although alternative production methods have been proposed, their yields are lower than the current, conventional production methods. The current processes will therefore continue to be used while attempts to improve them are ongoing.

18

CELLULAR DEATH

In the past it has generally been assumed that certain cell-culture conditions, such as interactions with specific hormones and deprivation of basic nutrients and growth factors, result in physical damage or metabolic collapse of the cell, leading to a passive, or "necrotic," death. It was thought that the cell was at the complete mercy of its environment and had no control over its fate. However, it is now clear that the cell does not always respond in such a simple and passive manner. Instead, a range of factors can trigger a highly complex and genetically regulated cellular response during which specific "death proteins" are activated, a phenomenon that has been named *apoptosis*. It is these proteins that are responsible, ultimately, for the death and destruction of the cell. Indeed, in many cases, the cell may have sustained only low levels of damage. Clearly, under these conditions apoptotic death will be premature and is often referred to as *cellular suicide*. *Apoptosis* is now acknowledged as a fundamentally important process that plays an essential role in embryogenesis and in the maintenance and functionality of the highly ordered cell populations that constitute higher organisms.

Indeed, there are now few aspects of biomedical research upon which apoptosis has not had an impact. It is as a consequence of its fundamental nature that any failure in its regulatory mechanisms leads to many of the diseases that pose the greatest challenges to medical science. This has resulted in an explosion of research into the genetic basis of apoptosis, the objective of which is to develop novel therapeutics for a wide range of disorders, including cancer, AIDS, and ischemic injury. Consequently a growing number of proteins have been identified that are involved in the induction suppression and execution of the apoptotic program. It is now clear that many of the cell lines used in the biotechnology industry for the production of therapeutics undergo apoptotic death. Obviously, the advances made in the characterization of the apoptotic pathway may be applied to the suppression of the apoptotic response in industrial culture processes. This should provide the biotechnologist with a new route to the optimization of culture performance. We begin by describing the biochemical basis of apoptosis, the morphological changes that accompany it, and some of the techniques that have been used to identify apoptotic cells. This is followed by a general discussion of the regulation and induction of apoptosis. We then describe studies that have investigated this phenomenon from a biotechnological perspective. Particular reference is made to the conditions that elicit an apoptotic response, the cell lines that are susceptible, and the effect of *bcl-2* over-expression on cell survival and productivity in the bioreactor environment.

MORPHOLOGICAL CHANGES

Apoptosis is defined by its highly characteristic morphology. Early changes include a reduction in cell volume and loss of surface microvilli. Cytoskeletal changes result in the formation of protrusions on the surface of the cell, which are referred to as blebbs. These may break away as intact vesicular structures, giving rise to “apoptotic bodies”. One of the most striking changes during apoptosis occurs within the nucleus. Fluorescence microscopy of a typical viable cell following staining with a DNA stain such as acridine orange, reveals a large spherical nucleus that may constitute most of the cellular volume. Often, the chromatin is highly diffuse, although occasionally it may be possible to see condensed chromosomes in mitotic cells.

During apoptosis, the nucleus undergoes major changes. Initially, the chromatin condenses and marginates to the nuclear membrane, forming crescent or ring-shaped structures that exhibit intense fluorescence. Eventually, the chromatin collapses into two or more particles, which are often highly spherical. Again, the remainder of the cell will be devoid of chromatin and will be almost transparent in appearance. The shape of the cell also undergoes a highly characteristic change. While viable cells tend to be highly irregular in shape, on entry into apoptosis they become smooth-surfaced and in many cases almost spherical. The changes in nuclear morphology coincide with the activation of a nuclease enzyme that cleaves chromatin first into 300 and/or 50-kbp fragments, and then ultimately into multiples of 200 bp.

The first stage is believed to be responsible for the morphological changes described above. The second stage represents the cleavage of chromatin at the internucleosomal linker regions. This generates a striking ladder like pattern when DNA samples from apoptotic cells are subjected to DNA gel electrophoresis. Together with condensation of chromatin, this so-called DNA ladder has become one of the hallmarks of apoptotic death. Many cell types express a further enzymatic activity—that of the transglutaminase enzyme. This cross-links proteins within the cell, generating a protein scaffold that is believed to hold the dead cell together, thus explaining the relative robustness of dead apoptotic cells as compared with cells that have undergone necrotic death. These highly controlled changes appear to have a specific task: to provide a very clean and rapid method of eliminating dead cells, which is a vital consideration when one considers the extent of apoptotic cell death during, for example, embryogenesis.

The cleavage of chromatin does not appear to be responsible, in itself, for the death of the cell. Instead, it has been argued that this ensures the complete destruction of the genetic material. It is suggested that this reduces the possibility of malignant transformation of surrounding cells by DNA from dying cells. The stabilization of the dead cell by the transglutaminase enzyme minimizes the probability of leakage of its contents onto surrounding cells. In vivo, surrounding cells and phagocytes engulf the dead cell before it can cause damage to surrounding tissue. Thus, inflammation of tissue as usually results from necrotic death is avoided. In vitro, in the absence of phagocytes, the apoptotic cell will eventually enter a degenerative phase called secondary necrosis.

Searching Symptoms

The search for simple techniques that allow for the identification of apoptotic cells has attracted a considerable amount of interest. As already stated, visualization of nuclease-mediated cleavage of DNA has been widely used to identify the presence of apoptotic cells. However, the technique can produce variable quality results that are of a qualitative, rather than a quantitative, nature. Perhaps the simplest techniques for the study of apoptosis are based on the identification of the morphological features of cell death. For example, fluorescence microscopic analysis of nuclear morphology is a highly effective method for identification and quantification of apoptosis. There are two ways in which this may be done. If samples cannot be analyzed immediately, cells maybe fixed in formaldehyde and stored at 4°C. Analysis involves staining with acridine orange, which reveals the condensation of chromatin in apoptotic cells. However, this technique has one major drawback—an inexperienced operator may confuse early necrotic cells with viable cells, which have a very similar nuclear morphology.

In order to avoid this difficulty, cells may be analyzed immediately while still in their culture medium by using the acridine orange–propidium iodide dual-staining technique. All cells are permeable to acridine orange, which stains chromatin green. Only membrane-damaged cells take up propidium iodide, and as a result exhibit red fluorescence. This technique therefore provides information regarding plasma membrane integrity, as well as nuclear morphology, and can consequently be used to simplify identification of necrotic cells. Additionally, the classification of apoptotic cells into early apoptotic and membrane-damaged apoptotic (sometimes referred to as secondary necrotic) cells gives an indication of the cell growth and death kinetics under the influence of a variety of environmental conditions during the cultivation process. The early-membrane-intact phase of death is relatively brief and, therefore, the presence of a large proportion of cells at this stage indicates that the rate of cell death is very high. Microscopic techniques have two major drawbacks—they are subjective and time consuming. In theory, the development of flow cytometric methods should overcome these difficulties and provide a powerful tool for the study of the biochemical features of apoptosis in heterogenous cell populations. At present, there are a number of techniques that are available, and some of the most commonly used are discussed further.

Cytometric method

The simplest flow cytometric method is based on the changes in light scattering properties that accompany cell death. Cells can be studied without the need for pretreatment, with the decrease in cell size producing a decrease in forward-scattered light. The increased granularity caused by nuclear condensation produces an increase in orthogonal light scatter. However, the latter is a transient stage, and eventually, a reduction in orthoganol light scatter is observed. The technique can also be used for the identification of necrotic cells (at least in their early stages). When necrosis is induced by a permeablizing agent such as saponin, a reduction in forward scatter is observed, but the increase in side scatter that occurs during apoptosis is not seen.

Molecular changes

When stained with a DNA-specific stain such as propidium iodide (PI), apoptotic cells exhibit a characteristically low DNA content that appears as a sub-G_1 peak (i.e., it appears below the position of the G_1 peak of the cell cycle of viable cells). This is believed to result from the leakage of cleaved DNA from apoptotic cells. The technique provides a rather good correlation with the fluorescence microscopic technique already described. However, necrotic cells undergoing degradation may also exhibit a reduced DNA content, although the passive and asynchronous nature of this process means that a clear "peak" is not usually observed.

Changes in plasma membrane

In viable cells, phosphatidyl serine (PS) is located on the inner leaflet of the plasma membrane. During apoptosis, one of the earliest changes is the loss of this asymmetrical distribution. Annexin V has a very high affinity for PS. Conjugation of annexin V to a fluorescent tag, such as FITC, enables the use of this interaction to identify cells that have lost PS asymmetry. By combining this method with PI staining, it is also possible to classify apoptotic cells into two subpopulations: early (membrane intact and therefore PI negative) and late (membrane damaged and therefore PI-positive).

The technique has been found to give a good correlation with levels of apoptosis during hybridoma batch cultures, during which apoptosis accounts for around 90% of cell deaths, as revealed by the fluorescence microscopic method described earlier. However, as with other flow cytometric techniques, necrotic cells can also give a false-positive result. This is because damage to the plasma membrane allows the annexin to enter the dead cell and bind to PS residues on the inner surface of the plasma membrane. Thus, the technique is reliable only when used to analyze early apoptotic cells, which can be seen as Annexin V positive and PI negative.

Identifying characters

Each of the techniques we described has advantages and drawbacks. When designing experiments to identify and quantify apoptosis, a number of points should be taken into consideration. First, there can be significant variations in the morphology of apoptosis from one cell type to the next, including, for example, absence of chromatin condensation. It is therefore recommended that several methods be used simultaneously to identify the mechanism of cell death. However, quantification of actual levels of apoptosis can vary significantly depending upon the technique used. Thus, in order to draw valid conclusions, it is essential that comparison of quantitative data collected using one technique is not made with data collected using a second technique.

Begining of cellular death

The list of factors that induce apoptosis has grown steadily over the last few years. In the present section factors that have been the center of purely biological studies will be considered, although where necessary, implications for animal cell technology will be highlighted. Factors that are specifically of interest to the process biotechnologist will be explored under "Apoptosis and Animal Cell Biotechnology."

Receptor–ligand interactions

The absence of certain hormones can result in the induction of apoptosis in certain cell types. This has led to the suggestion that apoptosis may play a pivotal role in the maintenance of tissue organisation in vivo. It is thought that all cells are primed to undergo apoptosis and are prevented from doing so through constant stimulation by parakrine survival factors. If a cell is removed from its physiologically correct location, the absence of the appropriate survival signal will lead to the induction of apoptosis. This role of survival factors as regulators of cellular distribution in vivo may also have an impact on the development of serum-free media preparations for industrial-scale cell-culture processes, as described later. The presence of receptor–ligand interactions can also lead to the induction of apoptosis.

For instance, in the Fas– FasL system, binding of the Fas ligand to the Fas receptor can trigger apoptosis. This mechanism is responsible for the regulation of the immune system. Autoreactive B cells undergoing maturation and autoreactive mature T cells are eliminated by the Fas-mediated induction of apoptosis. Fas also acts as the "off" switch for the immune system by inducing apoptosis in antigen-activated B and T cells. Molecular dissection of the Fas–FasL system has provided important insights into the early stages of the signaling cascade that leads to the expression of the death pathway.

Viral infection

A number of viruses have been shown to interact with the cellular apoptotic machinery. An apoptotic response to viral infection would appear to act as a protective mechanism that prevents viral replication by triggering the suicide of the infected cell. However, a number of virally encoded antiapoptotic genes have now been identified that suppress the expression of the death program, thereby providing the virus with the opportunity to propagate itself. Perhaps one of the most interesting examples of this anti-death mechanism from a biotechnology perspective is that of baculovirus, which has been synthesized at production scale by infection of insect cell lines, and has applications as a biological pesticide and, more recently, for the expression of recombinant proteins. A mutant was identified that induced high levels of apoptosis during infection of Sf21 insect cells. This was attributed to a mutation in the p35 viral gene that, in its wild-type form, acts as an antiapoptosis gene. Arguably, the most important example of virus-induced apoptosis is that mediated by HIV.

A number of reports have indicated that the binding of the viral gp 120 protein to the $CD4^+$ receptor of T cells triggers the induction of apoptosis, thus leading to the depletion of this class of cells during HIV infection. Interestingly, HIV infection of $CD4^+$ cells appears to provide protection from apoptosis. The viral Nef protein down-regulates the expression of $CD4^+$, thus preventing the induction of apoptosis. As a result, virus propagation in the infected cell is not prevented. There have also been reports of virus-induced suicide of bacterial cells. Until recently, it was assumed that altruistic cell death was not possible in single-celled organisms, simply because the genes involved in such a phenomenon would be lost when the cell concerned dies. However, bacteriophage infection in bacterial colonies has been reported to induce a suicide response, thus preventing the spread of the infection to the remainder of the colony. The genes that mediate such a response are propagated by other clones in the colony, thus preserving the altruistic nature of cellular suicide. Bacteria and viral expression systems have been used for the production of recombinant proteins. Clearly, the possibility that bacterial cells in such systems exhibit an apoptosis-like response needs to be investigated.

Cellular damage

Free-radical-mediated cellular damage has become an important area of study. In recent years there have been numerous reports that cell death following oxidative stress occur by apoptosis. Moreover, generation of free radicals has been postulated as being a universal triggering event in the induction of apoptosis. Indeed, for a period in the early 1990s, it was suggested that the widely studied antiapoptosis gene *bcl-2* functioned as an antioxidant. However, this theory has been challenged by the demonstration that anoxia-induced apoptosis can also be suppressed by Bcl-2 in the absence of measurable levels of free radicals.

Therapeutic agents

Exposure to high levels of toxic chemicals results in necrotic death of the cell. However, long-term exposure at a low level can trigger an apoptotic response. Many of the agents used in chemotherapy exert their affect by inducing apoptosis in tumor and normal cells. Overexpression of genes such as *bcl-2* has been linked to resistance to chemotherapy. Clearly, establishing the mechanism of induction of apoptosis in response to such agents and providing strategies that minimize the affect of antiapoptotic genes should provide novel and more effective chemotherapeutic strategies. Exposure of cells to ionizing radiation induces high levels of apoptosis in many normal tissues. Particularly susceptible are those cells from tissues that undergo rapid proliferation, such as spermatogonia and lymphocytes.

Such tissue would be especially prone to malignant transformation, and consequently the induction of apoptosis following DNA damage minimizes the likelihood of such an event. Irradiation of tumors can also lead to the induction of apoptosis. Notably, tumors that respond least favorably to irradiation exhibit the lowest level of apoptosis under such conditions. The induction of apoptosis following DNA damage is regulated by the product of the p53 gene, which has been referred to as the gaurdian of the genome because of its central role in preventing the propagation of cells that may have sustained genetic damage and thus, potentially, malignant transformation.

Genetic basis of Cellular Death

The study of the genetic basis of apoptosis has become a highly complex and fast-moving area of research. Consequently, this section will only deal with some of the most important aspects of this subject in order to provide a basis for the later discussion on genetic manipulation of the apoptotic pathway in commercially important cell lines. The genes involved in apoptotic death may be classified into three groups. The first group consists of the modulators of the apoptotic pathway that suppress or induce death. The second group comprises the components of the cell death pathway that mediate the cell death signal. The final component is the group of effector enzymes that is responsible for the death and destruction of the cell. Of the modulators of the apoptotic pathway, one group of closely related proteins, the Bcl-2 family, has attracted particular attention. The *bcl-2* gene, the first and best characterized member, was identified at the t(14;18) (q32↔1)breakpoint found in human follicular lymphoma. It encodes a 24-kDa protein that is located on the outer mitochondrial membrane, the cytosolic face of the nuclear membrane, and endoplasmic reticulum. Numerous studies have demonstrated the ability of this protein to suppress apoptosis in response to a wide variety of inducers. Until recently, studies of the molecular basis of cancer were directed at the identification and characterization of genes involved in the regulation of cellular proliferation. However, it is now evident that the failure of cells to undergo apoptosis at the correct time and location is also an important step in malignant transformation.

Table 18.1 Examples of genes involved in apoptosis

Inducers/Promoters	*Suppressors*	*Mediators*
bax	*bcl-2*	ICE family
bak	*bcl-xL*	Trans-glutaminase
bad	*mcl-l*	NUC-18
bcl-xS	*ced-9*	DNase-I
FasL/Fas	*p35*	DNase-II
	bhfrl	

Indeed, studies of *bcl-2* in this context have been instrumental in establishing the role of apoptosis in cancer. Mutations that result in the overexpression of *bcl-2* lead to the accumulation of cells due to life span extension. These cells then undergo further mutation, most notably involving the *c-myc* gene, which leads to the formation of high-grade tumors that combine the characteristics of high cellular survival with a high rate of proliferation. Clearly, these are characteristics that would be desirable in the ideal candidate host for the expression of recombinant proteins. One of the most common mutations identified in tumors is that of the p53 gene, mentioned earlier. In response to DNA damage, p53 causes cell cycle arrest, allowing the cell to repair damaged DNA, thus ensuring that any potentially carcinogenic mutation is not propagated. However, in some cell types, p53 triggers the induction of apoptosis, again in order to minimize the risk of transformation.

The well-characterized pattern of development of the nematode *Caenorhabtis elegans,* which includes the induction of apoptosis in specific cell types, has provided an important insight into the genetic basis of apoptosis. Most notable among the genes identified are *ced 9* and *ced 3*. The former is a *bcl-2* homologue that blocks the ability of *ced 3* to induce cell death. The *ced 3* gene is a cysteine protease that shares extensive homology with the mammalian protein called interleukin 1β converting enzyme (ICE). As the name suggests, this enzyme cleaves the active precursor pro–interleukin 1β to generate the active molecule interleukin 1β, and a number of studies have now implicated it in the induction of apoptosis. However, apoptosis can also be induced in macrophages and thymocytes of ICE-negative mice, indicating that ICE is not a universal mediator of apoptosis. Indeed, several ICE-related enzymes have been identified in recent years. The molecular mechanism of apoptosis induced by the interaction of the Fas-ligand with its receptor centers on the activity of ICE-related proteases. Indeed, important progress has now been made in deciphering the earliest events in the signal cascade that transduces the initial death stimulus to the death machinery of the cell. Activation of the Fas receptor by its ligand or agonist antibodies leads to the binding of the adapter protein MORT1/FADD (Fas-associating protein with death domain). This, in turn, binds to the ICE homologue FLICE (FADD-like ICE) or MACH (MORT1-associated ced-3 homologue). The targets of this enzyme are still under investigation.

BIOPHARMACEUTICAL ROLE

At present, there are around 100 new biopharmaceuticals in phase I, II, and III clinical trials. Production of a biologically active therapeutic will, in many cases, necessitate expression of the protein in mammalian cell lines. As a result of intense commercial pressure, tried and tested production processes are often adopted in order to ensure that the product reaches the market in the shortest possible time. Initially, such production processes were relatively inefficient and there was tremendous scope for process optimization. Advances in cell biology, in addition to improvements in process monitoring, control, and optimization, and the development of novel bioreactor designs, are making the transition from the research laboratory to industry standard practice. Clearly, this will lead to a reduction in the cost and complexity of the process of recombination-protein production using mammalian cell lines by placing the technology on a firmer scientific footing. In order to optimize the viable cell number and protein productivity, a detailed understanding of the factors that lead to cell death in the bioreactor is required. These factors may be classified into three groups:

1. The hydrodynamic environment of the cell
2. The accumulation of toxic metabolites
3. The exhaustion or local limitation of nutrients and oxygen

Although each of these areas has now been investigated to varying degrees, recent studies have demonstrated that at least some of the cell lines used in bioreactors undergo apoptotic death rather than necrosis, indicating that a reassessment of the subject is required. As described later, the greater understanding of the mechanism of cell death in commercial cultures should provide new routes to culture optimization. The resultant enhancement in culture efficiency would be expected to manifest itself in three forms:

1. First, the nutrients and culture time invested in generating a viable cell will be wasted if that cell should die prematurely. If the survival time of the cell can be enhanced, the proportion of culture resources utilized for production of the biopharmaceutical of interest can be increased by eliminating the need for regeneration of cellular biomass.
2. Second, as a cell dies, it releases proteolytic enzymes into the culture medium, which can lead to degradation of the product. Thus, product stability should be enhanced by minimization of cell death.
3. Finally, high levels of cellular debris in the culture medium can complicate the recovery of the target protein. This will add to the cost of downstream processing and lead to a reduction in the efficiency of target protein recovery.

Work Done

The first suggestion that apoptosis may account for cell death during the cultivation of hybridoma cells came from an electron microscopic study conducted by Al-Rubeai et al. Further studies by Franek and Dolnikova demonstrated the accumulation of nucleosomal DNA fragments in culture medium during the death phase of batch hybridoma cultures. Based on this observation, they estimated that around 30% of cells had undergone apoptotic death. Further evidence of apoptosis during hybridoma cultures was provided by the studies of Mercille and Massie and Singh et al. Upon DNA gel electrophoresis, both studies revealed the laddering pattern that, as mentioned earlier, is characteristic of apoptosis. Morphological analysis of the nuclei of the cells indicated that apoptosis accounted for 90% of the dead cells. Both groups also found high levels of apoptosis during the cultivation of murine plasmacytoma cell lines (sometimes incorrectly referred to as myeloma cells).

Furthermore, Singh et al. reported an absence of apoptosis during the death phase of CHO and Sf-9 batch cultures. However, studies by Moore et al. indicated significant levels of apoptosis during the death phase of serum-free batch cultures of CHO cells. Recent studies in our laboratory on a CHO 320 cell line overexpressing interferon indicate that this cell line may also be susceptible to apoptosis, although the morphology was not typical, and the frequency was much lower than that seen during hybridoma cultures under comparable conditions. Studies are now required to give an indication of variability in susceptibility to apoptosis between clones of the same cell type, and between different unrelated cell lines. The objective will be to produce a correlation between susceptibility to apoptosis and general cell robustness. Clearly, such a correlation would provide a simple and easily identifiable predictor of robustness following exposure of cells to a range of stresses, thus simplifying the process of cell-line selection.

Regulation

The features of the bioreactor environment that may result in cell death were outlined in the introduction. In the present section, studies that have considered these factors in terms of their ability to induce apoptosis are described and the implications of suppression of apoptosis are discussed.

Limited supply of nutrients

The nutrient limitations encountered by cells in the bioreactor may be classed into two groups: cycling or terminal limitation. The former may be encountered at all stages of large-scale or intensive culture systems. For example, in large-scale stirred tank reactors, the cells may be exposed to fluctuating nutrient levels because of inhomogeneity due to poor mixing. In intensive culture systems, the high cell densities reached will result in low local-nutrient concentrations. As a result, the level of cell death in such systems is relatively high. Terminal nutrient limitations will occur at the end of batch cultivation and, as a result, will become more extreme with time, invariably leading to cell death. In the case of hybridoma batch cultures, the first nutrient to become limiting is glutamine, and its exhaustion coincides with the onset of the death phase of the culture.

As stated above, cell death under these conditions is almost exclusively by apoptosis. Two studies have reported on the effect of *bcl-2* overexpression on cell survival during the death phase of hybridoma cultures. Itoh et al. found that bcl-2 over-expression significantly extended the duration of the culture by reducing the rate of cell death. Moreover, they reported a fourfold increase in the Mab (monoclonal antibody) titre in the culture medium. Simpson et al. have also reported an extension in culture duration, although there was only a 40% improvement in Mab titre. Necrosis became the predominant mechanism of cell death in the *bcl-2*-transfected cell line, indicating near-complete suppression of apoptosis under batch culture conditions.

More recently, Suzuki et al. have reported that transfection of COS-1 cells with *bcl-2* and then with the vector pcDNA-λ carrying the immunoglobulin λ gene for transient expression of λ protein have resulted in higher expression of the protein when compared to the control transfectant (i.e., *bcl-2* negative). In the same study, the mouse plasmacytoma p3-X63-Ag.8.653, which is used as a fusion partner in the generation of hybridomas, and the hybridoma cell line 2E3 were transfected with the human *bcl-2* gene. In both cases an extension of batch culture duration was reported. The *bcl-2*-transfected 2E3 cells survived 2 to 4 days longer in culture, producing a 1.5- to 4-fold larger amount of antibody in comparison with the control vector transfectants. A further enhancement in survival and antibody production in hybridoma 2E3 cultures was observed when cells were co-transfected with *bcl-2* and *bag-1*. In contrast with these promising studies, Murray et al. found that *bcl-2* transfection of the murine plasmacytoma NS0 failed to provide any protection from apoptosis. Although there was no endogenous Bcl-2 expression, they did report expression of Bcl-xL, a functional homologue of bcl-2. Thus, they suggested that Bcl-2 may be functionally redundant in this cell line.

Deficiency of amino acids and glucose

The link between the onset of apoptosis and exhaustion of glutamine during batch cultures of hybridoma cells has prompted systematic studies of the role of the various nutrients used in culture medium. Initial studies indicated that deprivation of glucose, serum, glutamine, cysteine, and methionine could all individually induce high levels of apoptosis. Moreover, recent studies in our laboratory indicate that this is not a feature of these particular nutrients alone. Deprivation of each amino acid individually from the commonly used RPMI 1640 culture medium was found to result in the induction of apoptosis, with particularly high levels observed following deprivation of essential amino acids. How might the deprivation of nutrients trigger an apoptotic response? Perreault and Lemieux have found that hybridoma cells undergo apoptosis when their protein biosynthetic machinery is compromised. It may be that deprivation of nutrients such as amino acids has the same effect, possibly resulting in the failure of the synthesis of a critical regulatory protein required to keep the apoptotic pathway in check.

Vaux and Strasser suggest that some agents or treatments lead to a reduction in the cellular ATP pool, and that this may be a trigger of apoptosis. They propose that such changes may be interpreted by the cell as being a consequence of viral infection, and the cell responds by inducing the apoptotic pathway. Overexpression of Bcl-2 was found to offer a high degree of protection following deprivation of each individual amino acid, with two exceptions, glutamine and threonine, which exhibited relatively less protection. This may indicate that these two amino acids either play a particularly important role in cellular metabolism and biosynthesis, or that they are essential components of the mechanism by which Bcl-2 protects the cell. The survival of *bcl-2*-transfected cells, even in the absence of supposedly essential amino acids, suggests a reduction in amino acid utilization due to down-regulation of nonessential cellular functions. Indeed, metabolic arrest has been reported following interleukin 3 (IL-3) withdrawal from the IL-3-dependent murine cell line Bo, which consequently underwent apoptosis. This state was stabilized by *bcl-2* transfection of the cells, thus extending survival time by 300%.

Presumably, Bcl-2 also stabilizes the metabolic arrest caused by amino acid starvation in murine hybridoma cells, possibly by maintaining the ATP pool above a threshold level. Moreover, studies by Simpson et al. would suggest that this state may be reversible by feeding the cells with fresh medium. Clearly, such a characteristic is far more desirable than a rapid, and obviously nonreversible, entry into apoptosis that occurs in apoptosis-susceptible cell lines.

Role of serum

The role of serum in the suppression of apoptosis has been well documented, and consequently, it was of no surprise when it was reported that commercially important hybridoma and plasmacytoma cell lines also undergo apoptosis on withdrawal of serum. This would be expected to have important consequences for the development of new serum-free media formulations. Previously, the rational behind the design of such media was not particularly scientific and often involved the inclusion of chemicals that were identified by empirical studies. However, demonstration of the role of serum in the regulation of apoptosis may provide a new avenue of research for the development of novel, and perhaps cheaper, serum-free media. The Bcl-2-mediated suppression of apoptosis following serum withdrawal was the earliest demonstration of the antiapoptosis activity of this gene. A Burkitt's lymphoma cell line transfected with *bcl-2* has been reported to grow better than control vector–transfected cells in commercially available serum-free media without the need for adaptation. Similar results were obtained using a *bcl-2* murine hybridoma cell line.

Aerobic Effect

In large-scale and intensive culture systems, effective aeration of the culture is a major difficulty. Thus, oxygen limitation is often the major limiting factor determining maximum cell number. Studies by Mercille and Massie demonstrated that deprivation of oxygen can induce apoptosis. Subsequently, it has been shown that Bcl-2 overexpression protects hybridoma and Burkitt's lymphoma cells from apoptosis. Clearly this will provide the cells with a considerable advantage in oxygen-limited intensive culture systems.

Hydrodynamic Environment

During the very early days of mammalian cell culture, it was generally assumed that cell lines would be far more sensitive to shear damage due to the absence of a cell wall. Considerable progress has been made in our understanding of the exact interactions that result in the greatest damage to the cells. It is now clear that it is not the shear forces generated at the impeller tip that are responsible for cell death induced by the hydrodynamic environment of the reactor. Far more damage is caused by the events that take place at the gas head space–liquid interface during bubble disengagement. Despite the importance of the hydrodynamic environment, the biochemical response of the cell to this component of the bioreactor environment has been rather neglected.

To address this issue Al-Rubeai et al. investigated the mechanism of cell death following exposure of cells to very high agitation levels.

Flow cytometric and morphological analysis indicated that cell death occurred mostly by apoptosis, although levels of necrosis were also significant. Singh et al. found that a Burkitt's lymphoma cell line that had been routinely passaged in stationary cultures underwent apoptosis when attempts were made to grow the cells in suspension. *Bcl-2* transfection of this cell line was found to allow much better cell growth in suspension without the need for adaptation. Simpson et al. reported similar behavior of a hybridoma cell line. Physiological studies have also investigated the relationship between apoptosis and shear stress. Dimmeler et al. have found that shear stress actually prevents induction of apoptosis in endothelial cells in the presence of the inducer tumor necrosis factor *a* or following growth factor withdrawal. Clearly, the possibility that a sublethal level of shear stress in the bioreactor protects the cells from apoptosis-inducing agents needs to be investigated.

Growth Rate and Productivity

Studies of p53 and *c-myc* have revealed a close relationship between cell proliferation and apoptosis. These studies may have important implications for process optimization strategies that have centered on the control of cellular proliferation. Such an approach allows for improvements in specific recombinant protein productivity during cultures that exhibit a negative correlation between growth rate and productivity. Furthermore, by controlling maximum cell number at an optimal level, cell death due to limitation of nutrients and oxygen should be minimized, thus simplifying medium clarification during downstream processing. However, such studies have had one major drawback: when attempts were made to control cellular proliferation, the cultures rapidly lost viability, and it would appear that, at least in the case of hybridoma cultures, this was due to the induction of apoptosis. Thus strategies designed to control cellular proliferation must also incorporate methods that minimize apoptosis. Initial results would appear to suggest that such an approach does work. Simpson et al. found that Bcl-2 overexpression delays hybridoma cell death following cell cycle arrest induced by thymidine treatment. Similar results have been reported during Burkitt's lymphoma cultures.

CONCLUDING REMARKS

The study of apoptosis from an animal-cell-technology standpoint is still in its infancy. However, even at this early stage, significant progress has been made, and it would appear that suppression of apoptosis can lead to improvements in culture viability and, indeed, productivity. It is now necessary to study the prevalence of apoptosis in other culture systems, and then to examine the effect of the suppression of bcl-2. overexpression on cell survival. Studies of a variety of high-density perfusion systems are currently underway. It is also necessary to establish the incidence of apoptosis in other industrially important mammalian cell lines, and to investigate the level of protection afforded by antiapoptosis genes. If the results presented here are found to be reproducible in many different industrial culture systems and cell lines, the study of apoptosis will provide the process biotechnologist with a further strategy in the quest for the optimal production process.

19

INTERACTION BETWEEN SUBSTRATES AND CELLS

The development and maintenance of tissues in vivo is acutely dependent on the ability of cells to form connections with substrates and neighboring cells. The formation of intimate contacts with various substrates provides sites of attachment and the scaffolding necessary for cells to communicate and develop cooperative function. These substrates, or attachment factors, include extracellular matrix (ECM) and basement membrane molecules such as collagen, laminin, fibronectin, and vitronectin. Additional substrates include proteins expressed on the surface of cells that function as attachment factors and mediate cell–cell interactions. Cell culture provides a powerful tool with which to identify attachment factors involved in cellular adhesion and to define the molecular mechanisms regulating cell adhesion. Additionally, in vitro approaches play a key role in the identification and cloning of specific attachment factor receptors. The cloning and characterization of attachment factor receptors has forced us to redefine the role of attachment factors in cellular functions. Growing in vitro evidence, combined with in vivo studies, make it clear that attachment factors serve not only as molecular scaffolds to support cell attachment and movement, but also as signaling molecules that play pivotal roles in regulating numerous cellular functions including growth, activation, differentiation, and death. We review here the ECM molecules that are widely used as attachment factors in culture and illustrate ways in which attachment factors regulate cellular function. Additionally, cell surface molecules that serve as attachment factors and facilitate cell-cell interactions in vivo are also reviewed.

MOLECULES OF EXTRACELLULAR ENVIRONMENT

Glycoproteins

Specific glycoproteins also called *fibronectins* consisting of two disulfide-linked subunits with a molecular weight of 220,000–250,000 kDa. Fibronectin was originally purified to homogeneity in 1970 by Mosesson and Umfleet from plasma cryoprecipitate and termed cold- insoluble globulin

(Clg). Later, fibronectin was detected as a cell surface protein on fibroblasts whose expression was lost upon transformation. In addition to existing as a disulfide-linked dimer, cell surface–associated fibronectin also exists as higher molecular weight disulfide- linked aggregates that are secreted by cells and accumulate as a major component of the basement membrane and ECM in vivo. The primary structure of fibronectin consists of a series of three different types of modular units that combine to generate the functional domains of the protein. The two subunits of fibronectin are connected by a pair of inter- chain disulfide linkages at the carboxy-terminal end. The intact dimer appears as an extended, flexible molecule consisting of two linear arrays of globular domains. The identification and characterization of the individual globular domains has resulted largely from examination of proteolytic fragments of fibronectin, many of which retain biological function.

The N-terminal domain of fibronectin binds fibrin, heparin, *Staphylococcus aureus*, thrombospondin, factor XIIIa transglutaminase, and IgG. Adjacent to the amino-terminal domain of fibronectin is a domain that binds gelatin and collagen. This interaction is important for localization of fibronectin to basement membranes and ECMs. The collagen-binding domain also binds to the C1q component of complement via the collagenous tail of C1q. Adjacent to the gelatin-binding domain of fibronectin is the cell-binding domain. Pierschbacher et al. showed that the cell-binding activity of fibronectin resided within a 12-kDa proteolytic fragment that contains the tetrapeptide sequence Arg-Gly-Asp-Ser (RGDS). Peptides containing soluble synthetic Arg-Gly-Asp (RGD) were shown to detach cultured cells and platelets from fibronectin substrates, implicating the RGDS sequence in cell binding. In addition to binding collagen, fibronectin has an affinity for several other components of the ECM, including the glycosaminoglycans heparin, heparan sulfate and chondroitin sulfate proteoglycan, and hyaluronic acid.

The interaction of fibronectin with these various components of the ECM is thought to play a role in the formation and maintenance of the ECM. The high-affinity cell surface receptors for fibronectin are the integrins, which consist of heterodimeric glycoprotein α- and β-subunits of approximately 140 and 120 kDa that bind to the RGD sequence of fibronectin. The extracellular domains mediate cell adhesion via binding to fibronectin and other ECM proteins. The cytoplasmic domains are responsible for transmitting signals to the cell interior following ligand engagement (outside-in signaling), and for modulating the affinity of the integrin for its ligand by interacting with intracellular signaling molecules (inside-out signaling). The cytoplasmic domains also interact with the actin-binding proteins filamin and α-actinin, thereby connecting the cell membrane with the cytoskeleton. These interactions play a role in the formation of focal adhesions and in phagocytosis. Fibronectin plays a central role in vertebrate development. Fibronectin matrices are utilized by migrating cells during embryogenesis. The migration of avian, amphibian, and mammalian neural crest cells is blocked by RGD-containing peptides and by antibodies to fibronectin or its cell surface receptor. Furthermore, inactivation of the fibronectin gene in mice leads to early embryonic lethality. Its role in development and in maintenance of the adult organism is due to its diverse influences on cell growth, adhesion, migration, and differentiation.

Cellular attachment

Fibronectin promotes the attachment of multiple cell types, including epithelial cells, follicular cells, ovary cells, neuronal cells, fibroblasts, keratinocytes, and adult liver cells. As previously mentioned, fibronectin is not solely an attachment factor for cells. In many cases, cell attachment to fibronectin is required for cell survival and promotion of cell growth. Fibronectin promotes cell division of numerous cell types, including follicular cells, primary chick fibroblasts, murine and human fibroblasts, neuroblastoma cells, keratinocytes, rat granulosa cells, and lymphocytes. Fibronectin also promotes DNA synthesis by serum-deprived quiescent normal hamster fibroblasts. In addition to regulating cell division, fibronectin has been shown to modulate the differentiation of cells in culture. Fibronectin coating of culture surfaces inhibits morphological changes and biosynthesis of lipogenic enzymes associated with the differentiation of preadipocyte fibroblasts.

Fibronectin also attenuates hormone-induced expression of cytoskeletal proteins associated with granulosa cell differentiation. Additionally, exogenously added fibronectin inhibits both spontaneous and chemically induced differentiation of cultured normal human keratinocytes in vitro. Alternatively, fibronectin induces the differentiation of neural crest cells, neuroepithelial embryonal carcinoma cells (F9), and neuroblastoma cells in vitro, consistent with observations showing that receptors for fibronectin are found on neural crest cell derivitives colocalized with fibronectin in vivo. Fibronectin also promotes the differentiation of primary human hepatocytes, as measured by the expression of albumin and the liver-specific enzyme cytochrome P450. It has been suggested that modulation of cell growth and differentiation by fibronectin is due in part to its role in regulating the cytoskeletal architecture of cells. The recent cloning of fibronectin receptors and identification of signaling pathways coupled to these receptors has provided evidence that specific signaling events play an important role in controlling cell growth and differentiation.

Collagen Fibers

Collagen is the most abundant component of the ECM and has been used extensively as an attachment factor for cultured cells. Collagen is composed of three helical α-chains that are wrapped around each other in a triple helical arrangement. In some forms of collagen the three chains are identical, whereas in other forms the chains differ from one another. The α-chains are composed of a series of triplet Gly-X-Y residues, where X and Y may be any amino acid. Frequently, X is a proline and Y is a hydroxyproline. Alternatively, hydroxylysine may also occupy the X or Y position and contribute to interchain cross-linking of the collagen α-chains. Collagen is secreted from cells as procollagen, with each of the α-chains containing large amino- and carboxy-terminal non-helical domains. Although some collagens exist in vivo as intact procollagen, many forms of collagen undergo proteolytic removal of the amino- and carboxy-terminal nonhelical domains.

To date, more than 14 collagen types have been described, which are generally subgrouped into six classes defined by the structural forms in which they assemble within the ECM. Several forms of collagen display ordered fibrous structures. For example types I, II, III, V, and XI collagen form fibrillar structures that contribute to connective tissues such as cartilage, skin, bone, and tendon, whereas type IX and XII collagens are fibril associated and interact with type I and II collagen fibrils. Type IV collagen is a nonfibrillar collagen and is the principle collagen type found in basement membrane. Type IV collagen differs from the fibrillar collagens by its existence in tissues in an unprocessed procollagen form. Other collagen types include filamentous collagen (type VI), short-chain collagens (types VIII and X), and long-chain collagen (type VII). Collagen matrices are associated with many cell attachment proteins in vivo, including fibronectin and laminin. Thus the role of collagen as an attachment factor may be divided into two types: (i) a direct mechanism whereby collagen serves as the substrate for cell attachment (VLA-2; collagen receptor); and (ii) an indirect mechanism whereby a secondary attachment factor, through its interaction with collagen, serves as the substrate for cell attachment.

Collagens as attachment factors

The principle collagens used as attachment factors in cell culture are the fibrillar collagens and the nonfibrillar type IV collagen. Fibrillar collagen isolated from connective tissues was the first collagen used in cell culture. A mouse tumor source of type IV collagen was subsequently identified, which produced large amounts of this collagen type. Collagen substrates serve as an attachment and growth-promoting factor for a variety of cells in culture. Collagen promotes the sustained growth of epithelial cells from various tissues including mammary epithelial cells, endometrial epithelial cells, vaginal epithelial cells, esophageal epithelial cells, and liver epithelial cells. Collagen substrates also promote the extended viability of primary chondrocytes.

Chondrocytes exhibit three-dimensional growth within collagen gels, deposit ECM, and incorporate or rearrange exogenous collagen into newly formed matrix, suggesting the formation of tissue-like structures. In addition to providing attachment and growth- promoting activity, collagen also modulates the differentiation state of cells in culture. Collagen induces morphological and biochemical characteristics associated with the differentiated state of various epithelial cell types including human endometrial cells, trachial epithelial cells, uterine epithelial cells, mammary epithelial cells, and intestinal epithelial cells. Collagen also promotes the differentiation of adrenocortical cells, osteoblasts, keratinocytes, hepatocytes, liver dendritic cell progenitors, thyroid follicle cells, smooth muscle cells, and Sertoli cells. Regulation of growth and differentiation, which is accompanied by alterations in gene expression, implicate collagen as an active signal-transducing molecule. Candidate receptors for collagen that may play a role in cell signaling include the integrins VLA-1, VLA-2, and VLA-3.

Role of Laminin in Organization

Laminin is the first extracellular protein expressed during development and plays a critical role in cellular development and tissue organization through modulation of cellular attachment, motility, growth, and differentiation. Laminin is a large glycoprotein that is a major component of basement membranes. The prototypical laminin molecule is composed of three subunits: a 440-kDa A-chain, a 200-kDa B1-chain, and a 220-kDa B2-chain. These are joined together by disulfide linkages to form a cruciform-shaped structure possessing multiple globular and rodlike domains that are responsible for its cell-binding activity and its interaction with a variety of ECM-associated molecules. The terminal end of the long arm, which contains a large, multilobed globular domain and forms the base of the cruciform, is responsible for receptor-mediated cell attachment, promotion of neurite outgrowth, and heparin binding. Regions within the short B-chain arms provide a second cell attachment site and also play a role in cell signaling and binding of the glycoprotein nidogen/entactin. Similar to fibronectin, laminin also binds to bacteria and glycosaminoglycans.

Laminin promotes cell attachment through interaction with specific cell surface receptors. The most extensively studied laminin receptors are the integrins. Laminin binds to at least four VLA integrins: $\alpha1\beta1$, $\alpha2\beta1$, $\alpha3\beta1$, and $\alpha6\beta1$. Laminin also binds to the $\alpha V\beta3$ integrin and $\alpha6\beta4$ integrin. The interaction of laminin with integrins is believed to occur through a non-RGD-mediated mechanism, although RGD-containing peptides can inhibit cell adhesion to the second cell adhesion site located within the central portion of the laminin cruciform. Other recently identified receptors for laminin include the 67-kDa laminin receptor, which has been implicated in tumor metastasis, and a 110-kDa laminin-binding protein from brain, which appears to be immunologically and biologically related to the β-amyloid precursor protein (APP) family. The interaction of the 110-kDa laminin-binding protein with the Ile-Lys-Val-Ala-Val (IKVAV) residues of the laminin A-chain plays a role in neurite outgrowth. A distinct 110-kDa protein, termed nucleolin, was co-purified with the APP-related 110-kDa protein and has also been identified as a laminin-binding protein that may play a role in ECM signaling. Laminin also binds to the actin-binding cell surface protein connectin and the 67-kDa elastin/laminin receptor.

Laminin in attachment

Laminin plays an important role in the attachment, growth, and differentiation of many cell types in culture, including neural and epithelial cells. When added to the culture medium, laminin promotes Schwann cell attachment, growth, and the formation of a stellate, process-bearing morphology and regulates glycolipid synthesis. Laminin promotes attachment and neurite outgrowth of sensory neurons and induces the differentiation of neuroblastoma cells. Antibodies directed against the heparin-binding domain of laminin reduce neuronal viability and inhibit neurite outgrowth in vitro, indicating that the heparin-binding domain is responsible for the effects of laminin on neurite outgrowth and neuronal survival. Laminin also modulates the proliferation of cultured astrocytes and oligodendrocytes.

Laminin promotes the differentiation of cultured endothelial cells to form capillary-like structures. Laminin also induces the differentiation of various epithelial cells, as measured by β-casein production by cultured mammary epithelial cells, the formation of cord-like structures by Sertoli cells, and the expression of alkaline phosphatase and lactase activity by intestinal epithelial cells. Laminin promotes the attachment and survival of F9 embryonal carcinoma cells, while it potentiates the differentiation of PCC4uva embryonal carcinoma cells to neurons following treatment with retinoic acid and dibutyryl cyclic AMP. Laminin also promotes the differentiation of granulosa cells and hepatocytes.

Serum Spreading Factor

Vitronectin, or *serum-spreading factor*, is a glycoprotein that is present in the blood and in other tissues. Vitronectin was initially described as a cell-spreading and growth-promoting $\alpha1$-glycoprotein enriched in a fraction of human serum isolated by glass bead chromatography. The spreading–promoting activity of this fraction was further purified from this fraction by Barnes et al. and was shown to be biochemically and immunologically distinct from fibronectin and laminin. Vitronectin exists in two distinct forms in human serum: a 75-kDa single-chain form and a 65-kDa two-chain form. The 65-kDa form is the product of proteolytic processing and lacks the carboxy-terminal 10-kDa fragment. Sensitivity to proteolytic processing may result from allelic differences in the vitronectin gene.

The gene encoding vitronectin was identified by expression cloning, and sequence analysis revealed a 1,545-bp open reading frame corresponding to the full-length plasma vitronectin. Sequence analysis of an independently isolated serum protein associated with complement, termed S-protein, revealed it to be identical to vitronectin. The major source of circulating vitronectin is the liver, although it is also produced by various hematopoetic cell types including platelets, monocytes, and macrophages. Vitronectin is also associated with tissues. In situ hybridization analysis indicates that vitronectin mRNA is expressed early in mammalian development primarily in the liver and central nervous system, suggesting that it plays an important role in mammalian development. In the mouse, vitronectin mRNA is detected by day 10 in the liver and central nervous system.

Expression of vitronectin mRNA in the central nervous system is first detected in the floor plate; later, vitronectin mRNA can be observed in the meninges of the cortex and the spinal cord. Vitronectin mRNA is associated with the vasculature of the central nervous system but not with blood vessels of peripheral tissues, suggesting that vitronectin may play a specific role in vascular function in the central nervous system. Surprisingly, mice deficient in vitronectin gene expression demonstrate normal development and survival, indicating that the functional role of vitronectin in development may be compensated for by other molecules. The vitronectin gene is composed of eight exons and seven introns, which define the domain structure of the mature protein. The amino-terminal cysteine-rich 44-residue domain is identical to somatomedin B. The second domain contains an RGD sequence that, analogous to fibronectin, is responsible for cell attachment via a number of integrins and also contributes to collagen binding. The second domain is also implicated in binding to plasminogen-activator inhibitor-1, and consistent with its localization to platelets, this interaction plays an important role in blood clotting.

Vitronectin also binds to heparin, and a domain located toward the carboxy-terminal region of vitronectin is responsible for the heparin-binding activity of vitronectin, which is dependent on the conformational state of vitronectin. Although native plasma vitronectin binds heparin weakly, vitronectin denatured by 8 M urea or other denaturing agents has an increased affinity for heparin. Denaturation is hypothesized to expose a cryptic heparin- binding domain and suggests that the biological functions of vitronectin are dependent on its conformation. Recently a form of vitronectin distinct from plasma vitronectin has been identified within platelet α-granules, which is able to bind heparin and may play a role in platelet function at sites of vascular injury. Additionally, similar to collagen and fibronectin, vitronectin binds to bacteria, including *S. aureus, Escherichia coli, S. pneumoniae,* and *Neisseria gonorrhoeae,* implicating vitronectin in the adherence and phagocytosis of bacteria by host immune defense cells.

Vitronectin in attachment

Based on the tissue distribution of vitronectin, it is not surprising that vitronectin promotes the attachment and growth of a wide variety of cell types in vitro, including endothelial cells, lymphoid cells, and neural cells. Human endothelial cells attach to vitronectin, adopt a flattened morphology, and exhibit increased viability. Vitronectin promotes the attachment and proliferation of IL-3-dependent mast cells. However, vitronectin is unable to replace IL-3 as a survival factor, indicating that vitronectin is not itself a survival factor for bone marrow-derived mast cells but is able to augment the IL-3-dependent signal in mast cells. Vitronectin promotes the attachment of myeloblast cells through the αv-containing integrins and stimulates megakaryocyte proplatelet formation through interaction with αv/β3 integrin. Vitronectin also promotes the attachment and long-term survival of cultured glioma cells, and this interaction may be mediated by gangliosides. In addition, vitronectin mediates attachment and nerve growth factor-dependent neurite outgrowth of rat peochromoccytoma cells via an RGD-dependent mechanism.

Other Factors Helping in Attachment

In addition to collagen, fibronectin, laminin, and vitronectin, other components of the ECM and basement membrane also contribute to cell attachment. Thrombospondin is a large trimeric glycoprotein consisting of three identical 140-kDa subunits joined together through interchain disulfide linkages. Thrombospodin is released by activated platelets and is found associated with the ECM proteins heparin, collagen, laminin, and fibrinogen. Thrombospondin is synthesized by human long-term bone marrow cells in vitro, promotes the adhesion of hematopoietic progenitor cells, and promotes the attachment and migration of monocytes and neutrophils. Receptors for thrombospondin that play a role in its function as an attachment factor include membrane-bound heparan sulphate, platelet glycoprotein IV (CD36; GPIIIb), and αv/β3 integrin. Tenascin is a very large, star-shaped ECM glycoprotein composed of six related subunits joined through inter- chain disulfide linkages. The individual subunits contain 13 epidermal growth factor–like repeats, 8 or more fibronectin type III repeats, and a globular carboxy-terminal fibronectin-like domain.

Tenascin is associated with mesenchymal-epithelial interactions during development, tissue remodeling, and wound healing. Tenascin promotes the attachment of many cell types in vitro, including tumor cells, fibroblasts, and endothelial cells, in an RGD-dependent manner through interaction with multiple integrins. Janusin, an ECM protein with structural homology to tenascin, promotes attachment and neurite outgrowth of hippocampal neurons in vitro. Glycosaminoglycans are a major component of the ECM and, as already mentioned, bind to fibronectin, collagen, laminin, and vitronectin. Glycosaminoglycans consist of long polysaccharide chains made up of disaccharide subunits and, with the exception of hyaluronan (HA), are sulfated and covalently attach to core proteins through serine residues. In addition to playing a role in tissue formation during development and remodeling, glycosaminoglycans also contribute to the process of inflammation and tumorigenesis. One glycosaminoglycan that has been widely studied is HA. The role of HA in various cellular functions including cell proliferation, cell activation, and cell migration is dependent on its interaction with specific cell surface glycoprotein CD44.

Molecular cloning of the CD44 gene revealed the presence of multiple isoforms, which result from alternate splicing. Analysis of the expression pattern of the different isoforms suggests that regulation of expression of specific isoforms plays a critical role in cellular interactions during early development. The CD44–HA interaction is involved in endothelial cell adhesion and proliferation and T-cell adhesion and activation. CD44 has also been implicated in the regulation of tumor growth and metastasis, and this activity is dependent on its interaction with HA.

CELL SURFACE RECEPTORS

In addition to molecules associated with the ECM and the basement membrane, cell surface receptors also serve as attachment factors to promote cell–cell interactions. Cell surface receptors implicated in cell attachment include the cell adhesion molecule family, the selectin family, and the cadherin family.

Cell Adhesion Molecules (CAMs)

The *cell adhesion molecules,* or CAMs, comprise a family of cell surface receptors that are members of the immunoglobulin gene superfamily. Members of this gene superfamily are characterized by the presence of multiple tandem immunoglobulin fold domains. Vascular cell adhesion molecule 1 (VCAM-1) is present on most peripheral blood leukocytes, activated endothelial cells, macrophages, dendritic cells, bone marrow fibroblasts, and myoblasts and plays an important role in the recruitment of leukocytes to sites of inflammation. At sites of inflammation, activated endothelial cells express VCAM-1, which facilitates the adhesion and subsequent transmigration of circulating leukocytes. The $\beta1$ integrin VLA-4 (CD$\alpha4/\beta1$) and the $\beta7$ integrin (CD$\alpha4/\beta7$) are ligands for VCAM-1. Similar to VCAM-1, intercellular adhesion molecules (ICAM-1, -2, -3) are expressed on most circulating blood leukocytes, activated endothelial cells, fibroblasts, keratinocytes, chondrocytes, and epithelial cells and play a role in immune and inflammatory responses by mediating cell–cell interactions.

The ligands for ICAM-1 and ICAM-2 include the integrins LFA-1 (CD18/CD11a) and MAC-1 (CD18/CD11b), and CD43. Neural cell adhesion molecule (NCAM) is expressed on neurons, astrocytes, Schwann cells, myoblasts, natural killer (NK) cells, and activated T lymphocytes. NCAM is thought to play an important role in vertebrate development through modulation of cell–cell interations between neurons, astrocytes, oligodendrocytes, and myoblasts. In addition to homotypic interactions, NCAM binds to heparan sulfate and heparin glycosaminoglycans. Two related members, L1-CAM and Ng-CAM, are expressed on postmitotic, premigratory neurons and play a role in neuronal migration and neurite outgrowth. Ligands for L1-CAM include axonin 1 and $\alpha V\beta3$ integrin. Ng-CAM also binds to axonin 1. Neurite outgrowth is stimulated by the interactions of L1-CAM and Ng-CAM with axonin I and requires calcium influx through L- and N-type calcium channels.

The condroitin sulfate proteoglycan neurocan also binds to NCAM, Ng-CAM, and L1-CAM and inhibits neuronal adhesion and neurite outgrowth. Platelet endothelial cell adhesion molecule (PECAM-1, CD31) is expressed on platelets, endothelial cells, monocytes, granulocytes, and a subset of T cells. Exogenous expression of PECAM-1 on mouse L cells induces cell aggregation both in a PECAM-1-dependent, homophilic manner, and a calcium-dependent heterophilic manner through interaction with heparin and chondroitin sulfate. Recently, the $\alpha v/\beta3$ integrin has been demonstrated to be a ligand for PECAM-1.

Role of Selectins in Cell Adhesion

The *selectin* family of cell surface adhesion molecules function specifically in leukocyte–endothelial cell adhesion. Selectins have a multidomain structure consisting of an amino-terminal carbohydrate-binding lectin domain, an epidermal growth factor–homologous domain, and a number of cysteine-rich domains homologous to the consensus repeats contained within complement regulatory proteins. Three members of the selectin family have been cloned and characterized, to date. L-selectin (CD62L) is expressed by leukocytes, E-selectin (CD62E) is found on endothelial cells activated by inflammatory mediators, and P-selectin (CD62P) is found on platelets and activated endothelial cells.

The E- and P-selectins are implicated in the initial adhesion and rolling of leukocytes on activated vascular endothelial cells that line blood vessels. Studies showing that monoclonal antibodies directed against the lectin domain of selectins blocked cell adhesion have implicated carbohydrates as ligands for selectins. Currently, a large number of sialylated, sulfated, and/or fucosylated carbohydrates have been identified as ligands for selectins. In addition to their adhesive properties, selectins activate leukocytes, causing intracellular calcium release, induction of IL-8 and TNF-α *mRNA transcription,* and activation of β2 integrin–dependent strong adhesion.

Role of Cadherins in Adhesion

The cadherin family of cell adhesion molecules are calcium-dependent membrane glycoproteins that are grouped into nine types based on structural and functional differences. Type I, or classical cadherins (E-, P-, N-, B-, R-, EP-cadherin), are the only cadherins that have been conclusively shown to promote cell adhesion. Type I cadherins contain five extracellular domains of approximately 110 amino acids each, a single transmembrane domain, and two cytoplasmic domains. The extracellular domains contain Asp-X-Gln-Asp-Gln-Asp and Asp-X-Asp amino acid motifs, which are responsible for calcium binding. The amino-terminal extracellular domain contains the cell adhesion recognition sequence His-Ala-Val (HAV), which mediates calcium-dependent homotypic and heterotypic interactions between cadherins.

The cytoplasmic domains of type I cadherins interact with a family of intracellular proteins termed catenins. The catenins in turn bind to microfilaments and serve to connect the cadherins to the cytoskeleton. The interaction of catenins with the cytoplasmic domain of cadherins is thought to play an important role in cadherin adhesive function. Cadherin interactions regulate the formation of adherens junctions and tight junctions between cells. In addition to mediating cell–cell contact, cadherins also regulate cell morphology and differentiation. E-cadherin, which has been classically associated with the formation, differentiation, and polarization of developing epithelia, has recently been shown to regulate erythropoietin-mediated differentiation of bone marrow mononuclear cells. Neural cadherin, or N-cadherin, promotes neurite outgrowth, which can be blocked by HAV-containing peptides. N-cadherin also regulates the adhesion of O-2A-lineage glial progenitor cells in culture.

CONCLUDING REMARKS

It has become clear that cell–cell and cell–substrate interactions mediated through adhesion molecules and their corresponding receptors play critical roles in cellular functions including growth, activation, differentiation, and death. One of the challenges facing researchers will be to define the molecular mechanisms triggered by the interaction between attachment factors and cell surface receptors that regulate cellular function. Recent advances in molecular biology have provided powerful tools by which to identify both extracellular and intracellular components of these complex signaling systems. Soluble, chimeric proteins containing the extracellular regions of attachment factors and cell surface receptors have been used to identify and clone novel interacting partners, to define domains and individual residues that participate in binding, and to examine the biological role of the attachment factor–cell surface receptor interaction.

In addition to being a tool for studying the biochemistry and biology of attachment factor–cell surface interactions, soluble chimeric proteins are also being pursued as novel drug candidates. The ability of soluble, chimeric proteins to mimic their native counterparts may allow for the generation of effective biological-based drugs that can modulate protein–protein, or cell–cell interactions and their downstream signaling events. In addition to defining extracellular interations, the development of powerful molecular biology approaches has provided novel methods by which to identify protein–protein interactions that occur inside cells. The yeast two-hybrid system has been used to identify signaling proteins that interact with the cytoplasmic domains of various attachment factor cell surface receptors.

This system has been used to identify several proteins that play important roles in integrin function, including the cytoskeletal protein filamin, cytohesin 1, integrin-linked kinase, and β3 endonexin. This system will continue to provide a means of identifying molecules that interact with the cytoplasmic domain of attachment factor cell surface receptors and participate in attachment factor–cell surface receptor interactions. These approaches, together with the current explosion in the identification and cloning of novel attachment factor receptors through use of genomic-based bioinformatics, will provide an exciting opportunity for interface between attachment factor biologists, molecular and cell biologists investigating cell signaling, and bioinformaticians in the development of the biology of cell attachment.

INDEX

T

V

Z